GELS HANDBOOK

Volume 2

GELS HANDBOOK

Volume 2
Functions

Editors-in-Chief

Yoshihito Osada and Kanji Kajiwara

Associate Editors

Takao Fushimi, Okihiko Hirasa,
Yoshitsugu Hirokawa, Tsutomu Matsunaga,
Tadao Shimomura, and Lin Wang

Translated by

Hatsuo Ishida

ACADEMIC PRESS

A Harcourt Science and Technology Company

San Diego San Francisco New York Boston
London Sydney Tokyo

This book is printed on acid-free paper. ∞

ACADEMIC PRESS
A Harcourt Science and Technology Company
525 B Street, Suite 1900, San Diego, CA 92101-4495, USA
http://www.academicpress.com

Academic Press
Harcourt Place, 32 Jamestown Road, London, NW1 7BY, UK

Library of Congress Catalog Number: 00-107106
International Standard Book Number: 0-12-394690-5 (Set)

International Standard Book Number, Volume 2: 0-12-394962-9

Printed in the United States of America
00 01 02 03 04 IP 9 8 7 6 5 4 3 2 1

Contents

Chapter 2 Functions 15

Preface

The development, production, and application of superabsorbent gels is increasing at a remarkable pace. Research involving functional materials in such areas as medical care, medicine, foods, civil engineering, bioengineering, and sports is already widely documented. In the twenty-first century innovative research and development is growing ever more active. Gels are widely expected to be one of the essential solutions to various problems such as limited food resources, environmental preservation, and safeguarding human welfare.

In spite of the clear need for continued gel research and development, there have been no comprehensive references involving gels until now. In 1996, an editorial board led by the main members of the Association of Polymer Gel Research was organized with the primary goal of collecting a broad range of available information and organizing this information in such a way that would be helpful for not only gels scientists, but also for researchers and engineers in other fields. The content covers all topics ranging from preparation methods, structure, and characteristics to applications, functions, and evaluation methods of gels. It consists of Volume 1, The Fundamentals; Volume 2, Functions; Volume 3, Applications; and Volume 4, Environment: Earth Environment and Gels, which consists of several appendices and an index on gel compounds.

Because we were fortunate enough to receive contributions from the leading researchers on gels in Japan and abroad, we offer this book with great confidence. We would like to thank the editors as well as the authors who willingly contributed despite their very busy schedules.

This handbook was initially proposed by Mr. Shi Matsunaga. It is, of course, due to the neverending effort by him and the editorial staff that this handbook was successfully completed. We would also like to express great appreciation to the enthusiasm and help of Mr. Takashi Yoshida and Ms. Masami Matsukaze of NTS Inc.

Yoshihito Osada
Kanji Kajiwara
November, 1997

Contributors

Editors-in-Chief

Yoshihito Osada, *Professor, Department of Scientific Research, Division of Biology at Hokkaido University Graduate School*

Kanji Kajiwara, *Professor, Department of Technical Art in Material Engineering at Kyoto University of Industrial Art and Textile*

Principal Editorial Members

Tadao Shimomura, *President, Japan Catalytic Polymer Molecule Research Center*

Okihiko Hirasa, *Professor, Department of Education and Domestic Science at Iwate University*

Yoshitsugu Hirokawa, *Technical Councilor, Science and Technology Promotional Office, Hashimoto Phase Separation Structure Project*

Takao Fushimi, *Examiner, Patent Office Third Examination Office at Ministry of International Trade and Industry*

Tsutomu Matsunaga, *Director, Chemistry Bio-Tsukuba*

Lin Wang, *Senior Scientist, P&G Product Development Headquarters*

Ito Takeshi, *Assistant Manager, Tokyo Office Sales and Development Division of Mitsubishi Chemical Co.*

Seigo Ouchi, *Head Researcher, Kanishi Test Farm at Agricultural Chemical Research Center of Sumitomo Chemical Co.*

Mitsuo Okano, *Professor, Tokyo Women's Medical College*

Masayoshi Watanabe, *Assistant Professor, Yokohama National University Department of Engineering, Division of Material Engineering*

Contributors

Aizo Yamauchi, *President, International Research Exchange Center of Japan Society of Promotion for Industrial Technology*

Yoshihito Osada, *Professor, Department of Scientific Research in Biology at Hokkaido University Graduate School*

Hidetaka Tobita, *Assistant Professor, Department of Engineering, Material Chemistry Division at Fukui University*

Yutaka Tanaka, *Research Associate, Department of Engineering, Material Chemistry Division at Fukui University*

Shunsuke Hirotsu, *Professor, Department of Life Sciences and Engineering, Division of Organism Structures at Tokyo Institute of Technology*

Mitsuhiro Shibayama, *Professor, Department of Textiles, Polymer Molecule Division at Kyoto University of Industrial Art and Textile*

Hidenori Okuzaki, *Assistant, Department of Chemistry and Biology, Division of Biological Engineering at Yamanashi University*

Kanji Kajiwara, *Professor, Department of Technical Art in Material Engineering at Kyoto University of Industrial Art and Textile*

Yukio Naito, *Head of Research, Biological Research Center for Kao*

(the late) Kobayashi Masamichi, *Honorary Professor, Department of Science, Division of Polymer Molecular Research at Osaka University Graduate School*

Hidetoshi Oikawa, *Assistant Professor, Emphasis of Research on Higher Order Structural Controls in Department of Reactive Controls at Reactive Chemistry Research Center at Tohoku University*

Yositsugu Hirokawa, *Technical Councilor, Science and Technology Promotional Office, Hashimoto Phase Separation Structure Project*

Makoto Suzuki, *Professor, Department of Engineering, Division of Metal Engineering at Tohoku University Graduate School*

Ken Nakajima, *Special Research, Division of Basic Science in International Frontier Research System Nano-organic Photonics Material Research Team at Physics and Chemistry Research Center*

Toshio Nishi, *Professor, Department of Engineering Research, Division of Physical Engineering at Tokyo University Graduate School*

Hidemitsu Kuroko, *Assistant Professor, Department of Life Environment, Division of Life Environment at Nara Women's University*

Shukei Yasunaga, *Assistant, Department of Technical Art in Material Engineering at Kyoto University of Industrial Art and Textile*

Mitsue Kobayashi, *Special Researcher, Tokyo Institute of Technology*

Hajime Saito, *Professor, Department of Science, Division of Life Sciences at Himeji Institute of Technology*

Hazime Ichijyo, *Manager of Planning Office, Industrial Engineering Research Center in Department of Industrial Engineering, Agency of Industrial Science and Technology at Ministry of International Trade and Industry*

Masayoshi Watanabe, *Assistant Professor, Yokohama National University Department of Engineering, Division of Material Engineering*

Kunio Nakamura, *Professor, Department of Agriculture, Division of Food Sciences at College of Dairy Agriculture*

Hideo Yamazaki, *Shial, Inc. (Temporarily transferred from Tonen Chemical Co.)*

Koshibe Shigeru, *Shial, Inc. (Temporarily transferred from Tonen Chemical Co.)*

Hirohisa Yoshida, *Assistant, Department of Engineering, Division of Industrial Chemistry at Tokyo Metropolitan University*

Yoshiro Tajitsu, *Professor, Department of Engineering at Yamagata University*

Hotaka Ito, *Instructor, Division of Material Engineering at National Hakodate Technical High School*

Toyoaki Matsuura, *Assistant, Department of Opthamology at Nara Prefectural Medical College*

Yoshihiko Masuda, *Lead Researcher, Third Research Division of Japan Catalytic Polymer Molecule Research Center*

Toshio Yanaki, *Researcher, Shiseido Printed Circuit Board Technology Research Center*

Yuzo Kaneko, *Department of Science, Division of Applied Chemistry at Waseda University*

Kiyotaka Sakai, *Professor, Department of Science, Division of Applied Chemistry at Waseda University*

Teruo Okano, *Professor, Medical Engineering Research Institute at Tokyo Women's Medical College*

Shuji Sakohara, *Professor, Department of Engineering, Chemical Engineering Seminar at Hiroshima University*

Jian-Ping Gong, *Assistant Professor, Department of Scientific Research, Division of Biology at Hokkaido University Graduate School*

Akihiko Kikuchi, *Assistant, Medical Engineering Research Institute at Tokyo Women's Medical College*

Shingo Matukawa, *Assistant, Department of Fisheries, Division of Food Production at Tokyo University of Fisheries*

Kenji Hanabusa, *Assistant Professor, Department of Textiles, Division of Functional Polymer Molecules at Shinshu University*

Ohhoh Shirai, *Professor, Department of Textiles, Division of Functional Polymer Molecules at Shinshu University*

Atushi Suzuki, *Assistant Professor, Department of Engineering Research, Division of Artificial Environment Systems at Yokohama National University Graduate School*

Junji Tanaka, *Department of Camera Products Technology, Division Production Engineering, Process Engineering Group at Optical Equipment Headquarters at Minolta, Inc.*

Eiji Nakanishi, *Assistant Professor, Department of Engineering, Division of Material Engineering at Nagoya Institute of Technology*

Ryoichi Kishi, *Department of Polymer Molecules, Functional Soft Material Group in Material Engineering Technology Research Center in Agency of Industrial Science and Technology at Ministry of International Trade and Industry*

Toshio Kurauchi, *Director, Toyota Central Research Center*

Tohru Shiga, *Head Researcher, LB Department of Toyota Central Research Center*

Keiichi Kaneto, *Professor, Department of Information Technology, Division of Electronic Information Technology at Kyushu Institute of Technology*

Kiyohito Koyama, *Professor, Department of Engineering, Material Engineering Division at Yamagata University*

Yoshinobu Asako, *Lead Researcher, Nippon Shokubai Co. Ltd., Tsukuba Research Center*

Tasuku Saito, *General Manager, Research and Development Headquarters, Development Division No. 2 of Bridgestone, Inc.*

Toshihiro Hirai, *Professor, Department of Textiles, Division of Raw Material Development at Shinshu University*

Keizo Ishii, *Manager, Synthetic Technology Research Center at Japan Paints, Inc.*

Yoshito Ikada, *Professor, Organism Medical Engineering Research Center at Kyoto University*

Lin Wang, *Senior Scientist, P&G Product Development Headquarters*

Rezai E., *P&G Product Development Headquarters*

Fumiaki Matsuzaki, *Group Leader, Department of Polymer Molecule Science Research, Shiseido Printed Circuit Board Technology Research Center*

Jian-Zhang (Kenchu) Yang, *Researcher, Beauty Care Product Division of P&G Product Development Headquarters*

Chun Lou Xiao, *Section Leader, Beauty Care Product Division of P&G Product Development Headquarters*

Yasunari Nakama, *Councilor, Shiseido Printed Circuit Board Technology Research Center*

Keisuke Sakuda, *Assistant Director, Fragrance Development Research Center at Ogawa Perfumes, Co.*

Akio Usui, *Thermofilm, Co.*

Mitsuharu Tominaga, *Executive Director, Fuji Light Technology, Inc.*

Takashi Naoi, *Head Researcher, Ashikaga Research Center of Fuji Film, Inc.*

Makoto Ichikawa, *Lion, Corp. Better Living Research Center*

Takamitsu Tamura, *Lion, Corp. Material Engineering Center*

Takao Fushimi, *Examiner, Patent Office Third Examination Office at Ministry of International Trade and Industry*

Kohichi Nakazato, *Integrated Culture Research Institute, Division of Life Environment (Chemistry) at Tokyo University Graduate School*

Masayuki Yamato, *Researcher, Doctor at Japan Society for the Promotion of Science, and Japan Medical Engineering Research Institute of Tokyo Women's Medical College*

Toshihiko Hayasi, *Professor, Integrated Culture Research Institute, Division of Life Environment (Chemistry) at Tokyo University Graduate School*

Naoki Negishi, *Assistant Professor, Department of Cosmetic Surgery at Tokyo Women's Medical College*

Mikihiro Nozaki, *Professor, Department of Cosmetic Surgery at Tokyo Women's Medical College*

Yoshiharu Machida, *Professor, Department of Medical Pharmacology Research at Hoshi College of Pharmacy*

Naoki Nagai, *Professor, Department of Pharmacology at Hoshi College of Pharmacy*

Kenji Sugibayashi, *Assistant Professor, Department of Pharmacology at Josai University*

Yohken Morimoto, *Department Chair Professor, Department of Pharmacology at Josai University*

Toshio Inaki, *Manager, Division of Formulation Research in Fuji Research Center of Kyowa, Inc.*

Seiichi Aiba, *Manager, Department of Organic Functional Materials, Division of Functional Polymer Molecule Research, Osaka Industrial Engineering Research Center of Agency of Industrial Science and Technology at Ministry of International Trade and Industry*

Masakatsu Yonese, *Professor, Department of Pharmacology, Division of Pharmacology Materials at Nagoya City University*

Etsuo Kokufuta, *Professor, Department of Applied Biology at Tsukuba University*

Hiroo Iwata, *Assistant Professor, Organism Medical Engineering Research Center at Kyoto University*

Seigo Ouchi, *Head Researcher, Agricultural Chemical Research Center at Sumitomo Chemical Engineering, Co.*

Ryoichi Oshiumi, *Former Engineering Manager, Nippon Shokubai Co. Ltd. Water-absorbent Resin Engineering Research Association*

Tatsuro Toyoda, *Nishikawa Rubber Engineering, Inc. Industrial Material Division*

Nobuyuki Harada, *Researcher, Third Research Division of Japan Catalytic Polymer Molecule Research Center*

Osamu Tanaka, *Engineering Manager, Ask Techno Construction, Inc.*

Mitsuharu Ohsawa, *Group Leader, Fire Resistance Systems Group of Kenzai Techno Research Center*

Takeshi Kawachi, *Office Manager, Chemical Research Division of Ohbayashi Engineering Research Center, Inc.*

Hiroaki Takayanagi, *Head Researcher, Functional Chemistry Research Center in Yokohama Research Center of Mitsubishi Chemical, Inc.*

Yuichi Mori, *Guest Professor, Department of Science and Engineering Research Center at Waseda University*

Tomoki Gomi, *Assistant Lead Researcher, Third Research Division of Japan Catalytic Polymer Molecule Research Center*

Katsumi Kuboshima, *President, Kuboshima Engineering Company*

Hiroyuki Kakiuchi, *Mitsubishi Chemical, Inc., Tsukuba Research Center*

Baba Yoshinobu, *Professor, Department of Pharmacology, Division of Pharmacological Sciences and Chemistry at Tokushima University*

Toshiyuki Osawa, *Acting Manager, Engineer, Thermal Division NA-PT at Shotsu Office of Ricoh, Inc.*

Kazuo Okuyama, *Assistant Councilor, Membrane Research Laboratory, Asahi Chemical Industry Co., Ltd.*

Takahiro Saito, *Yokohama National University Graduate School, Department of Engineering, Division of Engineering Research*

Yoshiro Sakai, *Professor, Department of Engineering, Division of Applied Chemistry at Ehime University*

Seisuke Tomita, *Managing Director, Development and Production Headquarters at Bridgestone Sports, Inc.*

Hiroshi Kasahara, *Taikisha, Inc. Environment System Office*

Shigeru Sato, *Head Researcher, Engineering Development Center at Kurita Engineering, Inc.*

Okihiko Hirasa, *Professor, Iwate University*

Seiro Nishio, *Former Member of Disposable Diaper Technology and Environment Group of Japan Sanitary Material Engineering Association*

VOLUME 2

Functions

CHAPTER 1

Review

Chapter contents

1.1 INTRODUCTION

In natural polymer gels, formation of networks (gel formation) is caused in many cases by the formation of intermolecular bonds as a result of temperature and pH changes and the presence of metallic ions. Thus, a reversible sol-gel transition takes place. On the other hand, synthetic gels consist of polymer chains with both ends connected by covalent bonds or other physical bonds. These structures typically lead to irreversible gel formation. Thus, the physical properties of these synthetic gels depend strongly on the degree of solvent incorporation into the gel networks. They show elastic rubber-like behavior while also possessing the plastic property to allow changes in shape in a nonlinear fashion. By controlling these mechanical properties, various applications have been developed, including those for food, paints, compounds for prevention of soil loss and vibration damping.

In an ionic gel, deep static potential wells exist along the polymer chain. At every crosslink point where these wells cross, deeper potential wells are formed [1]. High moisture absorption and the ability to adsorb the metallic ions of an ionic gel are related to these energy wells. These properties are widely used for sanitary products and ion exchange resins. The properties of gels are influenced by both networks and solvents. Solvents not only penetrate through the network chains but also determine the overall shapes of gels. The amount of solvent a gel can hold depends on the elasticity of the polymer networks and compatibility between the polymer network atoms and solvent. A gel swells when immersed in a good solvent. Swelling continues until the equilibrium between elasticity and amount of penetrated solvent is reached. Most gels swell and shrink proportionately according to the thermodynamic properties of the penetrating solvent. However, certain gels may suddenly change their size when the property of the solvent is changed slightly. For example, a polyacrylamide gel that is partially charged by copolymerization with acrylic acid gradually shrinks when its ethanol content is gradually increased in an ethanol-water mixed solvent after reaching equilibrium. However, in a certain compositional region, the volume of the gel suddenly reduces to several percent of the original volume with the addition of just a small amount of ethanol. This phenomenon is called the gel phase transition.

Another important property of gels is their ability to exhibit phase transition due to an open structure that interacts with the external environment. Gels can contain energy and information, such as pH, temperature, electric fields, and chemical compounds. They also change morphology. Their open system morphology allows gels to be used in the sustained release of substances; as well, they are able to control their own behavior by using external physical or chemical signals. The microenvironment of a gel allows it to be used for material exchange, separation, or active transfer, as well as for chemical reactions. Applications that use these properties include electrophoresis substrates, ion exchange resins, and culture substrates. Pattern formation and nonlinearity of gels are caused by the makeup of a particular microenvironment. Therefore, absorption, desorption and transport of materials are all possible because of the open system of gels. Table 1 summarizes the characteristics of gels and the function and application examples that employ those properties.

Gel properties and their functions will be reviewed here before the discussion of individual topics is pursued.

Table 1 Characteristics, functions and application examples of gels.

Water absorption, sustained water	Superabsorbent material, paper diapers, feminine products, oil-water separation materials
Moisture absorption	Drying agents, antifrost agents
Sustained release	Drug substrates, horticultural water holding agents, air fresheners
Absorption of materials, permeation	Impurity removal materials, selective separation membranes, chromatographic column materials, ion exchange resins
Stabilization of suspension, increased viscosity	Food materials, cosmetics, culture substrates
Transparency, light transmission	Artificial lens, optical lenses, display materials
Biocompatibility	Cell cultures, artificial skins, contact lenses
Chemomechanical responsiveness	Artificial muscles, actuators, switch elements, shape memory materials
Electrical properties	Sensors, electrodes
Vibration damping, sound absorption	Sound barrier walls, insulators, impact-damping materials, artificial joints
Flexibility, plastic deformation, elasticity, floatability, stretching	Shape-deformation materials, slurrying agents for soil, packing materials
Solidification of water at room temperature, freezing point depression	Digestive aids, fertilizer additives, cool pillows, coolants
Swelling	Toys, packing materials, ceiling materials

1.2 GEL FUNCTIONS

1.2.1 Water Absorption, Water Retention, and Moisture Absorption

From ancient times, cotton, pulp, and cloth materials were known to absorb water. These materials absorb water only by the capillary effect and suction power is weak. With pressure, water can be easily excluded and retention is also weak. In 1965 or thereabouts, poly(vinyl alcohol), poly(hydroxyethylmethacrylic acid), and poly(ethylene glycol) were developed and water retention improved.

The retention capability of gels depends on crosslink density and the types of monomers used. The lower the crosslink density, the greater the amount of water absorbed. Moreover, if there are polar groups in the monomer, water can be attracted due to static potential energy. Flory explained the water-retention capability of crosslinked polyelectrolytes by employing the ionic network theory [2]. There are two parameters that provide water absorptivity. One is compatibility between a polyelectrolyte and water; the other is the osmotic pressure generated by a high ionic concentration within the gel. The parameter that controls water absorption is rubber elasticity, which is based on the network structure. The ability to absorb water is determined by the balance of these parameters. While the readers should refer to monographs for detailed explanations, it can be briefly explained here.

$$\text{Ability to hold water} = (\text{osmotic pressure of ions} + \text{compatibility of polyelectrolyte and water})/\text{crosslink density}$$

According to this equation, it can be seen that the lower the osmotic pressure of ions (the higher the ionic concentration of electrolyte solution) the lower the ability to absorb water. When gels are used in feminine products or as water absorption agents, it is important to select application-specific gels. For feminine products, thinness and dryness are required. These requirements are accomplished by using absorbent layers [3–5].

1.2.2 Sustained Release

Along with recent developments of new drugs and the technology to create artificial organs, it has become important to minimize the side effects of these drugs or organs, both of which have been plagued by these sorts of problems. It is possible to reduce side effects if the right amount of drug can be administered at the right time and the right place. This is a much sought-after form of drug delivery system (DDS) and there has been much research in the areas of medicine, pharmacology, and engineering.

There are two methods of delivery of drugs that are contained in gels. In one the microenvironment is controlled and the gel's network size is also considered [6]; another method uses a stimuli-responsive gel based on phase transitions [7]. The former requires gels that have high-water content because any drug delivery system employed subcutaneously must take into account epidermal water content. This is what actually controls the release of the drug molecules through the reversible swelling and shrinking that is caused by changes in body temperature and pH. An intelligent drug system that can administer the appropriate amount of a drug at the appropriate times is being actively researched.

Solid air fresheners are an example of a sustained-release gel. Fragrance evaporation is strongly influenced by ambient temperature. In particular, the temperature inside a car in the sun can be extremely high, which causes problems for car air fresheners. Improved air fresheners have been developed using thermo-responsive gels [8]. This type of air freshener provides fragrance *only when it is necessary.*

Accordingly, in addition to the ability of gels to absorb materials and hold them, sustained release is also expected to be a factor that will be used in various applications.

1.2.3 Adsorption and Separation of Materials

The ability of polymer gels to adsorb is due to static force, hydrophobic interaction between polymer networks, adsorbent, hydrogen bonding, and van der Waals forces. In a polyelectrolyte gel, strong electric potential wells at the crosslink points of the polymer network pull the material and adsorption takes place. Upon adsorption, the depth of this potential well is compensated for gradually and eventually reaches equilibrium. Based on interactions between the material that passes through the gel and the gel material, certain materials adsorb and other pass. Thus, several materials can be separated. This characteristic of gels is already being used commercially for selective membrane purposes, and in chromatography column packing materials and ion exchange materials. Durability and safety are important issues here.

1.2.4 Transport and Permeation

Biomembranes perform simple and/or complex functions of filtration, permeation, and transport. These functions work either in series or parallel. They also function selectively. To imitate these functions, sensitivity in gels can be developed by varying crosslink density. Selective interaction such as compatibility or solubility with a penetrant is also required. An example of gels in which permeability is important is soft contact lenses [9]. The oxygen permeability of contact lenses is known to increase as the water content of the gel increases. For contact lenses, other parameters, such as biocompatibility, are important. For further details, readers are referred to Chapter 4, Volume 3, Applications.

1.2.5 Insolubility and Substrate Materials

Usually, oxygen exhibits activity as a solid state catalyst used in organic chemistry when it is included within a gel. By fixing oxygen on a substrate and allowing continuous use, recovery and reuse, it is possible to produce organic materials. This allows application to the fermentation and food industries and for chemical engineering purposes. Other uses, for example, to measure enzymes and antibodies, are also being explored. In biochemistry, fixing agarose, dextrin, or polyacrylamide to a gel along with an enzyme inhibitor or a co-enzyme is a standard way of separating and purifying enzymes. Because physical and chemical properties of the substrate itself affect the permeation of the matrix and the enzyme activities, it is necessary to consider the properties of the substrate (gel) and fixation method while also taking into account the properties of the fixing biocatalysts. It is also possible to produce useful materials, such as adenosine triphosphate (ATP) and steroids, by a multistage reaction system and a complex enzyme system. This can be done using not only enzymes but also by fixed microbe fungi and animal or plant cells as the catalyst in a bioreactor. In addition to the use of bioreactors, natural polysaccharide gels such as agar or agarose are also used as culture substrates for these cells. It recently became possible to grow liver and dermis cells using poly(N-isopropylacrylamide)(PNIPAAm) gel as the culture bed and then to detach these cells by using temperature changes and still maintain cellulal function [10]. Hence, a constructed cell structure that could not be detached using traditional enzyme treatment methods can now be recovered. This is a hopeful new transplantation method.

1.2.6 Viscosity Increase and Flow Properties

Internally crosslinked polymer microparticles called microgels can form colloidal suspensions [11]. When the microgel is colloidally dispersed in a good solvent, the suspension is a Newtonian fluid at low concentration, which then changes to a quasi-plastic fluid at high concentration. When concentration is further increased the system gels. In a poor solvent, the suspension is unstable and the microgels precipitate. Further, in a mixed solvent system, flow properties change, dependent upon solvent composition. The ability of microgels to adjust viscosity makes them useful in paints, inks, and adhesives. Both quality and functions of liquids are easily improved in this way. For details, readers are referred to Section 10 here.

1.2.7 Transparency

The only body part that has transparent materials in the human body is the eyeball. The cornea, lens and vitreous humor consist mainly of collagen and acidic mucopolysaccharides, which makes them gels. The attempt to use natural polymer gels as a cataract cure has met with limited success due to biocompatibility problems and long-term stability. On the other hand, if vitreous humor substances are replaced by artificial materials made of PVA hydrogel, the properties (transparency and refractive index) are very like those of the vitreous humor. Hence, it is ideal as a replacement material [12].

1.2.8 Biocompatibility

There are various requirements for medical polymers. Biocompatibility is one of them. Medical polymers must function immediately upon contact with internal organs and other systems without rejection reactions. Of the polymeric materials that have been used as artificial skin membranes, collagen is regarded as one of the most useful [13, 14]. Artificial skin requires proper moisture permeation. Collagen fulfills this requirement but is also soft, absorbs bodily fluid, and adheres well to scar tissue without causing inflammation or rejection. It also accelerates the growth of fiber cells. In many natural and synthetic gels, reduction in both the amount of adsorbed platelet protein and the number of adsorbed platelets, as well as formation of blood clots, have occurred. These phenomena are due not only to water content but also to the characteristics of constituent polymers. The sugar chains of the cell surface of these gels are several tens of nm in length [15]. A full understanding as to why these phenomena occur is not yet known, but it might be due to the small interfacial energy gap at the gel surface.

1.2.9 Conversion of Energy-Chemomechanical Materials

A chemomechanical system can be defined as one that is used to obtain macroscopic mechanical energy caused by microscopic deformation in response to changes in an external environment; it is also considered to be a system for obtaining large deformations effectively by using microscopic mechanical energy. Polymer gels can be functional polymers that possess complex system functions similar to those of biomaterials. Thus, they are potentially useful chemomechanical materials and various studies are underway today. Chemomechanical systems actuate by phase transition, oxidation-reduction, chelation, and formation of complexes between polymers. They are classified as follows:

1. ion formation, exchange [16, 17];
2. solvent exchange [17, 18];
3. thermal response [19–21];
4. electrical responses [22–28].

An ordinary actuator utilizes metal and its movement is awkward in comparison to its biomaterial counterpart. If an actuator is manufactured with a gel using one of the above-mentioned responses, an actuator that will not scratch its surroundings, and that will absorb energy and possess functions similar to those of a biomaterial can be expected. Various applications in the medical field are being considered because microscopic changes in position and shape caused by chemomechanical reactions can be used as control signals or switching elements. However, various problems need to be solved for actual application, including improving both efficiency and strength.

1.2.10 Electrical Properties, Magnetic Properties

Polyelectrolyte gels exhibit various anomalous phenomena, including volumetric phase transitional electrical shrinkage, and nonfrozen water [29–32]. Of these, the electrical properties of ionic gels (e.g., electrical shrinkage) have been extensively studied because of their important role in response and control by electricity. They can also be involved in the information transfer of biomaterials. The sol-gel transition using direct current electrical conductivity is also being studied [33]. It is extremely important to elucidate the electrical properties of gels in order to consider various applications.

Electrofluid is a suspension of solid particles in an electrically insulating liquid. The suspension normally exhibits good fluidity. However, when an electric field is applied, the viscosity suddenly increases. This is due to the dipole-dipole interaction of the polarized particles that became dipoles caused by the electric field. This phenomenon as a whole is similar to the transition from sol to gel. In addition, water that contains polymer particles is used as a suspension material. Therefore, these systems have been attracting the attention of gel researchers [34]. The properties of electrofluids are important in impact devices and for other uses.

Magnetic fluids are composed of magnetic particles suspended stably by a surfactant. Upon application of a magnetic field, the magnetic particles align themselves and the fluid looks as though it is a magnetic material. If this system is used gels with magnetic properties can be prepared.

1.2.11 Information Conversion Sensors

Sensors are being used to replace the sensing organs of humans and animals. They are essentially information devices that enhance the functions of natural sensing organs. Gels can be used by exploiting their ability to respond to changes in electrical conductivity, dielectric constant, or piezoelectricity [32] that result from changes in an external environment. However, many problems remain to be solved, among them reproducibility and stability.

1.2.12 Shape Memory

Some polymer gels possess shape memory. These polymer gels, which have side chains with strong crystallinity (long alkyl chains), possess stereoregularity and exhibit order-disorder transition when there are changes in solvent composition, temperature, or pH. For example, this type of gel can be swollen with water, maintain its shape if it is heated to above the phase transition temperature, deform upon melting of the crystalline portions, and then cool to recrystallize its side chains. Thus, even with heating it above the phase transition temperature, this gel will return to its original shape. This shape memory function is due to the suppression of the recovery force of the deformed gel network by the aggregation force of the side chains [35, 36].

1.3 FUTURE FUNCTIONAL MATERIALS

Thus far, some of the characteristics and functions of gels have been discussed. It may be noticed that gels have multiple properties, with myriad combinations of those properties possible.

Synthetic polymer gels are soft materials. Therefore it is possible to deform and maneuver them in ways not possible when rigid materials like metals are used. If a gel is used as an actuator, delicate materials can be picked up without harming them. As well, manufacture of a membrane that opens and closes micropores (chemical valves) is being attempted. An on-off drug delivery system that is triggered by specific symptoms, perhaps fever or blood sugar levels, is being developed. Applications to optical lenses or switching circuitry using the shape memory phenomenon are also possible. Furthermore, a touch sensor is possible that will exploit the capability of gels to change conductivity upon mechanical stimuli. However, there are many problems prior to developing actual applications, including improvements in response time, efficiency, and durability. It is important to control and understand the basic properties of polymer networks. Dynamic analysis of gel structure and mobility is also important.

REFERENCES

1 Gong, J.P., and Osada, Y. (1995). *Chem. Lett.*, **6**: 449.
2 Flory, P.J. (1953). *Principles of Polymer Chemistry*, Ithaca, NY: Cornell University Press.
3 Kobayashi, T. (1997). *Proc. 1st Symp. of Polym. Gel Study Group, Jpn.*
4 Masuda, Y. (1983). *Kogyo Zairyo*, **29**: 40.
5 Masuda, Y. (1982). *Polymer Digest*, **9**.
6 Kamiya, S., Hara, Y., Matsushima, S., Yamauch, Y., and Matsusawa, Y. (1978). *Ganki*, **29**: 420.
7 Okano, M., and Sakurai, Y. (1990). *Organic Polymer Gels*, Gakkai Publ. Center, p. 67.
8 Tokkyo Kaiho (1994). 122485.
9 Tokko Kaiho (1985). 6710.
10 Yamada, N., Okano, T., Sakai, H., Karikusa, F., Sawasaki, Y., and Sakurai, Y. (1990). *Makromol Chem., Rapid Commun.*, **11**: 571.

11 Kawaguchi, H. (1989), in *Polymer Gel Annual Review, 1998*, Polym. Gel Study Group, Jpn. (1989), p. 15.
12 Hara, Y., Hara, T., Hatanaka, O., Hirai, H., Ichiba, S., Kamiya, S., Nakao, S., Nishinobu, M., Hiasa, Y., and Yamauchi, A. (1984). *Ganki*, **35**: 1340.
13 Yoshisato, K. (1987). *Chem. Education*, **35**: 514.
14 Yoshisato, K. (1988). *Ensho*, **8**: 93.
15 Nosaka, A.Y., Ishikiriyama, K., Todoroki, M., and Tanzawa, H. (1989). *J. Bioactive Compatible Polym.*, **4**: 323.
16 Osada, Y. (1974) in *Functional Polymers*, vol. 9, Soc. Polym. Sci. Jpn., ed., Kyoritsu Publ.
17 Suzuki, M. (1986). *6th Int. Congress of Biorheology.*
18 Suzuki, M. (1987). *J. Soc. Rubber, Jpn*, **60**: 702.
19 Osada, Y. (1984). *Polymer Aggregates*, Polym. Complex Study Group, ed., Gakkai Publ. Center, p. 191.
20 Osada, Y. (1987). *Advances in Polymer Sci.*, vol. 1, *Conversion of Chemical into Mechanical Energy by Synthetic Polymers* (*Chemomechanical System*). Berlin: Springer Verlag.
21 Hirasa, K. (1986). *Kobunshi*, **35**: 1100.
22 DeRossi, *et al.* (1986). *Trans. Am. Soc., Artif. Intern. Organs*, **32**.
23 Grodzinsky, A.J. *et al.* (1980). *Biopolymer*, **19**: 241.
24 Kishi, R., and Osada, Y. (1989). *J. Chem. Soc., Faraday Trans.* 1 **85**(3): 655.
25 Osada, Y. *et al.* (1988). *Polym. Preprints Jpn.*, **37**: 291.
26 Osada, Y., Kishi, R., and Hasebe, M. (1987). *J. Polym. Sci., C. Polym. Lett.*, **25**: 481.
27 Osada, Y. (1991). *Advanced Materials*, **3**: 107.
28 Sawahata, K., Hara, M., Yasunaga, H., and Osada, Y. (1990). *J. Controlled Release*, **14**: 253.
29 Osada, Y., and Hasebe, M. (1985). *Chem. Lem. Lett.* **1285**.
30 Osada, Y., Umezawa, K., and Yamaguchi, A. (1989). *Bull. Chem. Soc. Jpn.*, **62**: 3232.
31 Miyano, M., and Osada, Y. (1991). *Macromolecules*, **24**: 4755.
32 Sawahata, K., Gong, J.P., and Osada, Y. (1995). *Macromol. Rapid Commun.* **16**: 713.
33 Chen, P., Adachi, K., and Kodaka, T. (1992). *Polymer*, **33**: 1813.
34 Tokkyo Kaiho (1990). 35933.
35 Osada, Y., and Matsuda, A. (1995). *Nature*, **376**: 219.
36 Tanaka, T., Kagamai, Y., Matsuda, A., and Osada, Y. (1995). *Macromolecules*, **28**: 2574.

CHAPTER 2

Functions

Section 1

Absorptivity of Water (Moisture Absorptivity and Retention of Water)

YOSHIHIKO MASUDA

2.1 SUPERABSORBENCY

2.1.1 Introduction

As superabsorbency is the topic in this section, superabsorbent polymers were chosen as representative of those gels with three-dimensional (3D) networks. In particular, we will consider both what constitutes *superabsorbency* and the appearance of the resin function, with special emphasis on acrylic superabsorbent polymers.

Hydrogels that are used for soft contact lenses are water absorbent polymers. However, these are highly crosslinked hydrophilic polymers that are designed to absorb at most the same weight of water (saline solution) as their own weight. It remains important to increase water absorption and maintain strength while also retaining oxygen permeability. *High water absorbing techniques*, however, are not synonymous with superabsorbency.

For example, if crosslink density is reduced gradually, water absorbency increases gradually and gel strength is reduced. These are the high water absorbing polymers. If crosslink density is further reduced, it becomes difficult to maintain the gel's shape and eventually all gels become water-soluble polymers. These changes as a function of crosslink density are continuous so that there is no maximum point *of high water absorption* for any of the hydrophilic polymers.

Perhaps 15 years ago when superabsorbent polymers were first used in disposable diapers, they were often described as "the superabsorbent polymer, which absorbs more than several hundred to a thousand times that of its own weight." Even today, superabsorbent polymers are often introduced as such in newspapers. However, in reality, polymers that absorb water thousands of times their own weight are seldom used in applications because a good balance among material properties is required, depending upon both the application and the uses of these products. These superabsorbent polymers form extremely weak gels and have poor water absorption characteristics (uneven morphology). For diaper applications, various developments have been reported to improve the balance among properties. For example, acrylic resins are often used to balance the properties of superabsorbency and gel strength.

As already described, superabsorbency criteria depend on the absorbent's properties, the absorption measurement method and use method, and the conditions under which it is used with other materials. Therefore in this section we will not define superabsorbency as the maximum absorption of water with respect to the weight of the absorbent. Rather, it will be defined as optimum water absorption based on the purpose of the application. Employing this definition, we will discuss the synthesis of superabsorbent polymers, measurement of water absorption, absorbency and its water absorption characteristics, change of the required properties for superabsorbency in the main areas of, health applications, other "superabsorbency" of industrial materials, and moisture absorption of general superabsorbent polymers.

2.1.2 Synthesis of Superabsorbent Polymers

The basic structure of superabsorbent polymers is simply "slightly crosslinked hydrophilic polymers." In order to form superabsorbent polymers, molecular weight should be higher and crosslink density should be lower. Here, the word "slightly" is quite qualitative. However, it is because highly crosslinked polymers show poor water absorption characteristics that they are no longer called superabsorbent polymers.

Generally speaking, hydrophilic polymers indicate those made of linear polymers with hydrophilic side chains or hydrophilic main chains. As shown in Table 1, there are natural and synthetic polymers. They are nonionic-, anionic-, cationic-, and betaine-type polymers. Depending on the type of polymer adopted, the water absorption characteristics differ greatly. The number of polymers actually produced and used is highly limited. In terms of production, alkali metal salt-crosslinked poly(acrylic acid) (sodium salt-crosslinked poly(acrylic acid)) dominates world production.

There are several reasons for its dominance. (1) As a raw material, the acrylic acid monomer is produced worldwide in large quantities as it is relatively inexpensive in comparison to other monomers. (2) Highly pure monomers are readily available and therefore high molecular weight polymers are easily produced. In addition, the molecular weight of the monomer is small and so the number of ionic species per gram is relatively high. Thus, polymers with large charge density can be obtained and these are readily rendered superabsorbent. (3) From the application point of view, the use of this material is mostly for disposable diapers. Although urine contains multivalent metallic ions, concentrations of such ions are not significant and the amount of liquid in contact with the

Table 1 Types and examples of hydrophilic polymers.

Natural materials
Nonionic-type: hydroxyethylcellulose (HEC) [1], starch [2]
Anionic-type:carboxymethylcellulose (CMC) [3], arginic acid [4, 5], hyaluronic acid [6], poly(glutamic acid) [7, 8]
Cationic-type: chitin [1], polylysine [8]
Synthetic materials
1. Those with hydrophilic side chains
Nonionic-type: poly(vinyl alcohol) [9], poly(acrylamide) [10], poly(vinyl pyrrolidone) [11], poly(hydroxyethyl acrylate), poly(vinyl methyl ether)
Anionic-type: partial alkali metal salt of poly(acrylic acid) (25–30), poly(isobutylene-maleic acid) [12, 13], poly(2-acrylamide-2-methylpropane-sulfonic acid) [14, 17], poly(acryloxypropane sulfonic acid) [15], poly(vinyl sulfonic acid) [16]
Cationic-type: poly(methacryloyloxyethyl quarternary ammonium chloride) [17, 18], poly(vinyl pyridine)
Betaine-type: N,N-dimethyl-N-(3-acrylamidepropyl)-N-(carboxymethyl) ammonium internal salt [19–21]
2. Those with hydrophilic main chains
Nonionic-type: poly(ethylene glycol) [9], poly(dioxirane) [22]
Cationic-type: poly(ethylene imine) [23]

polymer is limited. However, the crosslinked poly(acrylic acid) salt will not crosslink further with the multivalent metallic ions and shrink to the extent that the absorbed urine will be squeezed out of the polymer. Thus, the polymer serves its original purpose sufficiently (see Section 1.4).

There are several synthetic methods to produce this representative superabsorbent polymer (SAP). In general they are: (1) an aqueous solution polymerization method in which organic solvents are not used; and (2) a suspension polymerization method in which the aqueous solution of a monomer is suspended in an organic solvent (reverse phase suspension polymerization).

The synthetic route—from acrylic acid to crosslinked poly(acrylic acid)—is illustrated in Fig. 1. There are two methods. In one, acrylic acid or its sodium salt is polymerized and a linear polymer is obtained. Then the linear polymer is crosslinked by a crosslinking agent. In the other the acrylic acid or its sodium salt is polymerized simultaneously with a crosslinking agent and a crosslinked polymer is produced.

In this synthesis, factors requiring caution include the purity of the monomer, the type of initiator, concentration, temperature, selection of crosslinking agent (for the former method, the selection of the crosslinking agent that reacts after the polymer is formed, and for the latter the crosslinking agent that polymerizes simultaneously with monomer reaction), and the drying method and its conditions.

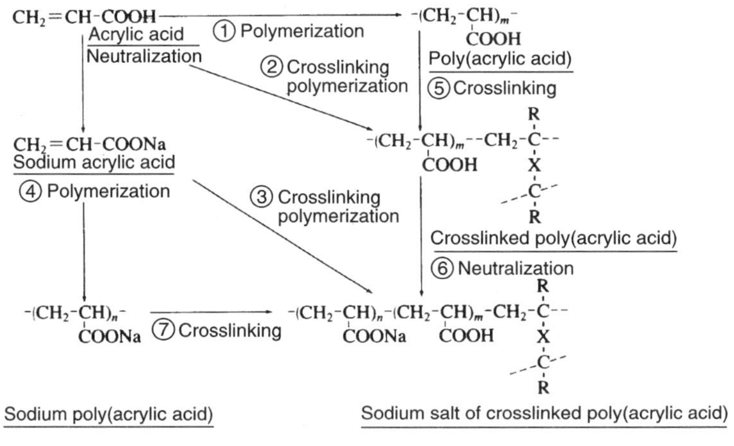

Fig. 1 Synthetic route of a superabsorbent polymer using acrylic acid as the raw material.

2.1.3 Measurement of Water Uptake

Water uptake varies greatly, depending on the measurement method and its conditions. The most desirable measurement method should reflect both the application and application method. The parameters that control water uptake can be divided into chemical and physical conditions. The former depends on the chemical structural difference of the absorbent (see Section 1.4), including the type of ions, concentration, and the content of the hydrophilic organic solvent in the aqueous solution; in the latter, parameters such as the temperature of the absorbate, measurement time, and the presence of pressure are important.

For general application purposes a measurement method for water uptake, JIS K 7223, was established on March 1, 1996. For more accurate information, readers are referred to the original document. In this method, the so-called tea bag method is adopted. This method utilizes a tea bag that contains a sample. The bag is immersed in deionized water or 0.9% sodium chloride solution for a specified time. Then, the increased weight of the bag is measured and water uptake is calculated. This method determines the amount of free swelling (the water uptake under no pressure conditions). Similar but differing in terms of how the water is squeezed are the filtration and centrifuge methods. The filtration method yields vastly differing results, depending on the type of resin and particle size, and thus requires caution. Other methods, which do not require squeezing water, include the blue dextrin method (measured by the high molecular weight pigment not absorbed into the gel) [24] or microscopic observation. In particular, for materials used in health applications, the demound wettability method (Fig. 2) (which measures water uptake under pressure) is available. In these methods, various measurement conditions such as pressure, measurement time, and type of absorbate can be used. In the same field but more advanced, a diffusive absorption ratio method [25] (Fig. 3) (which measures diffusivity and water uptake simultaneously) has been proposed.

Accordingly, the measurement method and conditions for water uptake vary greatly. Therefore, it is not possible to have a certain polymer show maximum water absorption under any method and conditions.

Thus, for specific applications, these types of water absorption measurements of superabsorbent polymers can be useful references to improve resin performance. It is for this reason that various measurement methods have been studied.

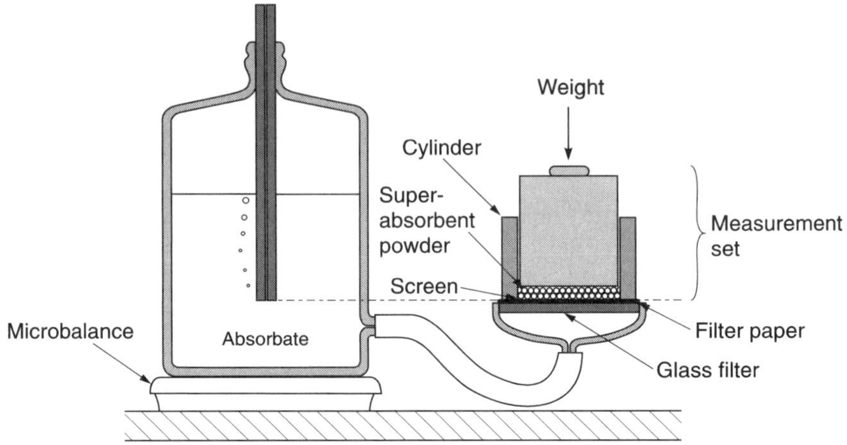

The measurement set is prepared by uniformly placing a certain amount of a superabsorbent polymer (W_0 *g*) on a fine mesh screen located at the bottom of the cylinder. This set is placed on a glass filter on which a wet filter paper is placed in such a way that there is no liquid pressure present. The weight ratio of the water absorbed per unit sample can be obtained as a function of time from the reduction of the absorbate (W_0 *g*) and W_0. In this case, the liquid is only absorbed vertically and this absorption is the so-called weight ratio under the applied pressure.

Fig. 2 Demound wettability measurement device.

2.1.4 Absorbate and Absorption

The water uptake of poly(acrylic acid)-type superabsorbent polymers differs markedly depending upon, in general, the chemical property differences of the absorbate, such as the type or concentration of ions in the absorbate, and the presence of a hydrophilic organic solvent in the aqueous solution.

Figure 4a shows the change in absorption ratio in the aqueous solutions of mono- and multivalent metallic ions using the tea bag method. The relationship between the monovalent ion and saturation absorption ratio is simple. However, when a small amount of multivalent metallic ions is present, the polymer first absorbs liquid until it is close to saturation swelling and then it gradually shrinks. This is due to the gradual formation of ionic crosslinks that connect carboxylic groups by multivalent metallic ions after the solution is absorbed. This leads to an increase in overall crosslink density. Figure 4b shows the time-dependent change in the

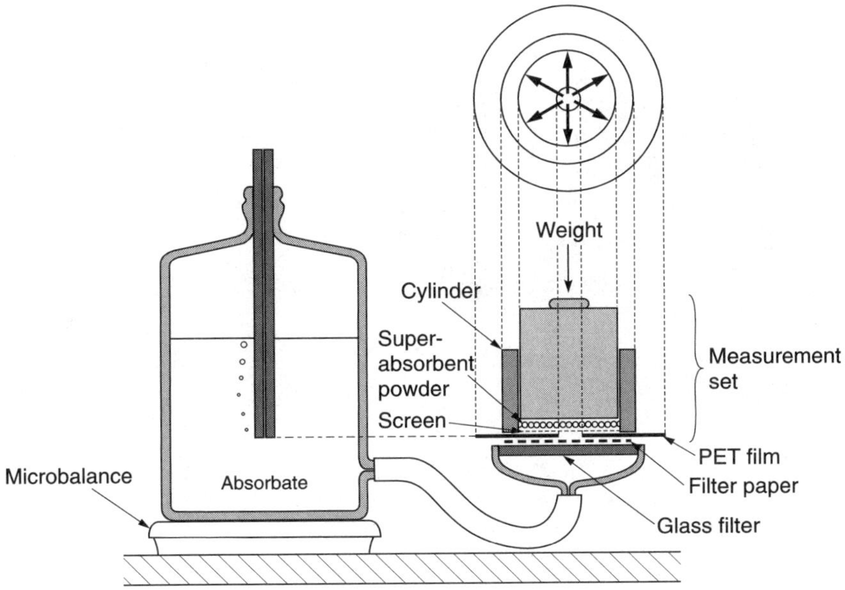

The measurement set is prepared by uniformly placing a certain amount of a superabsorbent polymer (W_0 g) on a fine mesh screen that is located at the bottom of a cylinder. This set is placed on a glass filter on which a wet filter paper and PET film with a small opening at the center are placed in such a way that there is no liquid pressure present. The superabsorbent polymer starts absorbing water at the center of the sample. Then, the water absorption gradually spreads horizontally by diffusive absorption. The weight ratio of the water absorbed per unit sample can be obtained as a function of time from the reduction of the absorbate (W_0 g) and W_0. The liquid is initially absorbed vertically through the PET hole and then further absorbed by diffusion. This allows measurement of the absorption ratio (W/W_0) taking into account the diffusivity.

Fig. 3 Measurement device using diffusive absorption ratio.

absorbent ratio. Figure 5 indicates absorption ratio change as a function of the composition of the hydrophilic organic solvent and the water mixture. As seen in the figure, a sudden change in absorption ratio and phase transition can be observed at a certain mixture concentration in the presence of an alcohol. As the water uptake varies greatly due to differences in the absorbate's properties, the approach used to increase water absorption changes, depending on the application and conditions of

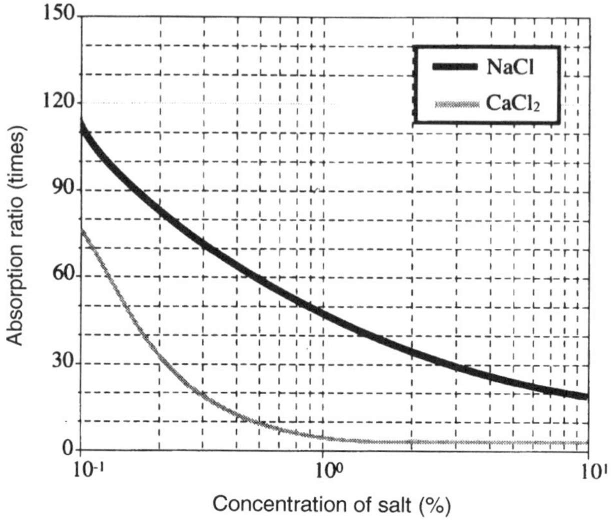

Fig. 4(a) Absorption ratio of the aqueous solution of electrolytes (the tea bag method, 30 min immersion).

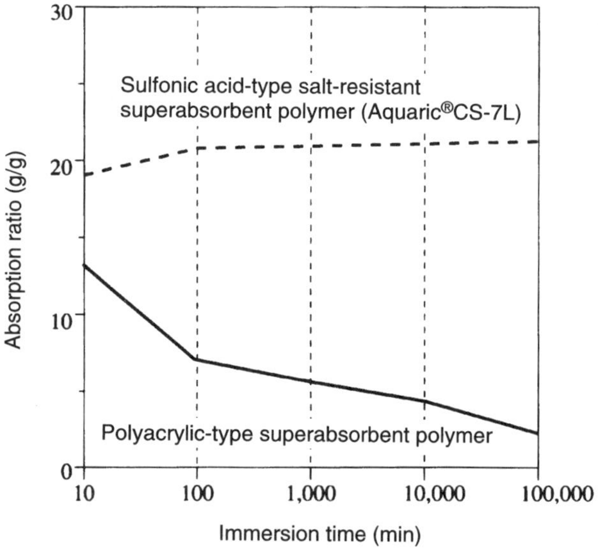

Fig. 4(b) Time-dependent change in absorption ratio (in simulated sea-water).

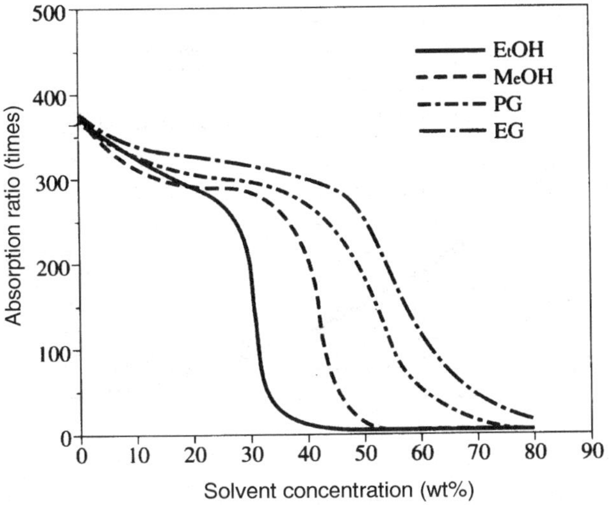

Fig. 5. Absorption ratio in hydrophilic organic solvent and water mixture.

use. Thus, there are occasions for which the type of primary polymer is more important than the crosslink density.

2.1.5 Superabsorbency in Hygiene Applications

Since this is currently the most significant application of superabsorbent polymers, many patent applications have been made. Particular effort has been expended to improve the "superabsorbency" and "high performance properties" of superabsorbent polymers.

For disposable diapers the absorbate is urine and so ionic strength is at most at the level of the saline solution, although urine does contain a small amount of multivalent metallic ions. The amount of absorbate per one diaper depends on the age of the person but in general it is around 200–300 ml including the absorption by the pulp. Under these conditions, the "superabsorbency" of superabsorbent polymers, which allows high performance of the diaper as a whole, has been pursued.

Initially, the amount of superabsorbent polymers used per diaper was minimal. Its use was only supplemental to the role of pulp fibers as the water absorbing material.

However, the amount of superabsorbent polymers used gradually increased and led to competition in diaper performance. This resulted in an increase in polymer use (from the supplemental nature of 1–2 g) to 3–4 g, 6–8 g, and eventually to today's mainstream, high-performance, thin diaper, which uses >10 g of superabsorbent polymer. During this transition, the total weight of diapers did not increase. Rather it decreased by reducing the amount of cotton-like pulp. Under such conditions of use, the "superabsorbency" requirements of the polymer gradually changed. These changes are listed in Table 2.

Various techniques have been developed during these historical changes, including a method to improve gel strength and water uptake under pressure while also maintaining the amount of free swelling. Surface treatment methods were also developed to prevent uneven swelling [26–28]. This can be achieved by further crosslinking the surface region of superabsorbent polymer particles. The interior has a low degree of crosslinking while near the surface there is a high degree of crosslinking.

Table 2 Change in application methods and the required properties of superabsorbent polymers used in disposable diapers.

Application method in diapers	Amount of polymer	Required properties for "superabsorbency"
1. The first generation Sandwich sheet and large amount of cotton-like pulp	Small amount (1–2 g)	(1) The amount of free swelling is large as measured by the tea bag method
2. The second generation a. Sandwich-type, a sandwich is used between medium-large amounts of cotton-like pulp, or	Medium amount (3–4 g)	(1) The amount of free swelling is large as measured by the tea bag method (2) High rate of absorption (3) High gel strength (4) Great suction power
b. Blend-type, a polymer is blended with medium-large amounts of pulp	Large amount (6–8 g)	(3) High gel strength (4) Great suction power (5) Large absorption ratio under pressure
3. The third generation a or b (high concentration core)	Large amount ($\geq$ 10 g)	(2) High rate of absorption (5) Large absorption ratio under pressure (6) High diffusion of liquid (gel permeability) (7) High performance of shape

This allows the outer shell to be mechanically strong without sacrificing the original superabsorbency of the slightly crosslinked polymer in the interior. This also allows prevention of uneven swelling due to the initial fast absorption of water and excessive swelling near the surface.

This technique, developed and commercialized in Japan, currently leads the superabsorbent polymers world market. On the other hand, companies in Western countries adopted polymers without surface treatment. As the trend for reducing thickness continued in the 1990s, use of a cotton-like pulp was reduced and the amount of superabsorbent polymers increased. Under such conditions, it became apparent that surface treatment was necessary to achieve "superabsorbency." As the polymer concentration increases, diffusion of urine throughout the diaper via the capillary effect of the pulp cannot be relied upon and uneven swelling results. It has become important to prevent this so as to achieve effective use of the entire diaper. Under the pressure of body weight, this phenomenon may be even further magnified. Due to such needs, Western companies also developed surface treatment technologies [29–31] and today most superabsorbent polymers for diaper application have surface-treated polymers.

However, both the make-up of diapers and method of evaluation differ depending on the company. Thus, the concept of "superabsorbency" also varies. As the design concept and diaper composition change in the future, the property requirements for superabsorbent polymers will also change. Thus, research and development on the properties of superabsorbent polymers will continue.

2.1.6 Superabsorbency of Other Industrial Superabsorbent Polymers

As the absorbate varies greatly for other industrial superabsorbent polymers, methods to achieve "superabsorbency" differ. Other than the application for coolants, most applications involve absorbates that will contain a high concentration of metallic ions, a large amount of multivalent metallic ions, or those that are expected to be in contact for a long period of time with a metallic ion solution. For chemical heater applications, relatively high concentrations of salt solution need to be contained. For soils the polymer is in constant contact with fertilizer solution or soil ground water. Sealing materials that swell with water are also often expected to be in contact with soils or seawater for lengthy periods of time. To prevent the destruction of concrete structures, the material used to absorb bleeding water will also be exposed to an absorbate that has a large amount of multivalent metallic ions.

Let us consider the example of water swelling of sealant materials. If the polymer is able to swell and stop water leakage when in contact with fresh water but it does not swell when in contact with sea water, then it will not have achieved its function as a sealant. It is desirable for these materials to swell to the same degree regardless of the type of water encountered. Therefore, the required property for this application to be "superabsorbent" is to swell similarly, independent of ion concentration and type in the absorbate. Furthermore, it must absorb as much water as possible (see Fig. 4b/sulfonic acid-type salt resistant superabsorbent polymer, Aquaric® CS-7L).

To cope with this situation, the superabsorbent polymers used for diaper application are insufficient. It is thus necessary to select the type of primary polymer in such a way that the hydrophilic group is salt resistant. As shown in Table 1, for this application various polymers have been studied. Useful polymers include those from nonionic-type, sulfonate-type, or betaine-type monomers.

As is obvious, when superabsorbency for various applications including diapers is considered, high performance under various conditions must also be evaluated.

2.1.7 Moisture Absorption of Superabsorbent Polymers (SAP)

Since a description of the moisture absorption of superabsorbent polymers is given in Section 2, Chapter 6, Part 3, Application, we will discuss only the fundamental aspects here.

Unlike the water absorption characteristics, in moisture absorption by superabsorbent polymers these characteristics do not appear because these polymers are crosslinked. The saturation moisture content is basically the same as for the hydrophilic polymers prior to crosslinking. However, due to the crosslinked structure, adhesion is minimized. Thus, even after moisture absorption/drying cycles, the surface area is kept relatively constant and the rate of moisture absorption/drying will also be kept relatively constant.

The saturation moisture content of poly(acrylic acid) salt-type superabsorbent polymers depends on the relative humidity as shown in Fig. 6.

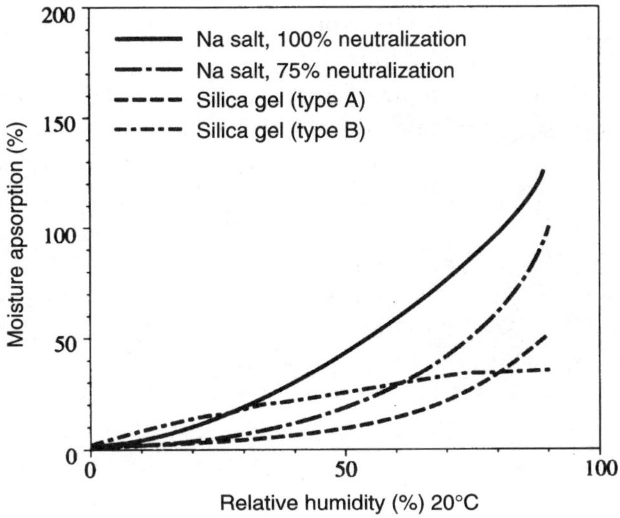

Fig. 6 Moisture absorption of a crosslinked poly(acrylic acid) salt.

The polymer's moisture absorption exceeds that of A-type silica gel at a relative humidity >60% if 75% of the acid is in salt form and at a relative humidity >20% if 100% of the acid is in salt form. At a relative humidity of 90%, its weight doubles and it absorbs 2 to 3 times the moisture of the silica gel.

2.2 HYALURONIC ACID GELS

TOSHIO YANAKI

2.2.1 Introduction

Hyaluronic acid (HA) is a linear polymer with N-acetyl-D-glucosamine residue and D-glucuronic acid residue. Its molecular weight varies greatly depending on the origin and purification method (from around several tens of thousands to several millions of gram molecules) and it is in the glucosaminoglycan family. Its secondary structure is thought to have the hydrogen-bond structure as shown in Fig. 1 because it is difficult to oxidize HA with iodide and also from nuclear magnetic resonance (NMR) results. Hyaluronic acid exists widely along with chondroitin sulfonic acid or dermatan sulfonic acid within the connective tissue of vertebrates. These substances retain water between cells, and form a jelly-like matrix by combining with proteins or other mucopolysaccharides. They are thought to facilitate a variety of biofunctions, including wound healing, prevention of infection, and lubrication. Concentrations of HA in skin decrease as one ages, which may be why the skin becomes dry. It is for this reason that HA obtained from chicken crowns is now added to some cosmetics. The role of HA as an effective moisturizer has now been widely recognized. This boom was further accelerated by a high-level HA-producing microbe and success in mass-producing HA by fermentation of this microbe.

Traditionally, HA has been manufactured by separating and purifying it from the crowns of chickens and from other animal organs. However, purification requires complex procedures to remove proteins, other mucopolysaccharides, and nucleic acids. As a result, production was

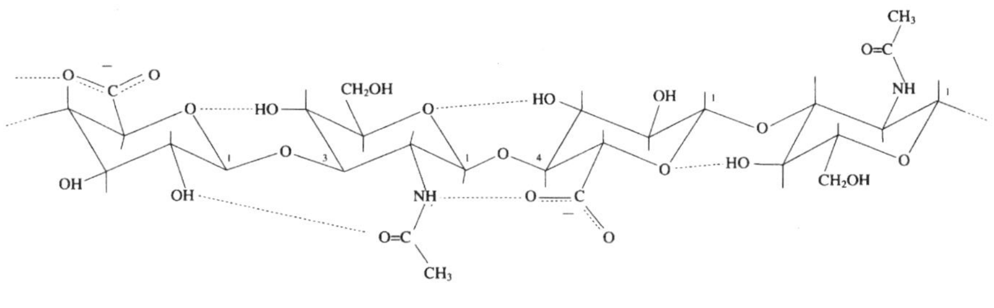

Fig. 1 Four kinds of hydrogen bonds formed among HA residues.

costly. At the same time, the supply was unstable as it was difficult to mass-produce high-quality materials. Although HA has been known to be produced by mucopolysaccharides without the use of bacteria in the streptococcal family, its production was extremely scarce, and industrial production was difficult. In the 1980s, a high-HA-producing organism, *S. zooepidemicus*, was discovered. After improvement in culture conditions and purification methods, the technology to obtain 5 g of high-quality HA from 1 liter of culture solution was established. From this point on, many companies entered successful commercialization activities as shown in Table 1. This success further accelerated the frequency of HA use. The first example of this was an anticipated application to the medical field as a result of its unusual viscoelasticity and highly safe profile. Some are used in eye surgery and in medicines that help with deformed joints. Today, these agents are in clinical use. Furthermore, their derivatives and cross-linked products have also been studied in the areas of cosmetics, medical devices, and pharmacology. They will be reported on as second-generation HA in the near future.

Table 1 Companies that mass-produce hyaluronic acid (HA).

Names of companies	Methods (microbe or raw material)	Cosmetic uses	Pharmacology uses
Shiseido	Fermentation method (*S. zoopidemicus*)	○	○
Denki Kagaku	Fermentation method (*S. equi*)	○	○
Kyowa Hakko	Fermentation method (Streptococcus microbes)	○	○
Cisso	Fermentation method (Streptococcus microbes)	○	
Kibun Foods	Fermentation method (*S. zooepidemicus*)	○	
Yakuruto	Fermentation method (*S. zooepidemicus*)	○	
Meiji Seika	Fermentation method (Streptococcus microbes)	○	○
Asahi Chemicals	Fermentation method	○	
Seikagaku Kogyo	Extraction method (chicken crowns)	○	○
Kyupi	Extraction method (chicken crowns)	○	○
Taiyo Gyogyo	Extraction method (chicken crowns)	○	

2.2.2 Production of HA by Fermentation Methods

The production of microbe polysaccharides including HA has been extensively studied and they are widely used in the food and pharmaceutical industries. In addition to HA, dextrin, xanthan gum, carrudrun, and pullulan are representative examples. As mucopolysaccharides from animals, heparin (a mucoitin glucoprotein) sulfate, and dermatan sulfate are known.

However, only HA is produced from a microbe, one of the variety of streptococcal strains from Lancefield groups A and C. However, the A group hemolytic *streptococci* must be separated and must not be included in industrial production for use by humans due to their disease-producing capability. However, among C group microbes, *streptococcus equi* and *streptococcus zooepidemicus* exist in the eyes and sinuses of animals, and as they are nonpathogenic for man, they are used to produce HA. This microbe is gram positive *S. lactis* and has complex nutrient requirements. Glucose is the best source for carbon nutrition for the culture at an optimum concentration of 4–6%. Yeasts and peptones are beneficial nitrogen sources. As the microbes multiply, HA and lactose are simultaneously produced and the pH of the culture decreases. It requires care to maintain a neutral pH as decreasing pH inhibits HA production. Figure 2 shows the culture process at an almost neutral pH, a temperature of 37°C, purge air rate of 1 vvm, and agitation rate of 450 rpm in a 30-liter jar fermenter. As time passes, HA is produced and the culture liquid gradually increases in viscosity. Thus, usually the culture process is halted after about 30–40 h. Highly pure HA is then extracted from this culture "soup" after a variety of techniques, including heat, chemicals, filtration and centrifuging, are used to remove microbes and other insoluble materials. The solution that remains is put through activated charcoal and alcohol to obtain the high-purity HA.

2.2.3 Dilute Solution Properties of HA

For polymers in general the relation between the molecular weight M and the intrinsic viscosity $[\eta]$ is expressed by the Mark-Houwink-Sakurada equation,

$$[\eta] = K \cdot M^a \text{ (}a\text{ and }K\text{ are constants)}$$

The constant a is a useful guide for molecular conformation of the isolated polymer chains in the solvent. For example, branched polymers show

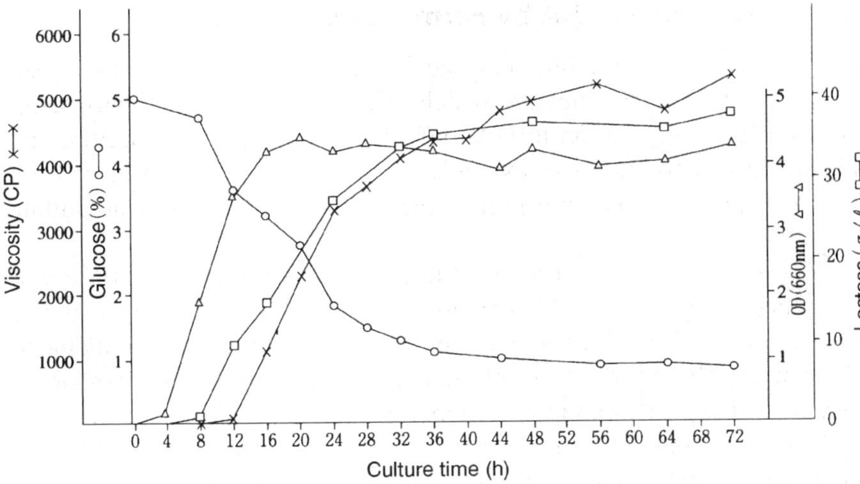

Fig. 2 The culture process of *S. zooepidemicus*.

$a < 0.5$, nonperturbed chains have $a = 0.5$, random coils that are spread by the excluded volume effect show $0.5 < a \leq 0.8$, partially bent polymer chains show $a \approx 1.0$, and rigid rod-like polymers show $a = 1.7$.

Table 2 shows the Mark-Houwink-Sakurada constants reported thus far for HA under physiological conditions. As can be seen in the table, the

Table 2 Mark-Houwink constants reported for HA.

Literature	Molecular weight ranges	$K \times 10^4$	a	Solvents	Measurement temperatures
Yanaki and Yamaguchi [44]	$40 \times 10^4 \leq M_w$	1.99	0.829	0.20 M NaCl	25°C
Bothner *et al.* [45]	$15 \times 10^4 \leq M_w$	3.46	0.779	0.15 M NaCl	25°C
Laurent *et al.* [46]	$10 \times 10^4 \leq M_w$	3.6	0.78	0.20 M NaCl	25°C
Fouissac *et al.* [47]	$13 \times 10^4 \leq M_w$	—	0.78	0.30 M NaCl	25°C
Cleland and Wang [18]; Cleland [49]	$40 \times 10^4 \leq M_w$	1.99	0.829	0.20 M NaCl	25°C
	$10 \times 10^4 < M_w$	2.28	0.816		
Shimada and Matsumura [50]	$M_w \leq 10 \times 10^4$	0.03	1.20	0.20 M Phosphoric acid buffer solution	37°C
	$20 \times 10^4 < M_w$	5.7	0.76		
Balazs *et al.* [51]	$10 \times 10^4 \leq M_w$	2.9	0.80	0.20 M NaCl	25°C

a value of HA with an average molecular weight M_w of 100,000 is 1.0–1.2, and M_w of 200,000 is 0.76–0.83.

This indicates that, under physiological conditions, the HA chain behaves like a partial bending chain in a low molecular weight region. However, in a high molecular weight region, it exists as a spread random coil due to the excluded volume effect. This effect can best be demonstrated this way—a short needle is difficult to bend but a long wire is easy to bend. This also suggests that, even with a partial bending chain of low molecular weight, HA can behave like a random coil at a high molecular weight. The continuous changes of molecular conformation as molecular weight increases have been supported by x-ray scattering and flow birefringence measurements.

Figure 3 shows the pH dependence of the intrinsic viscosity of HA, which is measured in a wide pH buffer solution. The quantity $[\eta]$ shows the highest value near neutral (a basic environment) and decreases in both acidic and alkaline environments. As shown in Fig. 1, the sudden decrease of $[\eta]$ in this basic environment is due possibly to the disruption of hydrogen bonds among various residues in this pH region and increased HA chain flexibility. On the other hand, the gradual decrease of $[\eta]$ in an acidic pH environment is due to the suppressed dissociation of carboxyl groups, resulting in decreased static repulsive force. Consequently, molecular size gradually decreases.

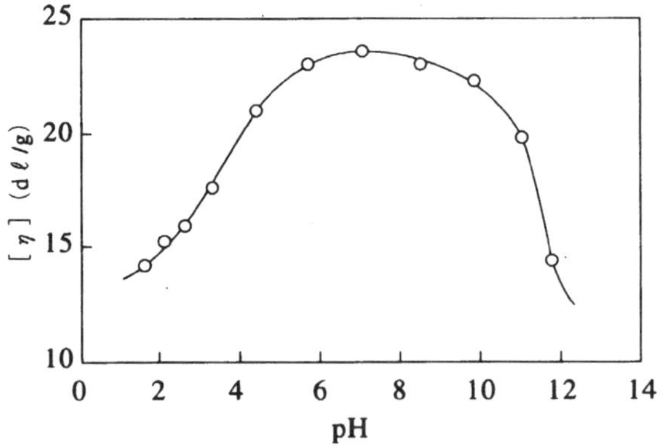

Fig. 3 The intrinsic viscosity $[\eta]$ of a high molecular weight (molecular weight approximately 1.20×10^6) in a broad buffer solution.

2.2.4 Crosslink Network Formation in Concentrated Solutions

When HA with a molecular weight of several million is dissolved in a saline solution at 1%, the solution shows zero shear viscosity of several hundred thousands to several million cp and obvious viscoelasticity. As can be seen from this, even 1% of HA must be treated as a sufficiently concentrated solution.

The viscoelastic properties of concentrated HA solutions have been studied by various groups using their knowledge of concentrated synthetic polymer solutions. Figure 4 shows the frequency dependence of the storage modulus $G'(\omega)$ and loss modulus $G''(\omega)$ of HA aqueous solutions under physiological conditions. In the figure, ω is the frequency and a_T is the shift factor. The G' and G'' curves cross in the vicinity of 2 s^{-1}. Below this region, the slopes of G' and G'' approach 2 and 1, respectively. This indicates that the HA solution flows in the low frequency region. However, above the crossover frequency, both moduli show both gradually constant values and the so-called rubbery plateau. In this region, HA molecules behave as though there are fixed crosslinked networks. From this result, a high molecular weight HA temporarily forms networks by molecular entanglement under physiological conditions. This is also the reason why

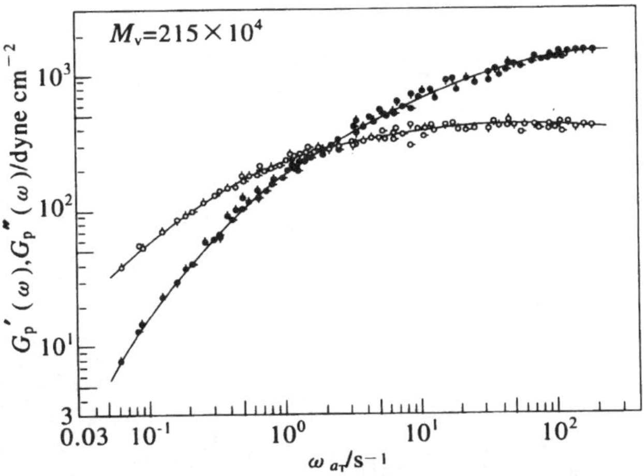

Fig. 4 The storage modulus G' (ω) and loss modulus G'' (ω), represented by ○, of a saline aqueous solution that contains 1% high molecular weight (the viscosity average molecular weight $M_v = 2.15 \times 10^6$).

HA shows significant viscoelasticity. In the case of a natural body, for slow movements these temporary networks behave like high viscosity fluids and upon impact it behaves like an elastic body that absorbs mechanical stress by deforming the HA chain.

The molecular weight and concentration where network formation starts can be determined from two parameters, that is, zero-shear viscosity η_0, and steady-state compliance. Figure 5 depicts the molecular weight dependence of η_0 of 1% HA aqueous solution under physiological conditions. The data in Fig. 5 seem to fit well the two straight lines, with slopes of about 0.9 and 3.8 and which show a breakpoint at a molecular weight of 400,000. This implies that HA shows molecular entanglement at >400,000 under physiological conditions and begins to form networks. Similar analysis is possible from the steady-state compliance.

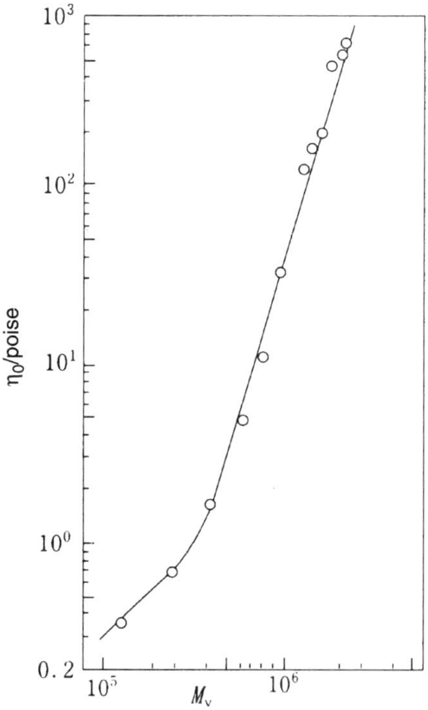

Fig. 5 The molecular weight dependence of the zero-shear viscosity η_0 of 1% HA aqueous solution under physiological conditions.

There are several interesting phenomena regarding pH dependence of the HA aqueous solution viscosity. The viscosity behavior of concentrated HA solutions under acidic conditions is of particular interest. Figure 6 shows the pH dependence of the apparent viscosity η_a that is observed at HA concentrations of $c = 0.1-1.0\%$. At an HA concentration of 0.1%, η_a monotonously increases as pH increases. However, at 0.3%, a shoulder appears near pH = 2.4, and at 0.6% a clear peak is observed. At 1.0%, the peak becomes even more pronounced. This anomalous viscosity behavior around pH = 2.4 was first studied by Ropes *et al.* [63] and

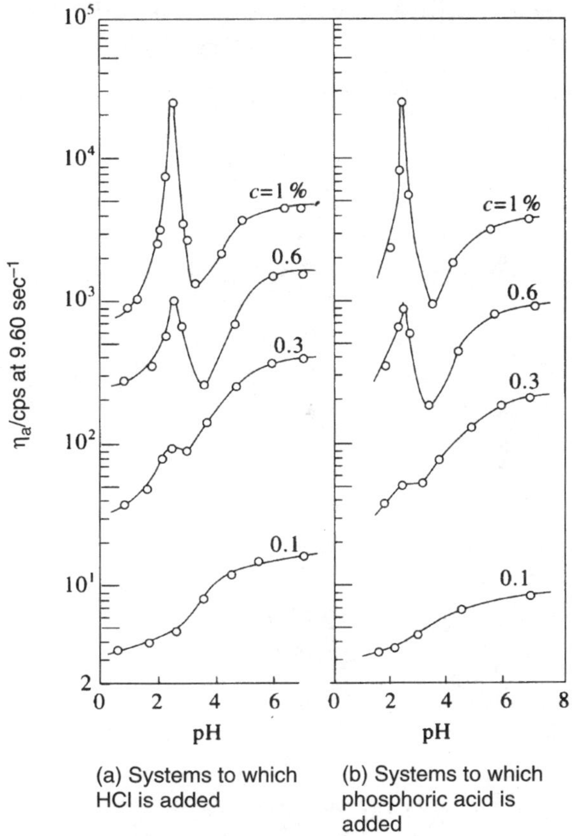

Fig. 6 The pH dependence of the apparent viscosity η_a that is observed at HA concentrations $c = 0.1–1.0\%$ [62].

Pigman *et al.* [64] and later Gibbs *et al.* [59] proposed the "molecular stiffening" concept to explain this phenomenon. However, this concept contradicts the results regarding intrinsic viscosity, and reports on the pH dependency of the diffusion coefficient and radius of gyration. At pH = 2.4, the dissociated carboxyl group and nondissociated carboxyl group exist in a 1 : 3 ratio. Thus, the repulsion by the dissociated carboxyl groups and attraction by hydrogen bonding or hydrophobic interaction achieve an ideal balance, which leads to the observed behavior.

2.2.5 Preparation and Application of HA Gels

2.2.5.1 Preparation and application of HA gels by noncovalent bonding

2.2.5.1.1 Gel-like concentrated HA solutions

When a high molecular weight HA dissolves at 1%, the solution becomes gel-like and exhibits significant viscoelastic behavior. As previously stated, this is because HA molecules form temporary network structures in solution. Using this property, concentrated HA solutions are used in ophthalmic operations or as knee joint lubrication improving agents. For the former, the use of HA solutions was accelerated because of improvements in corneal preservation methods and the increased number of corneal transplant operations that followed the introduction of antibody inhibitors. Due to an increasing population of elderly people, the many internal lens insertion operations being performed also contributed to increased use. In relation to the latter, if a concentrated HA solution is administered to a deformed joint, for example, caused by rheumatoid arthritis, the friction coefficient decreases in relation to the amount of injected HA. This method is now known to be effective. This material is now clinically used for both eyes and joints.

2.2.5.1.2 Acid-treated HA gels

As already described here, an HA aqueous solution suddenly increases the viscosity at pH = 2.4. When this solution is in contact with either ethanol or an acetone aqueous solution, it dehydrates, hardens, and forms a transparent gel. The modulus of this gel can be adjusted by the concentration and molecular weight of HA, and the concentration and amount of the dehydrating organic solvent. Figure 7 depicts an example of an HA gel that was prepared in a 50 ml beaker. The gel on the left is perfectly

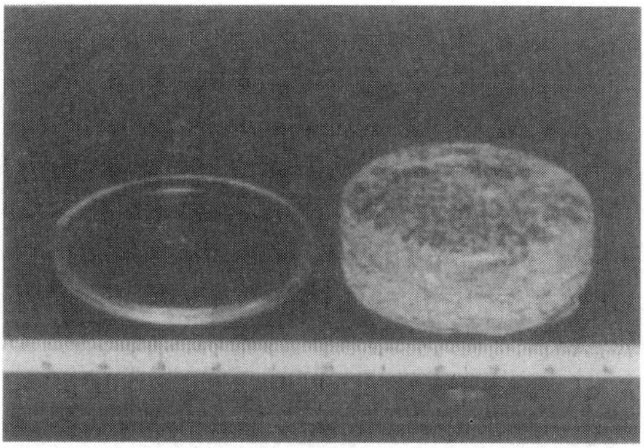

Fig. 7 Acid-treated HA gel prepared through contact with 80% ethanol aqueous solution.

transparent and the one on the right is a formed gel. A similar gel can be prepared by adding dehydrating organic solvent in a pH = 2.4 acidic buffer solution to a neutral HA aqueous solution. This method is quite useful for controlling the gel form. For example, spherical HA gels with radii of several tens to several hundreds of micrometers can be obtained by agitating a HA aqueous solution in a silicone oil at a very fast speed to form oily droplets.

Figure 8 shows the oil-containing microcapsule obtained with the method described here. In order to prepare this, an oil/water emulsion with HA aqueous solution in the outer shell is prepared. Then, using the aforementioned method, it is gelled in silicone oil and finally neutralized. If a material that is easily oxidized such as vitamin A is used as the oil-containing inner material, it is useful for stabilization. Another application involves use of a hard capsule that contains water-soluble drugs. For preparation, an ethanol aqueous solution at pH = 2.4 and with a cationic polymer is placed in contact with the HA aqueous solution. Then a water insoluble polyelectrolyte complex is formed at the interface. This technique is useful for the controlled release of the enclosed drug. Its release rate can be controlled by the particle size of the capsule and the membrane thickness of the polyelectrolytes. It is known that a polyelectrolyte complex of Oidrugit® and HA releases the drug according to the law of diffusion.

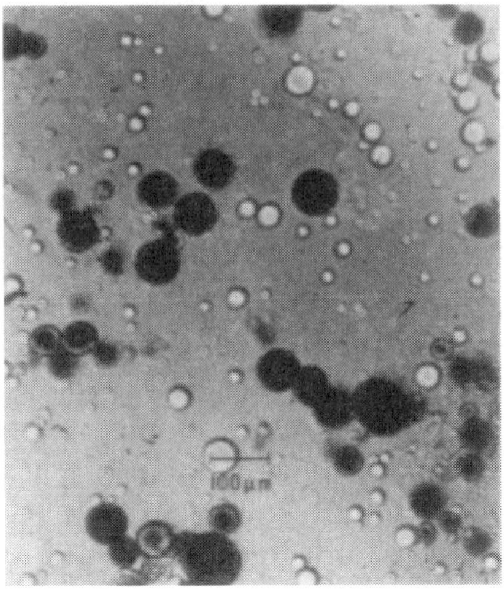

Fig. 8 Oil-containing microcapsule in which an O/W-type emulsion is fixed by HA gel [62] (the white droplets are made of a silicon oil, which is used as the suspension agent).

2.2.5.2 Preparation and application of HA gels made of covalent bonds

2.2.5.2.1 Gels crosslinked among hydroxyl groups

To obtain HA gels, that are crosslinked among hydroxyl groups, it is necessary to crosslink under some basic conditions. As crosslinking agents, 1,2,3,4-diepoxybutane, formaldehyde, dimethylol urea, dimethylolethylene urea, ethylene oxide, polyaziridine, polyisocyanate, and divinyl sulfone can be used. The properties of the obtained gels are determined by the type and amount of the crosslinking agent and the molecular weight and concentration of HA. Regardless of the properties, however, high biocompatibility will be maintained. In particular, Balazs *et al.* [51] have extensively studied HA gels that use divinyl sulfone as a crosslinker. They evaluated the applications of these gels for cosmetic materials such as moisturizers and medical drugs and devices. Research by the group at Seikagaku Inc. aims for similar water insoluble but swelling gels by using epichlorohydrin as a crosslinking agent.

Among similar approaches to crosslinking of hydroxyl groups, Yui *et al.* studied HA gels crosslinked by polyglycidyl ether for use as inflammatory response gels [70]. This approach uses the fact that HA decomposes in active oxygen, which is produced during inflammation. By administering the drug encapsulated in HA (one of the so-called intelligent drugs) it can be released as needed—when there is pain from inflammation. Figure 9 displays the results of *in vitro* tests. By forcibly generating hydroxy radicals using a divalent iron ion and hydroxyperoxide, crosslinked HA can be decomposed. This has also been confirmed by an *in vitro* test that used an embedded intelligent drug on the back of a mouse. Since HA gels will not decompose in a natural body under normal (noninflammatory) circumstances, this type of delivery system is perfect for delivery of drugs only as needed.

2.2.5.2.2 Gels crosslinked among carboxyl groups

When HA is kept with a crosslinking agent that is epoxydized at both chain ends at 50–90°C with quarternary ammonium salts, phosphorus salts, or imidazole compounds, a film that is crosslinked at the carboxyl groups can be obtained. This film swells slightly in water and becomes a gelled film. The decomposition rate of HA in the presence of hyaluronidase enzyme depends on the degree of crosslinking. Figure 10 shows an example where it can be seen that the higher the degree of crosslinking the slower the decomposition rate. A similar result has been observed during *in vivo* tests. Such a property has also been observed with the aforemen-

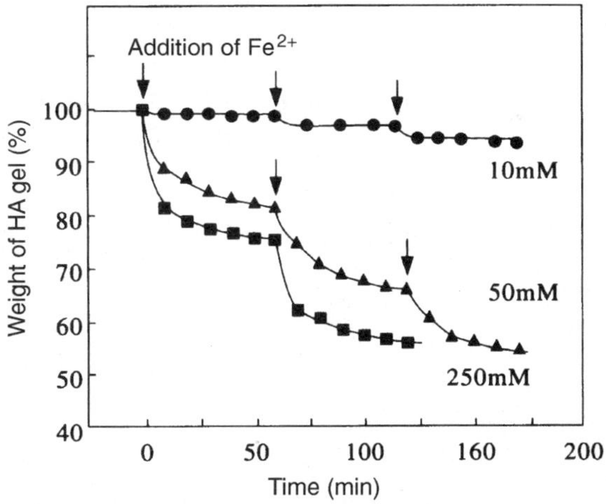

Fig. 9 Weight reduction of crosslinked HA gel accompanying the formation of hydroxy radicals [70] (free radicals are produced by 5 mM H_2O_2 and Fe^{2+} at various concentrations).

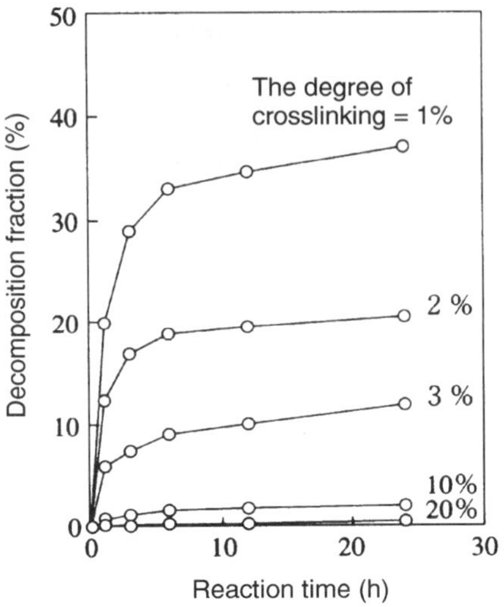

Fig. 10 Decomposition of crosslinked HA by hyaluronidase (50°C, pH6).

tioned hydroxyl-crosslinked HA gels. However, this tendency is more apparent in carboxyl-crosslinked HA gels.

Since a carboxyl-crosslinked HA film has a low degree of swelling and its rate of decomposition by enzymes can be controlled by the degree of crosslinking, applications for healing of surgical wounds and incisions have been evaluated. Organ adhesion after an operation can cause catastrophic damage. These HA gel films, which exhibit superior biocompatibility, bioabsorption, wound healing acceleration effects, and prevention of cell adhesion are extremely promising materials for prevention of adhesions. Powder from HA film has been evaluated as a lung cancer curing agent due to its ability to block blood supply. This treatment, which reduces blood supply to cancerous cells, is being clinically used. As this treatment allows administration of a small amount of drug at the same time blood supply is blocked, it is an effective treatment method. However, if the blood supply is halted for a long time, healthy cells will also die. Thus, it is designed to decompose in about half a day in the body by adjusting the degree of crosslinking.

2.2.5.2.3 Photocrosslinked gels

Cinnamic acid and thymine are known to dimerize upon irradiation with ultraviolet (UV) radiation. Miwa *et al.* [72] developed a technique by

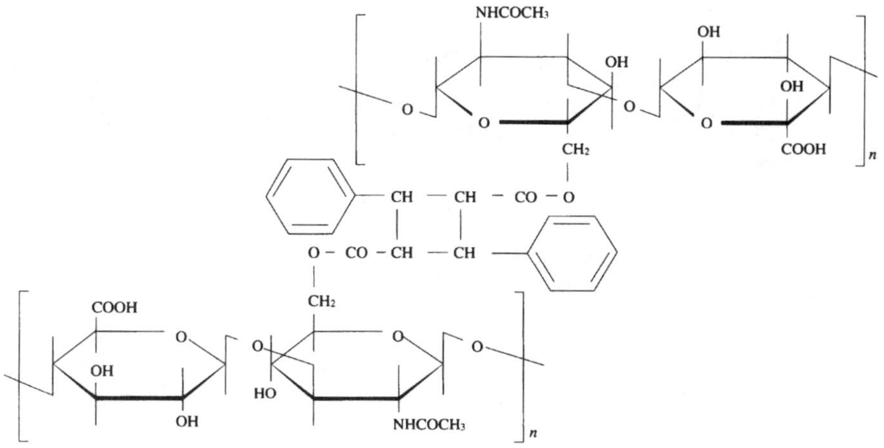

Fig. 11 The structure of the HA chain when it is crosslinked by ultraviolet radiation [41].

introducing cinnamic acid into HA and used this property to crosslink HA by UV radiation (see Fig. 11). They considered applying this technique to adhesion prevention, performed during *in vivo* tests on mice, and demonstrated that photocrosslinked HA shows release and biodegradation characteristics. They also confirmed that the developed material is useful in prevention of adhesions. Since the material can be crosslinked by UV radiation, it is hoped that this technique will someday be useful for other than just adhesion prevention.

2.2.5.3 The other HA related gels

Research on HA derivatives is currently very active. In particular, researchers at Fidia have been working diligently on ester derivatives. They have already developed a heparin-like blood-clotting agent using a sulfonic ester of HA benzylester. Among these, there are now materials that gel by absorbing bodily fluids. These materials have become commercially available as wound healing agents. This company has also been developing biodegradable microcapsules and surface coverings for artificial organs.

REFERENCES

1 Tokkyo Kaiho (1990). 145602.
2 WO9104278.

3 Tokkyo Kaiho (1988). 37143.
4 Tokkyo Kaiho (1993). 105701.
5 EP0459733.
6 Tokkyo Kaiho (1992). 30961.
7 Tokkyo Kaiho (1994). 322358.
8 USP5284936.
9 Tokkyo Kaiho (1983). 1746.
10 Tokkyo Kaiho (1990). 24565.
11 Tokkyo Kaiho (1984). 82864.
12 Tokkyo Kaiho (1992). 41522.
13 Tokkyo Kaiho (1995). 292023.
14 Tokkyo Kaiho (1983). 2312.
15 Tokkyo Kaiho (1986). 36309.
16 Tokkyo Kaiho (1996). 92020.
17 USP5130391.
18 Tokkyo Kaiho (1991). 81310.
19 USP5225506.
20 Tokkyo Kaiho (1995). 242713.
21 Tokkyo Kaiho (1996). 276225.
22 Tokkyo Kaiho (1996). 20640.
23 Tokkyo Kaiho (1994). 248073.
24 Tokkyo Kaiho (1987). 54751.
25 Tokkyo Kaiho (1996). 57311.
26 Tokkyo Kaiho (1982). 44627.
27 Tokkyo Kaiho (1983). 180233.
28 Tokkyo Kaiho (1984). 189103.
29 EP514724.
30 EP536128.
31 WO9305080.
32 Morris, E.R., Rees, D.A., and Welsh, E.J. (1980). *J. Mol. Biol.*, **138**: 383.
33 Heatley, F., and Scott, J.E. (1988). *Biochem. J.*, **254**: 489.
34 Toffanin, R., Kvan, B.J., A. Flaibani, A., Atzori, M., Biviano, F., and Paoletti, S. (1993). *Carbohydr. Res.*, **245**: 113.
35 Scott, J.E., Heatley, F., and Hull, W.E. (1984). *Biochem. J.*, **220**: 197.
36 Atkins, E.D., Meader, D., and Scott, J.E. (1980). *Int. J. Biol. Macromol.*, **2**: 318.
37 Meyer, K. (1947). *Physiol. Rev.*, **27**: 335.
38 Kokai Tokkyo Koho (1983). Showa 58-56692.
39 Akasaka, H. *et al.* (1988). *J. Soc. Cosmet. Chem., Jpn.*, **22**: 35.
40 Akasaka, H., and Yamaguchi, T. (1986). *Fragrance J.*, **78**: 42.
41 Seastonem, C.V. (1939). *J. Expt. Med.* **70**: 361.
42 For example, van Holde, K.E. (1971). *Physical Biochemistry*, Tokyo: Prentice-Hall, p. 148.
43 Norisue, T. (1982). *Kobunshi*, **31**: 338.
44 Yanaki, T., and Yamaguchi, M. (1994). *Chem. Pharm. Bull.*, **42**: 1651.
45 Bothner, H., Waaler, T., and Wik, O. (1988). *Int. J. Biol. Macromol.*, **10**: 287.
46 Laurent, T.C., Ryan, M., and Pietruskiewicz, A. (1960). *Biochem. Biophys. Acta*, **42**: 476.
47 Fouissac, E., Milas, M., Rinaud, M., and Borsali, R. (1992). *Macromolecules*, **25**: 5613.
48 Cleland, R.L., and Wang, J.L. (1970). *Biopolymers*, **9**: 799.

49 Cleland, R.L. (1984). *Biopolymers*, **23**: 647.
50 Shimada, E., and Matsumura, G. (1975). *J. Biochem.*, **78**: 513.
51 Balazs, E.A., Briller, S.O., and Delinger, J.L. (1981). *Seminars in Arthritis and Rheumatism*, New York: Grune & Stratton, p. 141.
52 Cleland, R.L. (1977). *Arch. Biochem. Biophys.*, **180**: 57.
53 Trim, H.H., and Jennings, B.R. (1983). *Biochem. J.*, **213**: 671.
54 Yanaki, T. (1996). *Colloid Science III*, Chem. Soc., Jpn., ed., Tokyo: Tokyo Kagaku Dojin, p. 198.
55 Morris, E.R., Rees, D.A., and Welsh, E.J. (1980). *J. Mol. Biol.* **138**: 383.
56 Reed, C.E., Li, X., and Reed, W.F. (1989). *Biopolymers*, **28**: 1981.
57 Welsh, E.J., Rees, D.A., Morris, E.R., and Madden, J.K. (1980). *J. Mol. Biol.*, **138**: 375.
58 Fouissac, E., Milas, M., and Rinaudo, M. (1993). *Macromolecules*, **26**: 6945.
59 Gibbs, D.A., Merrill, E.W., Smith, K.A., and Balazs, E.A. (1968). *Biopolymers*, **6**: 777.
60 Yanaki, T., and Yamaguchi, T. (1980). *Biopolymers*, **30**: 415.
61 Coleman, B.D., and Markovitz, H. (1964). *J. Appl. Phys.*, **35**: 1.
62 Matsuzaki, F., Yanaki, T., and Yamaguchi, M. (1995). *Industrial Biotechnological Polymers*, Lancaster, Pennsylvania: Technomic, p. 159.
63 Ropes, M.W., Robertson, W.B., Rossmeisl, E.C., Peabody, R.B., and Bauer, W. (1947). *Acta Med. Scand., Suppl.*, **196**: 700.
64 Pigman, W., Hawkins, W., Gramling, E., Rizvi, S., and Holley, H.L. (1960). *Arch. Biochem. Biophys*, **89**: 184.
65 Balazs, E.A. (1966). *Fed. Proc.*, **25**: 1817.
66 Kohara, T. *et al.* (1998). *New Opthalmol. Jpn.* **10**: 1251.
67 Mabuchi, K. *et al.* (1994). *J. Biomed. Mat. Res.* **28**: 865.
68 Tokkyo Koho (1992). Heisei 4-30961.
69 Kokai Tokkyo Koho (1986). Showa 61-12701.
70 Yui, N. *et al.* (1993). *Polym. Preprints, Jap.*, **42**: 3186.
71 Patent Application Number (1994). Heisei 6-341157.
72 Kokai Tokkyo Koho (1995). Heisei 7-97401.
73 Miwa, H. *et al.* (1993). *Jinko Zoki*, **22**: 376.
74 Kokai Tokkyo Koho (1988). Showa 63-105003.

Section 2

Sustained Release (Water Absorption)—Drug Delivery System

YUZO KANEKO, KIYOTAKA SAKAI, AND MITSUO OKANO

2.1 APPLICATION OF HYDROGELS IN DDS

During illness, homeostasis is often a casualty of the illness. Drugs are often useful for restoring homeostasis. Drugs administered orally or by injection are usually carried to organs by the bloodstream and perform their job, then they are metabolized by the body and eliminated. Drugs function most efficiently at the locus of illness at a selectively high concentration. Accordingly, if both concentration and time at the target location can be controlled for the drug, it is expected to work more efficaciously and with fewer side effects. In recent years, high-potency drugs such as peptides (developed by gene splicing) have been produced. Thus, it is a pressing and important issue to develop techniques to effectively and selectively administer such powerful drugs to a targeted area. It is not possible to maintain drug concentration in the bloodstream for a long period of time via oral or injectable protocols. Moreover, the possibility of a drug affecting something other than the problem for which it was administered is a shortcoming. As a solution to this problem, targeted drug delivery systems (DDS) are now attracting much attention.

In this way, a drug can be administered effectively to a desired body location with minimal side effects.

Research on DDS began with the development of sustained release drugs that are able to maintain drug concentration in the body within the effective (desired) range for an extended period of time. In particular, techniques to release drugs at a constant rate from the matrix (0-th order release) have been evaluated using the shape of the drug [1], diffusion control by permeable membranes [2], and control of osmotic pressure [3]. As already described, studies on effective delivery of drugs to areas of illness (targeting), and control of drug delivery time whenever necessary (pulse delivery system) [4] have become very important. Targeting of delivery involves both monoclonal antigens and passive targeting, which avoids fine internal veins [4]. In the group of time controlled drug delivery systems, there is an insulin delivery system that responds to blood sugar levels and a drug delivery system that responds to external stimuli that include heat [5], electric fields [6], magnetic fields [7] and pH [8]. Intelligent drugs [4, 9] detect the signal generated by an illness, judge the drug amount to be released based on the degree of illness, and deliver the drug (Fig. 1). A double targeting system is a combination of spatial and time control systems, which delivers the collected drugs to the target using pulse drug delivery (on-off delivery) and spatial control. Development of such systems will produce revolutionary cures.

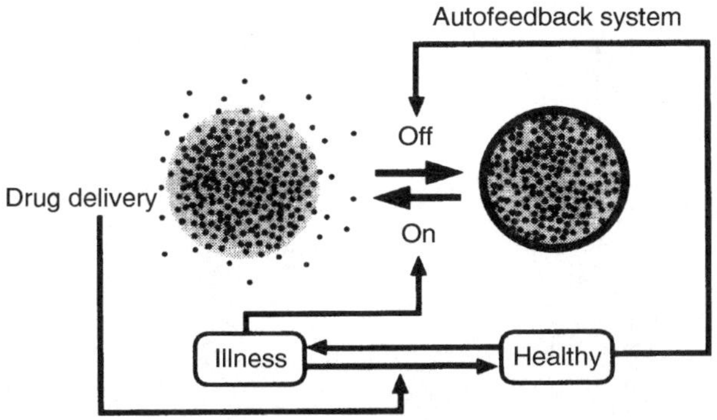

The drug itself detects a signal caused by illness and delivers the drug. Upon recovery, the drug actually detects this change and stops delivery.

Fig. 1 The concept of an intelligent drug.

Drug delivery that employs hydrogels is important because drug delivery systems that are able to control both concentration and distribution of the drug in the body with the aid of stimuli-responsive polymers are very important. In addition to quantitative control of drug delivery, studies related to timing control techniques, kinetic analysis based on the transfer phenomenon, on-off control of drug delivery, and autofeedback drug systems have been undertaken. In this section, drug delivery systems that utilize structural changes in stimuli-responsive polymer gels will be described. In particular, while dynamic mechanical analysis for the swelling and shrinking of gels and the mechanisms will be evaluated in detail, the effect on drug delivery from the viewpoint of rate control is also discussed here.

2.2 SWELLING AND SHRINKING OF POLYMER GELS

2.2.1 Swelling Mechanisms of Gels

In general, the process of a dried, glassy gel swelling after absorbing a solvent can be expressed by the three consecutive procedures illustrated in Fig. 2. They include step 1, the diffusion of solvent molecules into polymer networks; step 2, relaxation of polymer chains from a glassy

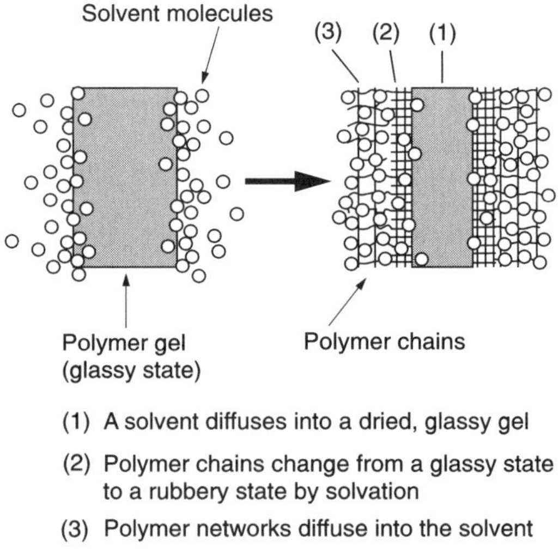

Fig. 2 The swelling process of polymer gels.

state to a rubbery state by solvation; and step 3, polymer networks diffuse into the solvent. Depending on which process is the limiting step, the swelling behavior of gels can be classified. When step 1 is the limiting factor, the amount of solvent absorbed by the gel is small and polymer chain relaxation seldom takes place. A further possibility occurs when the polymer chain relaxes quickly and in this way the swelling of the gel is controlled by the diffusion of solvent molecules into the networks (Fickian or case-I diffusion). When the relaxation of polymer chains contributes, that is, when step 2 is the limiting factor, the swelling behavior deviates from the Fickian diffusion (non-Fickian or anomalous diffusion). Finally, when step 3 is the limiting process, the swelling of the gel is controlled by the diffusion of polymer chains in the solvent. In this case, the swelling behavior of the gel is explained by cooperative diffusion [10].

As an index for the swelling of gels as Fickian or non-Fickian diffusion, Vrentas *et al.* [11] proposed the use of the Deborah number, $\mathrm{DEB} = \lambda/\theta$, where λ is the stress relaxation time and θ the diffusion time of the solvent. The values of these quantities vary as the gel absorbs solvent. If the DEB change is small during the absorption process, the average value can be regarded as the intrinsic value for the system. However, if the DEB value variation is large, then the smallest and largest values are used. In the beginning and end of the process, when DEB $\gg 1$, the larger the value the smaller the absorption of the solvent. In this case, the structural changes of the polymer can be regarded as rather small. On the other hand, when DEB $\ll 1$ at the beginning and end of the process, relaxation takes place quickly. In any case, relaxation of the polymer chains can be neglected and the solvent absorption of gels is controlled by the diffusion of the solvent.

When DEB is ≈ 1, the polymer chains do not relax immediately or, even if the chain relaxes immediately, polymer conformation is different from the equilibrium state. As indicated in Fig. 3, when DEB is ~ 1 in zone II, the swelling behavior is different from Fickian diffusion due to the influence of the relaxation of polymer chains, which is unlike those cases where DEB $\gg 1$ (Zone I) or DEB $\ll 1$ (Zone III).

2.2.2 Analysis of the Swelling Behavior of Gels

In the aforementioned step 1, the solvent diffuses into polymer networks (case-I diffusion). At this time, the solvent absorption behavior of the gel can be obtained by solving the diffusion equation (Fick's second law). In

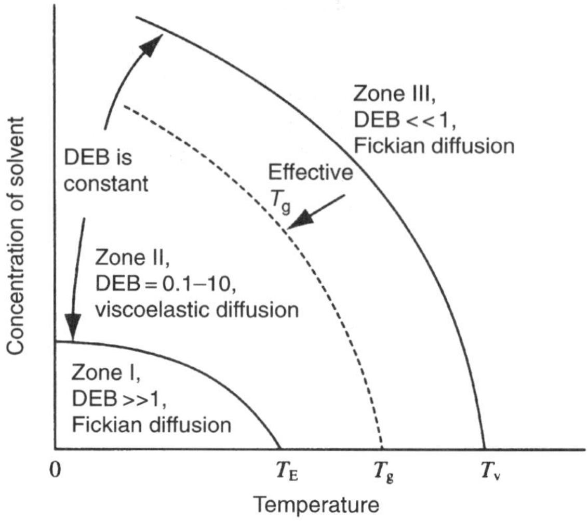

The concentration of the solvent in the gel is on the ordinate, the T_E on the abscissa is the temperature below which the polymer can be regarded as an elastic body, T_g is the glass transition temperature of the polymer, and T_v is the temperature below which the polymer acts as a viscous fluid.

Fig 3. Classification of the solvent absorption behavior of gels [11].

the case of a flat film, the amount of solvent absorbed increases in proportion to the $\frac{1}{2}$th power at the beginning as shown in the following equation [12]:

$$\frac{M_{st}}{M_{s\infty}} = 4\left(\frac{D_s t}{\pi l^2}\right)^{1/2} \left(0 \leq \frac{M_{st}}{M_{s\infty}} \leq 0.6\right)$$

where M_{st}, $M_{s\infty}$ are the total amount of solvent absorbed at time t and infinite time, respectively, l is the thickness of the film, and D_s is the diffusion coefficient of the solvent in the polymer. A different situation is observed in step 2, that is, relaxation of polymer chain influences (anomalous diffusion), in particular when the rate of polymer relaxation is slow in comparison to the rate of permeation of the solvent. In this case, there will be a clear separation into two layers between the swollen portion near the surface and the glassy, non-swollen part and with the relaxing region as the borderline. As shown in Fig. 4, the interphase between the two regions is called the swelling front. This front propagates into the gel

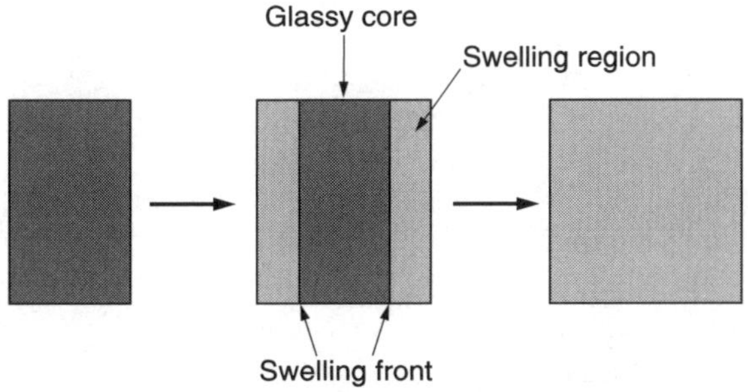

Fig. 4 The swelling process of gels by case-II transport.

as the solvent is absorbed (case-II diffusion) [13]. Under this condition, the swelling front propagates to the interior of the gel at a constant rate.

The amount of solvent absorbed by the gels of various shapes can be expressed by the following equation:

$$\frac{M_{\mathrm{st}}}{M_{\mathrm{s}\infty}} = 1 - \left(1 - \frac{k_0 t}{C_{\mathrm{s}} a}\right)^N$$

$$(\text{flat film: } N = 1, \ \text{rod: } N = 2, \text{ and sphere: } N = 3)$$

where k_0 is the relaxation constant, C_s is the concentration of solvent in the swollen region, and a is half the thickness of the flat film, or radius of a rod or sphere. In particular, for a flat film gel ($N = 1$), the amount of solvent absorbed proportionately increases with time. This case-II diffusion solvent absorption has been ordered in various solvent-polymer systems, such as polystyrene in hexane [14] and poly(methyl methacrylate) in methanol [15]. There are cases in the latter stage of case-II diffusion in which there is a sudden increase in the amount of absorption due to the change in film thickness or temperature. Hopfenberg observed this phenomenon in the absorption of *n*-hexane by polystyrene by changing the temperature or thickness of the film [14] (see Fig. 5). This accelerating phenomenon was defined as super case-II diffusion [16]. Siegel *et al.* [17] also found this kind of phenomenon in the absorption of water by poly(methyl methacrylate-co-N,N′-dimethyl ethyl methacrylate) and explained it using the swelling front model (see Fig. 6). There are

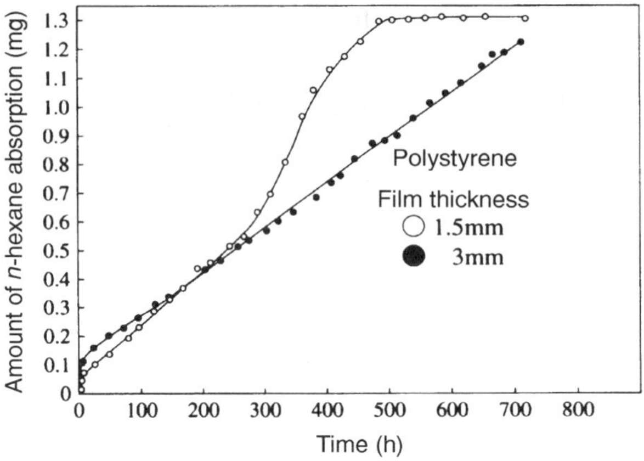

In the film with thickness of 1.5 mm, super case-II diffusion is observed.

Fig. 5 The absorption of *n*-hexane by polystyrene with film thicknesses of 1.5 mm and 3 mm.

several explanations: (1) a dried gel is in contact with a solvent; (2) as water is gradually absorbed, the swelling front propagates from both surfaces and the glassy core inhibits the swelling towards the radial direction; (3) the swelling fronts meet at the center of the film and the core disappears; and (4) the inhibiting effect disappears and the gel swells, resulting in both an increased rate of water absorption and achievement of

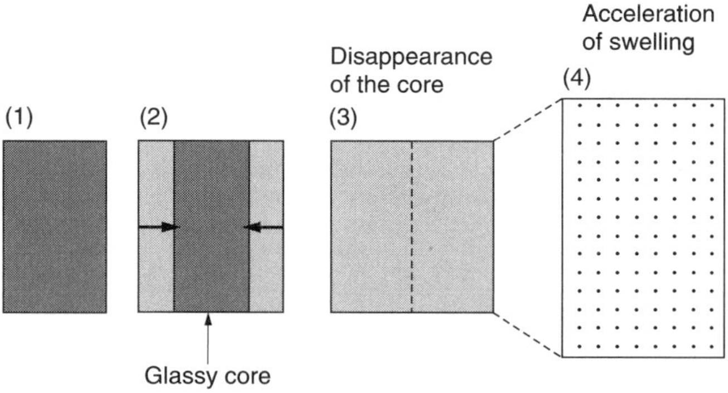

(1) The solvent diffuses into a dried gel; (2) the swelling fronts propagate into the inner part of the gel as it absorbs water; (3) the core disappears; and (4) the gel swells and the water absorption rate increases.

Fig. 6 Swelling front model [17].

equilibrium. In this case, the swelling exhibits sigmoidal behavior, which accelerates in the middle.

The author and others synthesized and measured the water absorption behavior of a gel from a copolymer of N-isopropylacrylamide (NIPAAm) and butylmethacrylate (BMA), which changes its degree of swelling significantly upon temperature change [18]. Since the equilibrium degree of swelling can be changed by temperature, the different behavior shown in the preceding explanation (1)–(4) can be generated only by changing the temperature without changing the gel structure. Gel with 5% BMA shows a phase transition around 25°C in a phosphate buffered saline solution (PBS) at a pH of 7.4. Above this temperature the gel expels water and shrinks and paradoxically, above this temperature the gel also can absorb water and swell significantly. At lower temperatures, the degree of swelling is larger. The degree of swelling of a disk-like gel from a dried gel is proportional to the $\frac{1}{2}$th power of time above the phase transition temperature. This behavior agrees with the diffusion theory. This is because hydration did not take place and thus water absorption was controlled only by the diffusion of water into the networks. On the other hand, when the gel was swollen at 20°C, which is below the phase transition temperature, the degree of swelling increased in proportion to the immersion time. This is because the water absorption followed case-II diffusion (see Fig. 7(b)). Furthermore, at 10°C, where the equilibrium swelling is greater, the swelling behavior is sigmoidal and accelerates in the middle of the diffusion process (see Fig. 8(b)). This is a kind of super case-II diffusion. When a gel accelerates diffusion, towards the radial direction the dimension increases and reduces in the thickness direction. Therefore, based on restriction by the glassy core, as is the case in Siegel's example, swelling behavior can be explained.

The swelling behavior of step 3 is caused by the cooperative diffusion of polymer networks.

Cooperative diffusion has been studied experimentally by Matsuo and Tanaka [10]. Traditionally, the swelling of gels was thought to be determined by solvent diffusion. On the contrary, they analyzed swelling based on the diffusion of polymer networks in the solvent. If the radius of the spherical gel is $\Delta R(t)$, its total radius change ΔR_0, the cooperative diffusion coefficient of the networks D_c, and the equilibrium

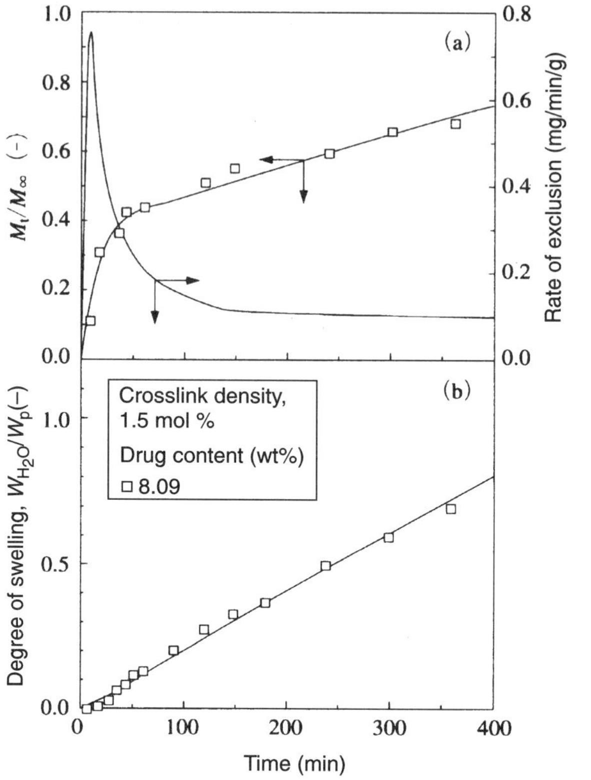

Fig. 7 (a) Exclusion pattern of indomethacin, and (b) swelling behavior of the NIPAAm-BMA copolymer gel at 20 °C.

radius R, radius changes that accompany swelling can be expressed as follows:

$$\Delta R(t) = \frac{6\Delta R_0}{\pi^2}\sum_{n=1}^{\infty}\frac{1}{n^2}\exp\left(-\frac{n^2\pi^2 D_c}{R^2}t\right)$$
$$\approx \frac{6\Delta R_0}{\pi^2}\exp\left(-\frac{\pi^2 D_c}{R^2}t\right)$$

As can be seen, the gel size changes exponentially and the cooperative diffusion coefficient of gels D_c can be expressed by the ratio ($D_c = E/f$) of the modulus of the networks E, and the friction coefficient of the solvent and networks by f [10]. In gels, cooperative movement of the networks is generated by crosslinking and the elasticity of the polymer

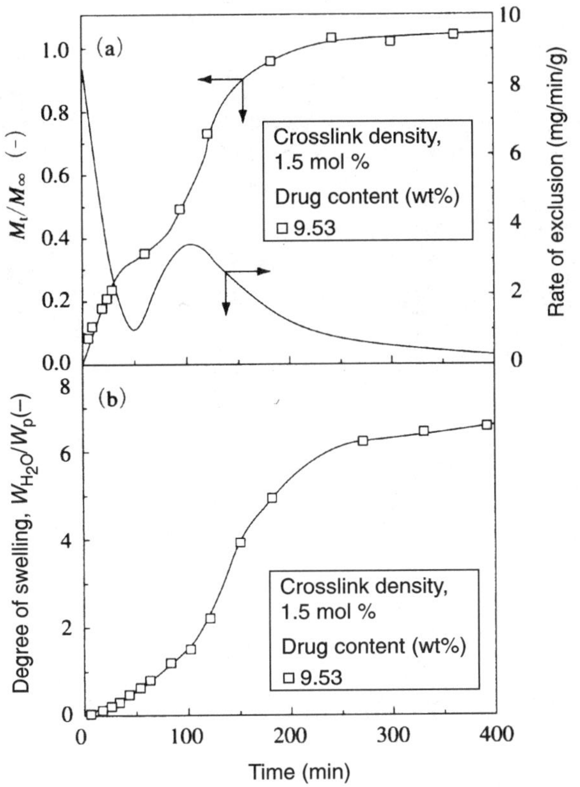

Fig. 8 (a) Exclusion pattern of indomethacin, and (b) swelling behavior of the NIPAAm-BMA copolymer gel at 10°C.

chains contributes to this movement. Accordingly, this diffusion differs from ordinary diffusion of molecules and is regarded as a unique phenomenon of gels.

2.2.3 Shrinking Mechanism of Gels

When gels shrink, the structural changes caused by aggregation of polymer chains influence shrinking behavior in a manner similar to the influence of polymer chain relaxation on swelling behavior. Polymer aggregation can be understood as the phase transition via desolvation from the swelling to the shrinking state. Gels shrink at the surface in response to changes in the external environment. When a crust is formed by the shrinking polymers at the surface of the gel, the shrinking process slows. This process is shown in Fig. 9. In Step 1 the shrinking layer (skin layer) contains very little solvent and it suppresses permeation of solvent from the interior of the gel, thus causing reduced shrinking of the gel as the process continues. In step 2 pressure gradually builds in the interior of the gel, leading to bubble formation near the gel surface. As a result, it is seen in step 3 that the solvent in the gel interior flows out because of pressure building up and the gel then continues to shrink. Accordingly, when polymer chains undergo phase transition and aggregation, the dynamics of a gel are strongly influenced by skin layer formation. On the other hand, when gels proceed from a swollen state to a smaller equilibrium state without the accompanying phase transition of polymer chains, the shrinking behavior of gels can be explained by the cooperative diffusion of polymer chains.

The author and others measured and analyzed the shrinking behavior of disk-shaped NIPAAm-BMA copolymers at various temperatures [19]. In a manner similar to swelling behavior, shrinking behavior showed marked differences above and below the phase transition temperature. Based on the cooperative diffusion theory of gels, the shrinking behavior of the gel was analyzed using the diffusion equation

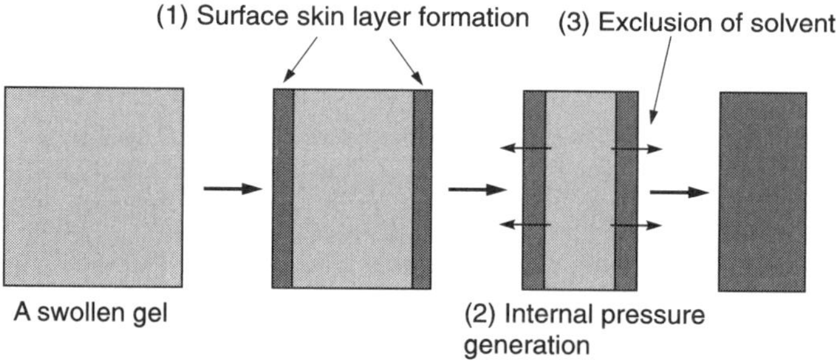

(1) The gel shrinks from the surface and the shrinking layer (skin layer) is formed at the surface of the gel. Solvent permeation through the skin layer is low.

(2) Following gel shrinkages, internal pressure builds.

(3) Due to the internal pressure, the solvent in the interior of the gel is pushed out and thus the gel continues to shrink.

Fig. 9 Shrinking process of polymer gels.

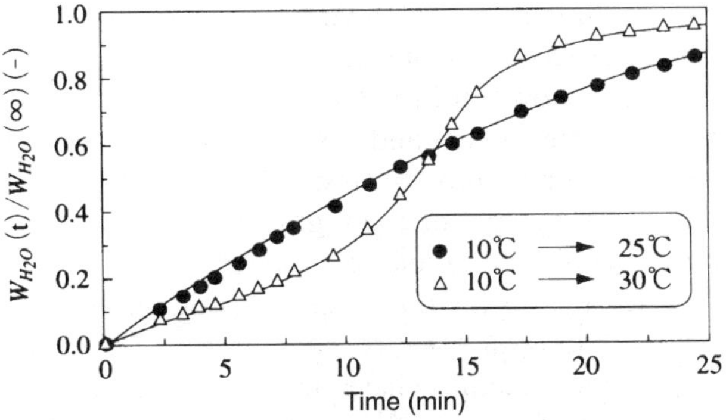

At 25°C, the exclusion of solvent is proportional to $t^{1/2}$.
At 30°C, the amount of water excluded is seen to slow down at the beginning but it later accelerates.

Fig. 10 The change in the amount of water excluded by a NIPAAm-BMA copolymer gel, that was swollen at 10 °C and shrunken from 25 °C to 30 °C.

(Fick's second law of diffusion). Figure 10 shows the amount of water that is excluded from the gel interior upon shrinking versus $t^{1/2}$. When the amount of water excluded from the gel increases in proportion to $t^{1/2}$ (ordinate values are from 0 to 0.6), the shrinking behavior of the gel is controlled by the cooperative diffusion of polymer networks. When the gel that is swollen at 10°C has shrunk at 25°C, which is below the phase transition temperature, the gel is seen to have shrunk during the cooperative diffusion dominated mode. On the other hand, when it was observed to have shrunk at 30°C, which is above the phase transition temperature, water exclusion was first slowed and later it was accelerated. This is thought to be caused by the retardation of shrinking by both the skin layer that formed at the beginning and the internal pressure build-up in the later stage, which increases the water exclusion.

When gel thickness is increased, the internal pressure increases as the skin layer forms. This is due to the increased surface area in comparison to the volume of the gel. When internal pressure increases, a sudden exclusion of water takes place. At this time, the skin layer experiences strong pressure and the gel structure is destroyed, which often leads to the formation of cracks. In this case, the swelling process was irreversible [19].

For those gels in which it is desired to move them back and forth between the swollen and shrunken stages, the use of porous sponge-like gels has been studied [20, 21]. In this case, the surface area of the gel is large and solvent exclusion takes place easily. As a result, even if the gel is large, fast shrinking is observed. This is the best method to prevent build-up of the internal pressure that is caused by the surface layers. The shrinking behavior of such gels follows a cooperative diffusion mechanism. The authors studied the rapid shrinking behavior using a new method [22, 23] involving the synthesis of a graft-type gel structure with a temperature-responsive NIPAAm. As the one end of the grafted chain is free, the motion of these chains is faster than the cooperative motion of the polymer chains with both ends crosslinked. Unlike the ordinary three-dimensional (3D) gels, a gel that has different molecular architecture but the same chemical composition and crosslink density was also synthesized (see Fig. 11). When the shrinking behaviors in response to temperature change are compared, the motion of the networks of the ordinary 3D gels indeed followed cooperative diffusion and the size changed slowly. In contrast, the free grafted chains are independent from the networks. Thus, the gel excluded water suddenly and the phase transition took place in response to the temperature change. These grafted chains lowered the phase transition temperature of the network main chains. At the same time, as the hydrophobic interaction among network chains increased, the cohesive force within the gel increased. The gel did not form the skin

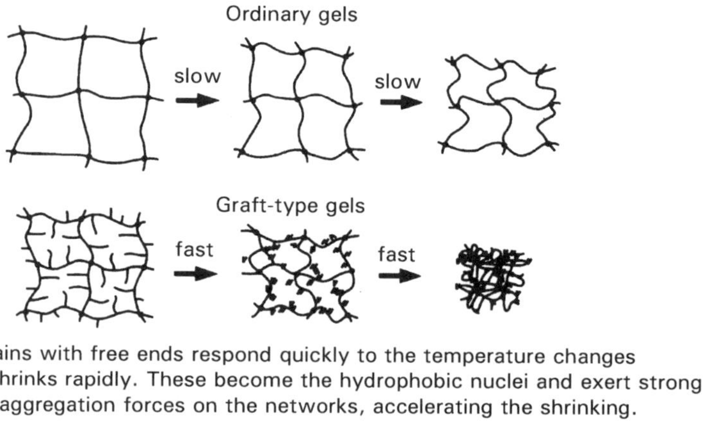

Fig. 11 The polymer gel made of NIPAAm with a graft-type molecular architecture and its rapid shrinking mechanism [22]

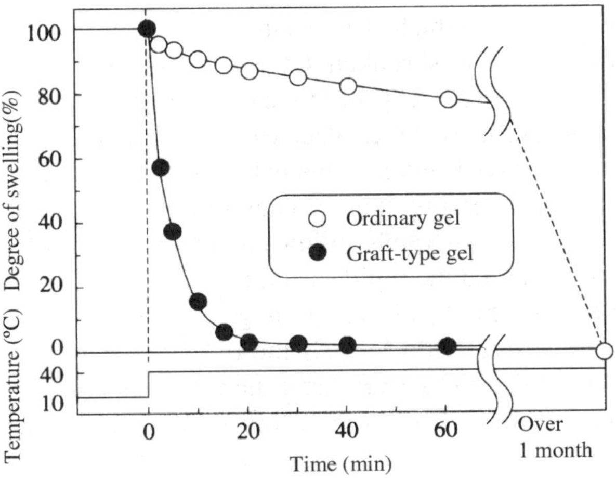

Fig. 12 Comparison of the shrinking behavior at 40°C of the ordinary and graft-type gels, which were swollen at 10°C.

layer. Thus, the internal water was pushed out and the gel shrank. The result is shown in Fig. 12. From this result, a new mechanism of gel shrinkage that is based on the elasticity of the phase transition of polymer chains has been proposed [22, 23]. This new mechanism is different from the cooperative diffusion of networks.

2.3 CHANGE OF SWELLING OF GELS AND ITS EFFECT ON DRUG DELIVERY

2.3.1 Free Volume Theory

Drug delivery that uses gels is strongly influenced by the degree of swelling of the gels. Therefore, when the rate of drug delivery is discussed, it is important to clarify the relationship between the degree of swelling and the diffusion of drugs. In this respect, Yasuda *et al.* proposed a free volume theory [24]. They applied the dissolution diffusion concept, which has been established for the diffusion phenomenon of polymer-gas systems in combination with the free volume concept [25] to swollen gel membranes. When a solute permeated a membrane, they considered the probability of finding a space (free volume) larger than the volume of the solute and correlated the diffusion phenomenon and free volume within the gel membrane. The diffusion of the solute within the membrane is explained as follows: (1) polymer chains and solvents are randomly mixed and there will be voids whose size and location are not specified; (2) the interaction of polymer chains with solvent or solute can be ignored; (3) solute can diffuse through the film only when there is a void larger than the solute molecule; and (4) the size of the free volume through which solute can diffuse is equivalent to the size of the free volume of the solvent in the film.

Based on these assumptions, the relationship between the diffusion coefficient and water content of the film can be expressed by the following equation [24]:

$$\frac{D_m}{D_w} = \Psi(q)\exp\left(-B\left(\frac{q}{V_w}\right)\left(\frac{1}{H} - 1\right)\right)$$

where q is the cross-sectional area of the solute, $\Psi(q)$ is the probability of finding the voids larger than q, B is a proportionality constant, and V_w is the free volume of water. As can be seen from this equation, the ratio between the diffusion coefficient of the solute in the swollen gel film D_m and in water D_w is proportional to the water content of the film H. Hence, the higher the water content of the gel (the degree of swelling), the greater the permeability of a material. Yasuda reported the diffusion coefficient of sodium chloride in the films of hydrogels of various polymers, cellulose, and acetate with water contents in the range 10–70%. As shown in Fig. 13, the permeation of sodium chloride in a swollen film followed the free volume theory [26].

2.3.2 Diffusion-Controlled Drug Delivery

If a gel does not show structural changes during drug delivery (change in the degree of swelling or degradation), then drug delivery behavior can be explained based solely on drug diffusion. This behavior is called diffusion-controlled drugs. Delivery of diffusion-controlled drugs depends on the gel structure used in the delivery device. These include reservoir devices in which a drug is coated with a polymer film, and monolithic devices in which a drug is dispersed into a polymer matrix. In reservoir-type devices the drugs are coated by polymer films, which is similar to microcapsules. The entrapped drug permeates the film slowly and is delivered. Various polymers, such as silicone rubber, ethylenevinyl acetate, polyurethane, and polyethylene can be used as polymer films. If the drug is delivered as a mixture of saturated solution and solid drug, the solution concentration is constantly saturated due to continuous dissolution of the solid drug into

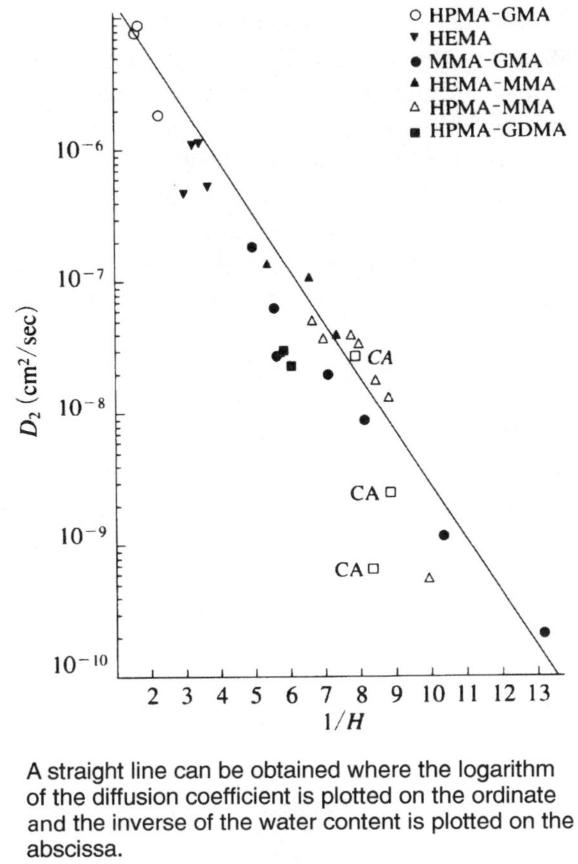

Fig. 13 Diffusion coefficient of sodium chloride in films with various water contents [26].

the solution. Hence, if the external concentration is zero, the delivery rate is constant regardless of the time (0th order delivery). However, if the drug concentration is less than the saturation, the drug concentration distribution changes and delivery rate is reduced. From Fick's first law of diffusion, the delivery rate from the device with film thickness l is expressed as follows:

$$\frac{dM_t}{dt} = \frac{ADKC}{l}$$

where M_t is the cumulative drug delivered until time t, A is the surface area of the device, K is the distribution coefficient, C is the concentration of the drug in solution, and D is the diffusion coefficient of the drug molecule in the polymer.

On the other hand, a monolithic device is made of a polymer matrix in which a drug is dispersed. Therefore, it is the preferred mode of drug delivery because it is easily manufactured and safe if the device is accidentally destroyed in the body. A monolithic device can be divided into a monolithic dispersion type, in which the initial drug concentration is greater than the saturation concentration, and a monolithic dissolution type in which the drug concentration is less than the saturation concentration. Drug delivery behaviors can be expressed by the Higuchi model [27], as shown in Fig. 14. The model assumes a quasisteady state expressed by the following assumptions: (i) that there are regions where the drug is dissolved and regions where the drug is dispersed after time t; (ii) as time proceeds, the interface between the two regions penetrates the material and the thickness of the dispersed zone decreases; (iii) the drug concentration distribution in the dissolved zone is linear; and (iv) the initial drug concentration C_0 is greater than the solubility of the drug in polymer C.

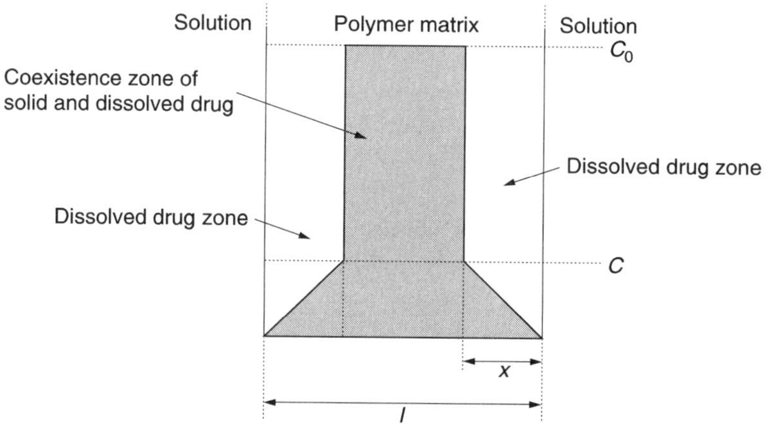

The dispersed drug in the matrix dissolves and diffuses to the surface to be delivered.

Fig. 14 The Higuchi model.

Accordingly, the following equation was obtained:

$$\begin{aligned} m_t &= A[DtC(2C_0 - C)]^{1/2} \\ &= A(2DtCC_0)^{1/2} \qquad (C_0 \gg C) \end{aligned}$$

The delivery rate is also expressed as

$$\frac{dM_t}{dt} = \frac{A}{2}\left(\frac{2DC_0C}{t}\right)^{1/2}$$

On the other hand, the drug delivery behavior of the monolithic dissolution type can be expressed by Fick's second law of diffusion. For a device with thickness t at both an early and later period of delivery the following equation is obtained [12]:

$$\frac{M_t}{M_\infty} = 4\left(\frac{Dt}{\pi l^2}\right)^{1/2} \qquad \left(0 \leq \frac{M_t}{M_\infty} \leq 0.6\right)$$

$$\frac{M_t}{M_\infty} = 1 \frac{8}{\pi^2}\exp\left(\frac{-\pi^2 Dt}{l^2}\right)$$

$$\left(0.4 \leq \frac{M_t}{M_\infty} \leq 1.0\right)$$

where M_∞ is the amount of cumulative drug to be delivered—namely the total amount of drug in the gel. Drug delivery from the monolithic dissolution type device is proportional to the $\frac{1}{2}$th power of the time at an early stage of delivery. At a later period, exponential behavior is observed due to the reduction of the drug concentration in the device.

2.3.3 Swelling-Controlled Drug Delivery

When a gel undergoes structural changes such as swelling or shrinking drug diffusivity changes as was previously explained in the free volume theory discussed in Section 2.3.1. Delivery characteristics also change as the gel swells. Hence, the drug delivery pattern will not be proportional to the $\frac{1}{2}$th power as it was with the diffusion-controlled type. The swelling-controlled type takes advantage of this property to design a more accurate, sustained drug delivery system. In particular, when a gel absorbs water as the drug is delivered, the delivery rate is a function of both water absorption and drug diffusion rate. If the water absorption rate is fast and drug diffusion becomes the limiting step, delivery behavior can be explained using the equations described in Section 2.3.2. On the other hand, when the swelling rate is much less than the diffusion rate, drug delivery follows the swelling behavior. As a parameter to judge whether the drug delivery is diffusion-controlled or swelling-controlled, Korsmeyer and Peppas proposed a swelling interface number (S_w) as follows [28]:

$$S_w = \frac{v\delta(t)}{D_i}$$

As shown in Fig. 15, v is the propagation rate of the interface between the rubbery state and glassy state assuming that they are clearly separated, $\delta(t)$ is the thickness of the swollen region at time t, and D_i is the diffusion coefficient of the drug in the swollen region.

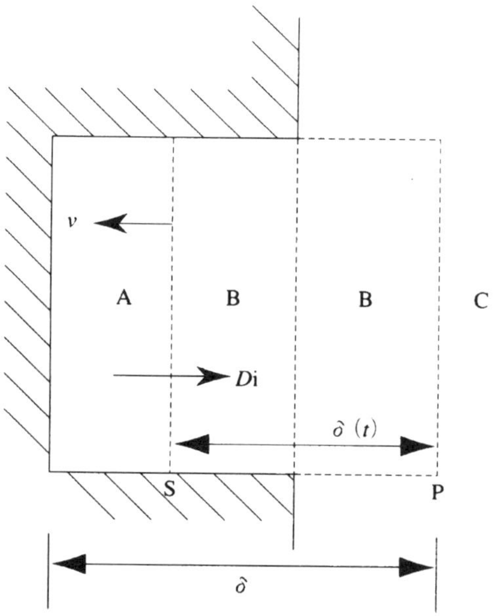

A: glassy region; B: rubbery region; C: solvent;
S: interphase; P: polymer-solvent interface.
The S advanced into the inner part of the gel
at a constant speed *v*.

Fig. 15 Drug delivery from a swelling-controlled polymer gel [28].

Segot-Chicq and Peppas and Korsmyer and Peppas investigated the relationship between solvent absorption and drug delivery of various swelling polymers such as copolymers of 2-hydroxyethyl mechacrylate and N-vinyl-2-pyrrolidone [15] and copolymers of ethylene and vinyl alcohol [29]. They quantified the drug delivery behavior with the following empirical equation:

$$\frac{M_t}{M_\infty} = kt^n$$

where k is a constant. As shown in Table 1, drug delivery behavior from the gel was classified by the corresponding Deborah number (DEB), which is the parameter used to evaluate the swelling behavior of gels and S_w. When the structural changes in the gel are small (DEB $\ll 1$ or $\gg 1$) and the advancement speed of the interface is much faster than the drug diffusion rate ($S_w \gg 1$), Fickian delivery with $n = 0.5$ results. In contrast, if the polymer chain relaxes (DEB is ≈ 1) and the drug diffusion rate in the swollen region is much faster than the advancement speed of the interface ($S_w \ll 1$), 0th delivery with $n = 1$ results (constant drug delivery rate). At this time, if S_w is ≈ 1, non-Fickian behavior with $n > 0.5$ will be observed. Accordingly, DEB and S_w are important parameters for evaluating the drug delivery behavior.

2.3.4 Changes of Drug Delivery Behavior Accompanying Swelling and Shrinking of Gels

The authors manufactured a monolithic device made of a NIPAAm-BMA copolymer gel with a film thickness of 0.5 mm and a diameter of 15 mm in which the nonsteroidal drug, indomethacin, was used as the model drug. The drug delivery behavior of this system was measured. After the initial delivery, 0th delivery was observed (Fig. 7(a)). The delivery rate accel-

Table 1 Classification of drug delivery behavior.

	$\frac{M_t}{M_\infty} = kt^n$	*DEB*	S_w
Fickian diffusion	$n = 0.5$	$\ll 1$ or $\gg 1$	$\gg 1$
Non-Fickian diffusion	$n > 0.5$	≈ 1	≈ 1
Case II transport	$n = 1$	≈ 1	$\ll 1$

DEB, S_w and drug delivery behavior (M_t/M_∞) are compared

erated at 10°C (Fig. 8(a)). The delivery at 20°C is due to case-II transport and the delivery behavior followed the swelling behavior. Because the degree of swelling at 10°C was 6–7 times greater than that at 20°C and super case-II transport was observed, drug delivery accelerated in the middle of the observation. Accordingly, the drug delivery pattern could be controlled by the swelling of the gel. There were also occasions where the hydrophilicity/hydrophobicity of the drug itself influenced the swelling behavior of the gel and influenced in a complex manner the drug delivery patterns. Swelling behavior and drug delivery are both quite complex and much has been reported on to date [30].

When a gel undergoes structural changes as it shrinks, drug delivery behavior will necessarily be influenced. The authors prepared disk-shaped NIPAAm-BMA copolymers and used a sodium salicylic acid aqueous solution at a low temperature for swelling them. They then measured the drug delivery rate by shrinking the gels above the phase transition temperature [19]. Skin layer formation and internal pressure build-up caused by gel shrinkage pushed out water. The drug inside the gel was also expelled along with this water. The drug delivery rate reduced once due to skin layer resistance. It then increased and exhibited a peak due to the internal pressure (see Fig. 16). The magnitude of this peak, the time to reach the peak, and the number of the peak can be controlled by

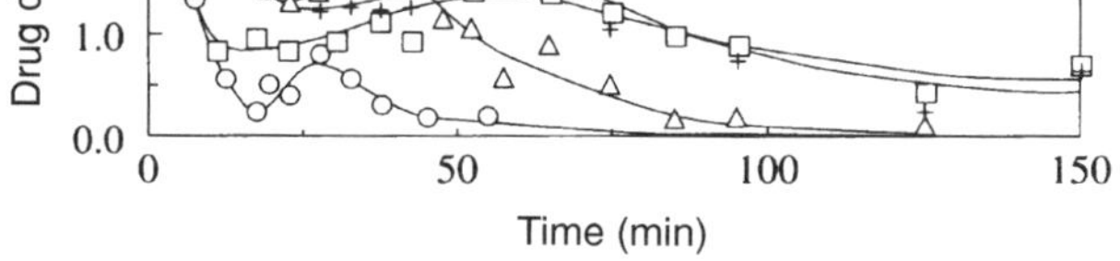

After the formation of a skin layer, a peak appears due to the internal pressure. The internal pressure is proportional to the increase in gel film thickness.

Fig. 16 Change of delivery rate of sodium salicylic acid upon shrinkage of a NIPAAm-BMA copolymer gel at 30°C.

temperature, thickness, and chemical composition of gels [31]. Additionally, because the hydrophilicity/hydrophobicity of any drug (as well as the swelling phenomenon) also influences this behavior, delivery of hydrophobic drugs where shrinkage also is involved will be described in relation to on-off drug delivery in Section 2.4.1. Hoffman *et al.* [32] measured the delivery pattern of a solute from a NIPAAm gel swollen in a methylene blue solution at low temperature as it underwent shrinkage when various elevated temperatures were imposed. The solute was delivered by diffusion below the phase transition temperature of 20°C. However, above the phase transition temperature of 50°C, delivery was influenced by the formation of the shrinking surface layer and accumulation of internal pressure.

As shown in Fig. 17, upon appearance of the initial quick delivery, a slow delivery continued, which resulted in a two-stage behavior. This is due to the inhibition of drug delivery by the formation of a shrinking surface layer. Drug delivery patterns are now better understood in terms of diffusion and the role of internal pressure [33].

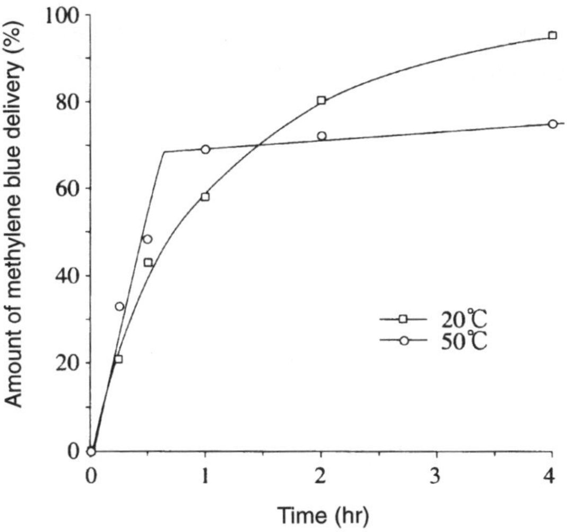

At 20°C, delivery due to diffusion is observed. However, at 50°C, following initial delivery, a two-stage delivery behavior is observed due to inhibition of delivery caused by the shrinking surface layer.

Fig. 17 Change of methylene blue delivery upon shrinkage of a NIPAAm polymer gel.

2.4 DRUG DELIVERY CONTROL USING INTERNAL STRUCTURAL CHANGES OF GELS

2.4.1 On-off Drug Delivery Control by Thermo-responsive Gels

The authors have been studying on-off drug delivery control of thermo-responsive NIPAAm polymer gels [4, 5, 34–38]. It is possible to inhibit drug delivery from a monolithic device using hydrophobic indomethacin dispersed in a gel. This system did not exhibit exclusion of the drug upon shrinkage of the gel [32]. Hence, this provides perfect on-off drug delivery, where delivery is halted where the skin has formed and resumes when the gel swells [34]. Figure 18 shows the delivery rate of indomethacin in PBS (pH 7.4) upon repeated temperature changes of between 10 and 30°C using a 0.5-mm-thick film of NIPAAm-BMA copolymer. The gel swelled at lower temperatures and drug diffusivity increased. Hence, the drug was delivered as swelling progressed. From this on-state to off-state at an elevated temperature, a squeezing effect is observed due to the large reduction in gel volume. At this point, a sharp peak in delivery was observed followed by inhibition when a nonpermeable skin layer was reached. Using such a surface shrinking layer, extremely rapid drug delivery control was achieved. Because the interior of the gel is still swollen even in its off-state, the drug can diffuse near the surface after a

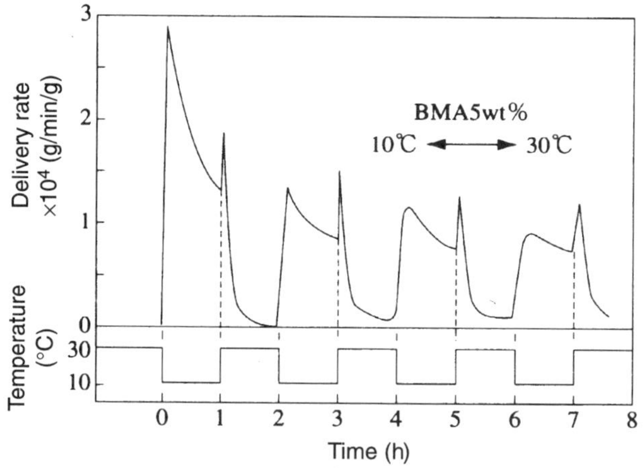

Fig. 18 Delivery rate changes of NIPAAm-BMA copolymer gel upon repeated temperature changes of between 10 and 30°C.

prolonged period. Thus, when the system is brought back to the on-state, the drug delivery rate was similar to that of the initial pulse-type delivery pattern. If the temperature range is narrowed to between 20 and 25°C, a delivery peak due to the delayed time caused by the accumulated internal pressure was observed upon temperature increase [31]. An even further narrowed temperature range showed a vibration-like delivery pattern in which drug delivery stops after pulsed delivery upon gel shrinkage. This mechanism has also been analyzed [33].

The authors have shown that the density and thickness of a skin layer as well as internal pressure can be handled by both the composition and volume of a gel and by the temperature. In other words, it has become possible to control the behaviour of gel in drug delivery systems by using temperature control. The authors have succeeded in on-off drug delivery with a small temperature variation of near body temperature by adjusting the phase transition temperature of a thermoresponsive gel in PBS [38]. This is achieved with a tercopolymer of NIPAAm, hydrophilic dimethylacrylamide (DMAAm) and hydrophobic BMA. Introduction of BMA will adjust the transition temperature so that it approaches body temperature and concomitantly it strengthens skin structure. A device in which indomethacin was dispersed in the gel achieved on-off control as shown in Fig. 19 over a narrow temperature range of between 36 and 38°C. In an actual application to control fever, the system must be in the off-state at a lower temperature and in the on-state at a higher temperature. The authors have developed hydrogels that swell at higher temperatures and shrink at

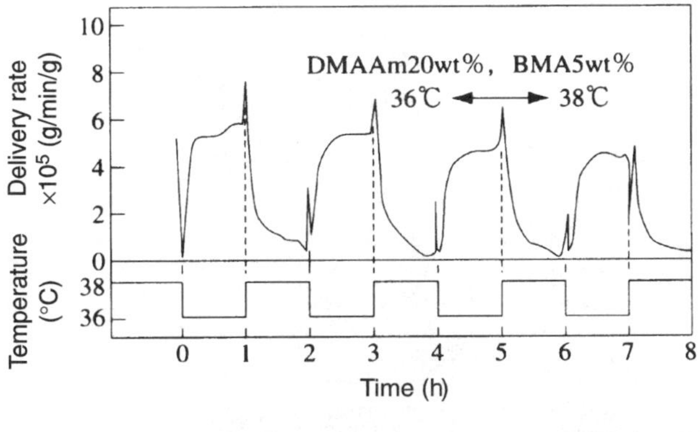

Fig. 19 Delivery behavior of indomethacin from a NIPAAm-DMAAm-BMA tercopolymer in response to temperature changes of between 36 and 38°C.

lower ones [39–41]. We also manufactured a new device (shown in Fig. 20) in which indomethacin was dispersed in a copolymer of NIPAAm and hydrophilic acrylamide (Aam) that does not form a dense skin layer even at high temperatures [42]. The device has drug delivery pores covered by a nonpermeable polymer. The gel continues to deliver drugs through these pores at high temperatures because the skin layer is not dense even if it shrinks. However, the gel swells and fills the capsule completely at low temperatures, which inhibits drug delivery. Accordingly, it has become

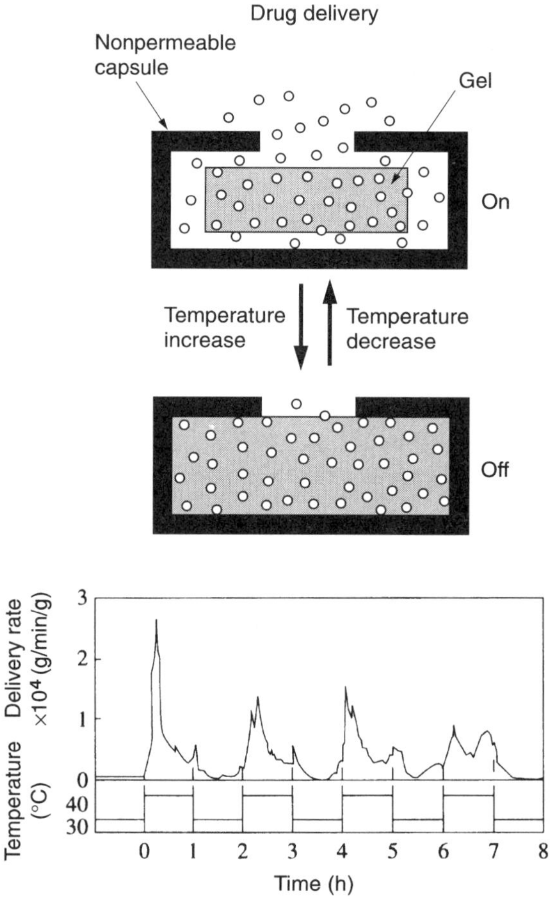

Fig. 20 Delivery behavior of indomethacin dispersed from a surface area-controlled NIPAAm-AAm copolymer gel upon repeated temperature changes of between 30 and 40°C.

possible to convert a negative thermoresponsive gel into a positive thermoresponsive drug delivery system simply by controlling the surface area.

2.4.2 Drug Delivery Control by Chemical Compound-Responsive Gels

A drug delivery system that responds to a certain chemical compound is typified by a system that can deliver insulin in response to blood sugar levels. Pancreatic malfunctions (of either type A or B diabetes mellitus) require that insulin be administered to prevent localized and systemic problems. Blood sugar equilibrium is then maintained using an external protocol, that is, manufactured insulin. These include systems that use molecular exchange reactions and enzyme-matrix reactions [44, 45]. Kost *et al.* [44] fixed glucose oxidase (GOD) in a gel film made of a copolymer of N,N-dimethylaminoethyl methacrylate and hydroxyethyl methacrylate. They controlled insulin delivery by gel swelling in response to pH changes vis-à-vis glucose oxidative reactions. Ishihara *et al.* [45] developed a composite film of GOD-fixed polyacrylamide film and an oxidation-reduction film that contains nicotide (niacin). When blood sugar levels rise, hydrogen peroxide is produced and this then oxidizes the nicotide in the oxidation-reduction film, resulting in the formation of a positive charge in the film. This leads to gel swelling and insulin delivery so as to equalize blood sugar.

Other than systems that employ enzymatic reactions, systems with a polymer containing boric acid group are also being evaluated [46, 47]. A water-soluble copolymer was synthesized using a vinyl monomer that contains phenyl boric acid and acrylamide. This copolymer forms a complex with poly(vinyl alcohol) via hydrogen bond formation. This complex dissociates through an exchange reaction in the presence of glucose. Hydrogel beads made of polyacrylamide and that contain boric acid as the side chain were prepared and polyhydroxyl-modified insulin was fixed onto the gel. These beads were packed into a column and buffer solutions of glucose of different concentrations were alternately passed. A pulse-like delivery behavior was observed by the exchange reaction of the modified insulin to the glucose concentration (see Fig. 21).

There are drug delivery systems that respond to the sorts of chemical compounds that are produced at inflammation-affected body sites [48–50]. Yui *et al.* [49, 50] evaluated drug delivery control using

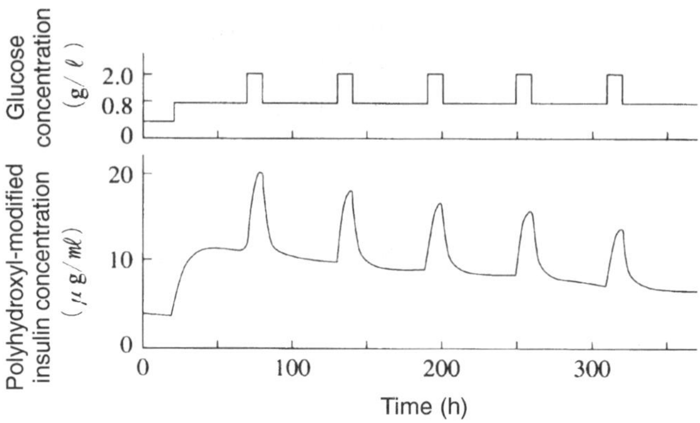

Fig. 21 Pulse-like drug delivery of a boric acid group-containing gel in response to the glucose concentration changes.

inflammation-responsive biodegradable gels. This system took advantage of the production of a hydroxyl radical by the inflamed cells. The hydroxyl radical then specifically decomposed hyaluronic acid (HA), which is a polysaccharide. They manufactured a heterogeneous degradable drug using a crosslinked HA gel in which drug-containing lipid microspheres were included. It was demonstrated in experiments using mice that this gel decomposes very little during healthy conditions. However, it rapidly and specifically decomposed when inflammation was present. The HA degradability in the presence of hydroxyl radicals can be controlled with the crosslinking method and degree of crosslinking. Such biodegradable polymers, which respond to biostimulation while in the body and are able to control drug delivery, have seldom been reported because it has been difficult to control degradation and discrete associated degradation behavior. It is also necessary to explore whether drug delivery is diffusion- or degradation controlled from a time management viewpoint.

2.4.3 Drug Delivery Control by pH-Responsive Gels

It is desirable for drugs like polypeptides, which are deactivated by a low stomach pH, and anti-inflammation drugs such as indomethacin, which severely affect the stomach, to bypass the stomach and be delivered and absorbed in the intestinal tract. It is possible to develop position selectivity that allows drug delivery only in the intestines using pH-responsive polymers because the pH of the stomach and intestine differ significantly. Hence, systems that respond to external pH changes have been investigated. Dong and Hoffman [8] synthesized tercopolymers of thermoresponsive NIPAAm, vinyl-terminated polysiloxane (VTPDMS) and an anionic monomer, acrylic acid (AAc). These tercopolymers show thermo- and pH-responsivity. In the stomach (at pH 1.4), the carboxylic acid groups of acrylic acid do not dissociate. However, they dissociate in the intestine at 6.8–7.4 pH and the gel swells due to the repulsive forces among charges, which causes drug delivery. The phase transition temperature depends on the pH. At 37°C, the gel reversibly shrank at lower temperatures and swelled at higher temperatures. The delivery behavior of indomethacin at pH levels of 1.4 and 7.4 was investigated. When the gel is in a swollen state at a pH of 7.4, the diffusivity of the drug in the gel is high and thus the drug is delivered (Fig. 22), whereas drug delivery was inhibited at a pH of 1.4. The drug delivery rate increased as the AAc content increased. The gel that had 10 mol (NSA-10) had a higher delivery rate than the one that had 2 mol (NSA-2). This is because the increased AAc concentration increases the water content of the gel and drug diffusivity increased.

Siegel and Pitt [51] and Baker and Siegel [52] attempted to control glucose permeability in response to pH by using copolymers of NIPAAm and methacrylic acid (MMA) where the gel was sandwiched in the diffusion cell. Glucose permeability was investigated when one side of the gel was maintained at pH 7.0 and the pH of the other side of the gel was varied. When the pH was reduced to 4.9–5.0, the dissociated carboxylic acid groups of methacrylic acid return to carboxylic acid. This, along with the hydrophobic interaction of NIPAAm, led to gel shrinkage, which resulted in controlled glucose permeation. If the pH is gradually increased, the gel swelled at pH 5.1–5.2 and glucose was passed again. As shown in Fig. 23, the gel reversibly swelled and shrank as the pH changes were repeated. This then led to on-off control of glucose permeation. Hysteresis (or lag) is observed for onset of pH during swelling and shrinking. This is probably due to the reduced diffusivity of solute molecules and ions in the shrunken gel, which may influence reaction rate. There are reports that the combination of pH hysteresis and glucose oxidase enzyme reactions are being used to develop pulse-type delivery systems [51, 52].

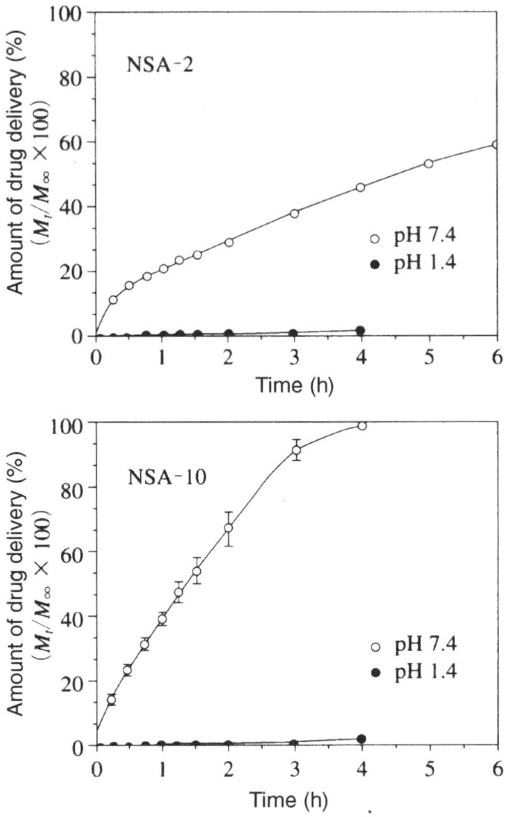

NSA-2 and NSA-10 are gels of AAc/NIPAAm = 2/100 and 10/100 mol%, respectively.

Fig. 22 Delivery behavior of indomethacin from NIPAAm-VTPDMS tercopolymers and an anionic mononer (AAc) in the presence of pH values of 1.4 and 7.4.

2.4.4 Drug Delivery Control by Physical Stimuli-Responsive Gels

To control drug delivery, external signals must penetrate the body. For this purpose, physical stimuli, such as electric fields, magnetic fields, or ultrasound, are effective. Along with recent developments in catheter technology, various light signals can also be used. The importance of this field is thus expected to grow. Sawahata *et al.* [6] attempted on-off control of drug delivery with electrical signals and a field-responsive gel to

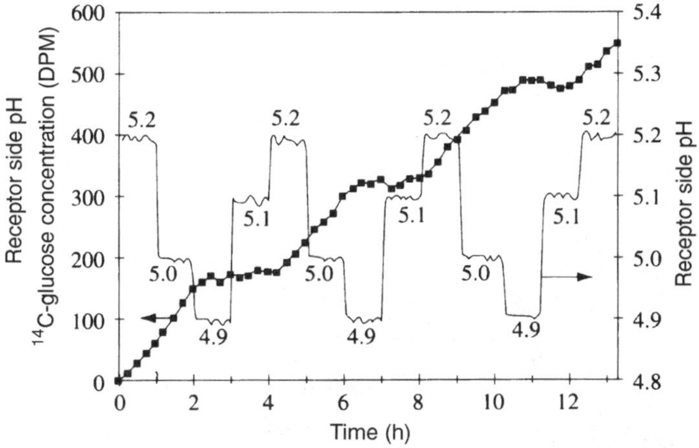

Fig. 23 On-off control of glucose permeation by NIPAAm-MMA copolymer gel film in response to pH changes.

which a drug was added. Insulin glucose and raffinose were mixed into gels made of poly(methacrylic acid), poly(dimethylaminopropylacrylamide) and poly(acrylic acid). Upon application of voltage at a constant interval, drug delivery was accelerated. Drug uptake had slowed or halted due to gel shrinkage. The amount of drug delivered can be controlled by the voltage.

Kwon *et al.* [53] manufactured a delivery device by ionically fixing a drug of positive charge to the sulfonic acid group of a copolymer that fell between 2-acrylamide-2-ethylpropane sulfonic acid and BMA. When voltage was applied to this device, the drug was delivered only then. This was due to the exchange reaction between the hydrogen ions produced at the anode and the drug. Furthermore, Kwon *et al.* [54] developed a device made of poly(ethyl oxazoline) and poly(methacrylic acid) polymer complex to which insulin was added. These polymers form a polymer complex via hydrogen bonding at low pH and dissolve in water >pH 5.4 by eliminating hydrogen bonds. Upon application of voltage to the device, which is fixed on a cathode, the hydroxyl ions increase the pH near the cathode. Hence, the polymer complex dissociates and on-off control becomes possible only when voltage is applied. This result is shown in Fig. 24.

Hsieh and Langer [7] manufactured a hemispherical device made of ethylene vinylacetate that was coated by a nonpermeable film. This device had a delivery hole in the center. By introducing a ring magnet in the

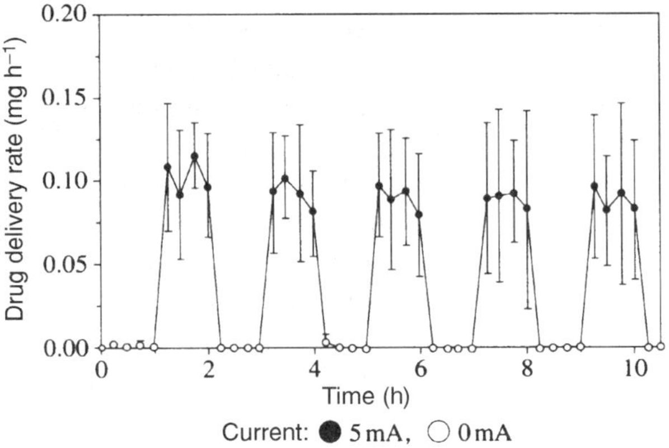

Fig. 24 Field responsivity of insulin delivery from poly(ethyl oxazoline) and polymethacrylic acid polymer complex.

matrix of this device and applying a magnetic field, pulse-like drug delivery was achieved. Furthermore, Negishi *et al.* [55] injected methotrexate, a chemotherapeutic drug, that had been covalently bonded to biodegradable poly(hydroxypropyl glutamine) into a tumor. As the drug is a polyelectrolyte, the irradiation of external microwave selectively heated the device and the heat was used to deliver the drug. As a result, they have demonstrated that external microwaves are an effective way to control drug delivery. Ishihara *et al.* [56] synthesized a copolymer made of an azobenzene monomer that responds to UV light by reversibly transferring from trans- to cis- and 2-hydroxyethyl methacrylate (PHEMA). Upon irradiation by UV light, the polymer changed from trans to cis. The dipoles in the cis-form interact with PHEMA and the hydration of PHEDMA is controlled, resulting in gel shrinkage. Utilizing a light-responsive gel that shrinks and swells reversibly, drug delivery control was attempted using ethyl-p-aminobenzoate as a model drug.

2.5 CONCLUSIONS

Systems that control both time and amounts of drug to be delivered by external stimuli are extremely useful in avoiding rejection by the body, developing intelligent drugs, and for various other pharmacological uses. The DDS that controls material transport by stimuli-responsive gels was discussed here. In particular, the mechanism of swelling and shrinking of polymer gels and drug delivery using such behaviors were summarized. Pulse-like drug delivery using reversible swelling-shrinking of gels was also discussed. It is possible to achieve control of the amount and time of drug delivery because these new stimuli-responsive gels change their structures and functions based on changes in external conditions. It is therefore indispensable to study the detailed mechanisms of dynamic structures and drug delivery behavior using a deductive approach that taps polymer science, pharmaceutical engineering, and chemical engineering. Major advances in DDS technology are expected in the near future.

REFERENCES

1 Lipper, R.A., and Higuchi, W.I. (1977). *J. Pharm. Sci.*, **66**: 163.
2 Okano, T., Miyajima, M., Kodama, F., Imanidis, G., Nishiyama, S., Kim, S.W., and Higuchi, W.I. (1987). *J. Controlled Release* **6**: 99.
3 Theeuwes, F.D., Swanson, D., Wong, P., Bonsen, P., Place, V., Heimlich, K., and Kwan, K.C. (1983). *J. Pharm. Sci.* **72**: 253.
4 Okano, T., Yui, N., Yokoyama, M., and Yoshida, R. (1994). *Advances in Polymeric Systems for Drug Delivery*, New York: Gordon and Breach Science Publishers.
5 Okano, T., Bae, Y.H., and Kim, S.W. (1990). *Pulsed and Self-regulated Drug Delivery*, J. Kost, ed., Boca Raton, FL: CRC Press, pp. 17–46.
6 Sawahata, K., Hara, M., Yasunaga, H., and Osada, Y. (1990). *J. Controlled Release* **14**: 253.
7 Hsieh, D.T., and Langer, R. (1982). In *Controlled Release Delivery System*, T.J. Roseman and S.Z. Masdorf, eds., New York: Marcel Dekker, p. 107.
8 Dong, L.C., and Hoffman, A.S. (1991). *J. Controlled Release* **15**: 141.
9 Yoshida, R., Sakai, K., Okano, T., and Sakurai, Y. (1993). *Adv. Drug. Delivery Rev.* **11**: 85.
10 Matsuo, E.S., and Tanaka, T. (1988). *J. Chem. Phys.* **89**: 1695.
11 Vrentas, J.S., Jarzebski, C.M., and Duda, J. L. (1975). *AIChE J.* **21**: 894.
12 Crank, J. (1975). *The Mathematics of Diffusion*, London: Oxford University Press.
13 Alfrey, T., Jr., Gurnee, E.F., and Lloyd, W.G. (1966). *J. Polym. Sci.* Part C **12**: 249.
14 Hopfenberg, H.B. (1978). *J. Membrane Sci.* **3**: 215.
15 Korsmeyer, R.W., and Peppas, N.A. (1984). *J. Controlled Release* **1**: 89.
16 Jacques, C.H.M., Hopfenberg, H.B., and Stannett, V.T. (1974). *Permeability of Plastic Films and Coatings to Gases, Vapors, and Liquids*, New York: Plenum Press.
17 Siegel, R.A., Falamarzian, M., Firestone, B.A. and Moxley, B.C. (1988). *J. Controlled Release* **8**: 179.
18 Okuyama, Y., Yoshida, R., Sakai, K., Okano, T., and Sakurai, Y. (1993). *J. Biomatr. Sci. Polym. Edn.* **4**: 545.
19 Kaneko, Y,. Yoshida, R., Sakai, K., Sakurai, Y., and Okano, T. (1995). *J. Membrane Sci.* **101**: 13.
20 Kabra, B.G., and Gehrke, S.H. (1991). *Polym. Commun.* **32**: 322.
21 Wu, H.S., Hoffman, A.S., and Yager, P. (1992). *J. Polym. Sci.* Part A **30**: 2121.

22 Yoshida, R., Uchida, K., Kaneko, Y., Sakai, K., Kikuchi, A., Sakurai, Y., and Okano, T. (1995). *Nature* **374**: 240.

23 Kaneko, Y., Sakai, K., Kikuchi, A., Yoshida, R., Sakurai, Y,. and Okano, T. (1995). *Macromolecules* **28**: 7717.

24 Yasuda, H., Peterlin, A., Colton, C.K., Smith, K.A., and Merrill, E.W. (1969). *Makromol. Chem.* **126**: 177.

25 Cohen, M.H., and Turnbull, D. (1959). *J. Chem. Phys.* **31**: 1164.

26 Yasuda, H., Lamaze, C.E., and Ikenberry, L.D. (1968). *Makromol. Chem.* **118**: 19.

27 Higuchi, T. (1961). *J. Pharm. Sci.* **50**: 874.

28 Korsmeyer, R.W., and Peppas, N.A. (1983). *Controlled Release Delivery Systems*, (1983). T.J. Roseman, and S.Z. Mansdorf, eds., New York: Marcel Dekker, pp. 77–90.

29 Segot-Chicq, S., and Peppas, N.A. (1986). *J. Controlled Release* **3**: 193.

30 Yoshida, R., and Okano, M. (1993). *Hyomen* **31**: 474.

31 Yoshida, R., Sakai, K., and Okano, M. (1992). *Jinko Zoki* **21**: 244.

32 Hoffman, A.S., Afrassiabi, A., and Dong, L.C. (1986). *J. Controlled Release* **4**: 213.

33 Yoshida, R., Sakai, K., Okano, T., and Sakurai, Y. (1992). *Ind. Eng. Chem. Res.* **31**: 2339.

34 Bae, Y.H., Okano, T., Hsu, R., and Kim, S.W. (1987): *Makromol. Chem. Rapid Commun.* **8**: 481.

35 Okano, T., Bae, Y.H., Jacobs, J., and Kim, S.W. (1990). *J. Controlled Release* **11**: 255.

36 Okano, T., Yoshida, R., Sakai, K., and Sakurai, Y. (1991), in *Polymer Gels*, D. DeRossi, ed., New York: Plenum Press, pp. 299–308.

37 Okano, T., and Yoshida, R. (1993), in *Biomedical Applications of Polymeric Materials*, T. Tsuruta, T. Hayashi, K. Kataoka, K. Ishihara, and Y. Kimura, eds., Boca Raton, FL: CRC Press, pp. 407–427.

38 Yoshida, R., Sakai, K., Okano, T., and Kimura, Y. (1994). *J. Biomater, Sci. Polym. Edn.* **6**: 585.

39 Katono, H., Maruyama, A., Sanui, K., Ogata, N., Okano, T., and Sakurai, Y. (1991). *J. Controlled Release* **16**: 215.

40 Katono, H., Sanui, K., Ogata, N., Okano, T., and Sakurai, Y. (1991). *Polymer J.* **23**: 1179.

41 Aoki, T., Kawashima, M., Katono, H., Sanui, K., Ogata, N., Okano, T., and Sakurai, Y. (1994). *Macromolecules* **27**: 947.

42 Yoshida, R., Kaneko, Y., Sakai, K., Okano, T., Sakurai, Y., Bae, Y.H., and Kim, S.W., (1994). *J. Controlled Release* **32**: 97.

43 Makino, K., Mack, E.J., Okano, T., and Kim, S.W. (1990). *J. Controlled Release* **12**: 235.

44 Kost, J., Horbett, T.A., Ratner, B.D., and Singh, M. (1985). *J. Biomed. Mater. Res.* **19**: 1117.

45 Ishihara, K., Kobayashi, K., and Shinohara, I. (1983). *Makromol. Chem. Rapid Commun.* **4**: 327.

46 Kitano, S., Kataoka, K., Koyama, Y., Okano, T., and Sakurai, Y. (1991). *Makromol. Chem. Rapid Commun.* **12**: 227.

47 Shiino, D., Murata, Y., Kataoka, K., Koyama, Y., Yokoyama, M., Okano, T., and Sakurai, Y. (1994). *Biomaterials* **15**: 121.

48 Heller, J. (1985). *J. Controlled Release* **2**: 167.

49 Yui, N., Okano, M., and Sakurai, Y. (1992). *Seitai Zairyo* **10**: 218.

50 Yui, N., Okano, T., and Sakurai, Y. (1992). *J. Controlled Release* **22**: 105.

51 Siegel, R.A., and Pitt, C.G. (1995). *J. Controlled Release* **33**: 173.

52 Baker, J.P., and Siegel, R.A., (1996). *Makromol. Chem. Rapid. Commun.* **17**: 409.

53 Kwon, I.C., Bae, Y.H., Okano, T., Berner, B., and Kim, S.W. (1990) *Makromol. Chem. Rapid Commun.* **33**: 265.
54 Kwon, I.C., Bae, Y.H., and Kim, S.W. (1991). *Nature* **354**: 291.
55 Negishi, N., Yoshida, H., and Kikuchi, S. (1988). *Jinko Zoki* **17**: 531.
56 Ishihara, K., Hamada, N., Kaot, S., and Shiohara, I. (1984). *J. Polym. Sci. Polym. Chem. Ed.* **22**: 881.

Section 3
Adsorption and Separation

SHUJI SAKOHARA

3.1 ABILITY TO CONCENTRATE SOLVENT BY GELS AND SEPARATION OF MIXED SOLVENT BY GEL MEMBRANES

3.1.1 Introduction

It is well known that control of chemical and crosslinking structures yields various functional polymer gels. These functions have been actively evaluated for engineering applications. Concentrating and separating of materials are application examples. As a concentration and separation operation, gel filtration is well known. Further development in the concentration and separation functions and their engineering applications through control of the chemical and crosslinking structures of gels is desired.

Concentration and separation are among the most important operations in the production process. Many engineering production methods have already been developed and commercialized and therefore concentration and separation with gels will be of no use unless these methods are superior to the traditional ones. Many traditional concentration and separation methods, such as distillation, require much energy and the materials to be separated may be exposed to severe conditions. On the other hand, polymer gels can be controlled by slight changes to the environment. Thus, concentration and separation can be achieved under better conditions and with energy conservation an added benefit.

In this section, concentration and separation of mixed solvents using polymer gels will be described. The concentration and separation characteristics of solvents as examples of engineering applications will be introduced. This will be done by controlling the chemical and crosslinking structures of gels and separation by gel membrane. In the concentration and separation of mixed solvents, separation on the molecular level will be required. These phenomena will be described based on the author's experience.

3.1.2 Concentration and Separation of Materials by Polymer Gels

One of the major concentration and separation functions accomplished by controlling the chemical and crosslinking structures of polymer gels is the sifting effect based on the size difference of the materials to be separated and distribution based on compatibility/incompatibility with the gel. Sometimes, these effects appear in tandem.

When molecular size is relatively large, the sifting effect can be achieved by gels. Gels have been applied to separate molecules [1, 2]. When a gel is swollen in either solution or suspension, the solvent penetrates the gel. When the size of the network (the effective pore size of the network) is controlled by changing the degree of crosslinking, the penetration of the relatively large solute or suspension can be inhibited and the gel exhibits a molecular separation function. As an example, in a concentrated waste sludge system where it is difficult to eliminate water, a proposal that uses poly(vinyl methyl ether) has been made [3, 4].

The concentration and separation function of gels using compatibility differences can be achieved on the molecular level. However, there are only a few examples at this point, including a report in which the amounts of water and organic solvent absorbed into the hydrophilic gel are different [5, 6]. This is caused by solvation of the hydrophilic group of the gel network. If the spatial structure of the gel is controlled by the crosslinking structure of the gel, improvement in selective concentration can be achieved and if the chemical structure of the gel is changed, a similar effect can be expected for other mixed solvent systems. This is the subject of this section.

3.1.3 Chemical and Crosslinking Structures of Polymer Gels and Their Selective Concentrations of Mixed Solvent

3.1.3.1 Chemical structures of polymer gels and their swelling characteristics for solvents

Studies on concentration and separation of polymer gels using differences in compatibility with the solvent are dominated by work on hydrophilic gels. Studies on amphoteric or hydrophobic gels have rarely been reported [7, 8]. To use a gel for concentrating media, it is essential to understand the chemical structure of the gel and its compatibility with solvents. In other words, swelling characteristics need to be examined. Some examples are provided.

Figure 1 shows the degree of swelling of an amphoteric dimethylacrylamide gel, which exhibits both hydrophilicity and hydrophobicity in various solvents. The composition is listed in Table 1. This gel swells in polar solvents like alcohol to the same extent as in water or even more. However, it does not swell as much in ketones and does not swell nearly at all in non-polar solvents that include benzene and cyclohexane. However, as the sizes of the solvent molecules are different, compatibility cannot be evaluated simply by differences in the degree of swelling. Judging from the number of moles of the absorbed solvent per 1 g of dry gel, the amount of absorbed solvent monotonously decreased as the polarity decreased. Thus, compatibility with the gel decreased according to this order. Many amphoteric gels likely show a similar trend.

Figure 2 illustrates the degree of swelling for various solvents for gels made of methacrylic acid derivatives in the ester group. It shows both ethyl methacrylate (ENA) gel and 2-dimethylaminoethyl methacrylate (DMAEMA) gel. The composition of the gels is listed in Table 2. The

Table 1 The composition of dimethylacrylamide gel.

Primary monomer	N,N′-dimethylacrylamide	1000 mol/m^3
Crosslinking agent	N,N′-methylenebisarylamide	80 mol/m^3
Catalyst	N,N,N′N′-tetramethylenediamine	10 mol/m^3

Initiator: ammonium sulfate (0.5 mol/m^3)
Solvent: water
Gelation temperature: 50°C

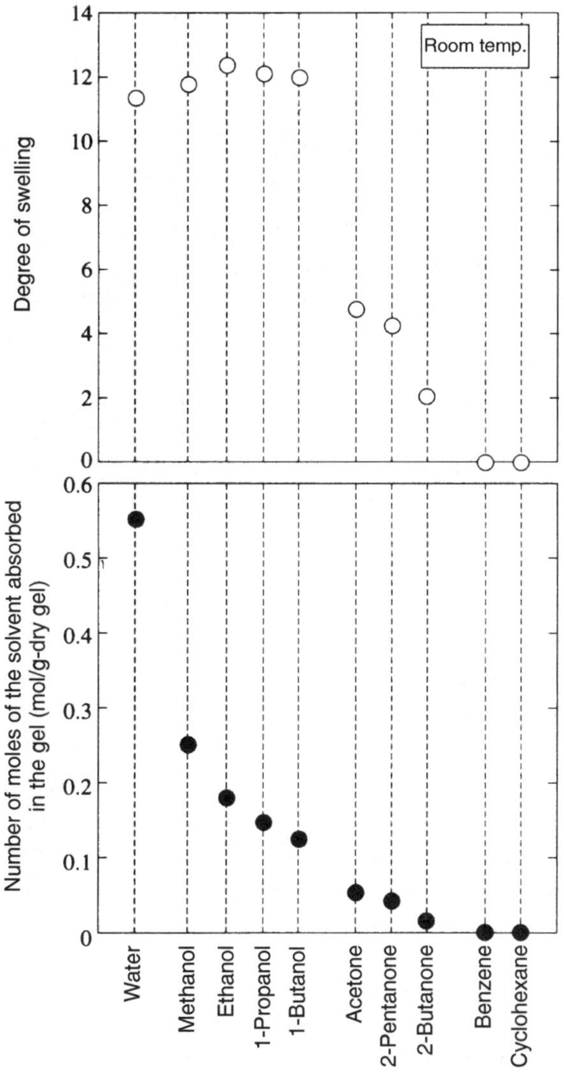

Fig. 1 The degree of swelling and the number of moles of the absorbed solvent for an amphoteric dimethylacrylamide gel.

two differ in that DMAEMA possesses tertiary amine. Thus, a DMAEMA gel shows the amphoteric property whereas the EMA gel exhibits hydrophobicity. Hence, the degree of swelling with water and alcohol differ markedly between the two. Both gels swell with aromatic or ketone

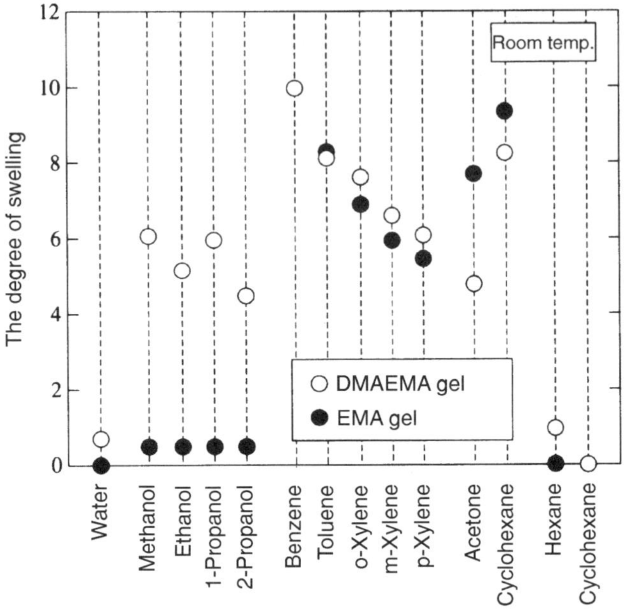

Fig. 2 The degree of swelling of dimethylaminoethyl methacrylate (DMAEMA) gel and ethyl methacrylate (EMA) gel in various solvents.

solvents. However, they do not swell in solvents such as paraffin. In addition, the degree of swelling among the xylene isomers is different. As a result, it is possible to synthesize to some extent the gels that swell selectively in the desired solvent if the chemical structure of the gel is properly chosen.

Table 2 The compositions of DMAEMA gel and EMA gel.

Primary monomer	2-Dimethylaminoethyl methacrylate (DMAEMA)	1000 mol/m^3
	Ethyl methacrylate (EMA)	
Crosslinking agent	Diethylene glycol dimethacrylate (DEGDMA)	200 mol/m^3
Catalyst	N,N,N′,N′-tetramethylethylenediamine (TEMED)	60 mol/m^3

Initator: α, α'-azoisobutylonitrile (AIBN) (100 mol/m^3)
Solvent: dimethylformamide (DMF)
Gelation temperature: 50°C

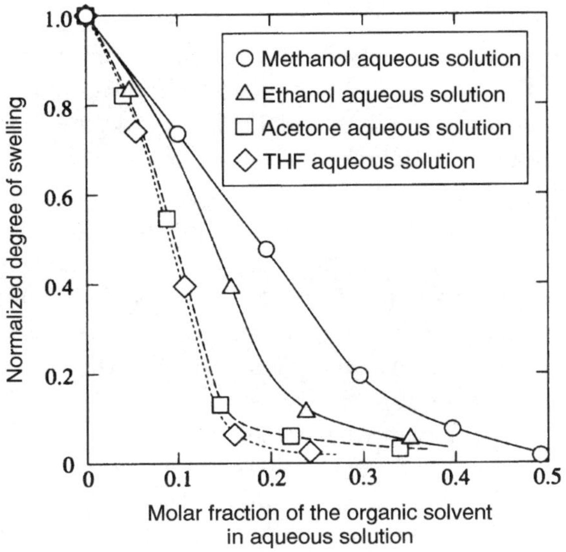

Fig. 3 The degree of swelling of an acrylamide gel in various organic solvent aqueous solutions.

3.1.3.2 Swelling characteristics and selective concentrations of solvent of polymer gels for mixed solvents

As a hydrophilic gel does not swell nearly at all in organic solvents, the degree of swelling of the gel in an organic aqueous solution decreases as the concentration of the organic solvent increases. The gel selectively absorbs water. (See References [7–10] for more information.) Figure 3 depicts the reduction in the degree of swelling of a typical hydrophilic gel of the acrylamide variety. This reduction is shown for various organic solvent aqueous solutions as a function of the concentration of the organic solvent. Gel composition is listed in Table 3.

Table 3 The composition of the acrylamide gel.

Primary monomer	Acrylamide	1000 mol/m^3
Crosslinking agent	N,N′-methylenebisacrylamide	40 mol/m^3
Catalyst	N,N,N′,N′-tetramethylethylenediamine	10 mol/m^3

Initiator: ammonium sulfate (0.5 mol/m^3)
Solvent: water
Gelation temperature: 50°C

Reduction in the degree of swelling depends on the type of organic solvent. The lower the polarity the lower the concentration at which the degree of swelling decreases. Judging from the absorption of water and organic solvents into the gel, as shown in Fig. 4, the amount of water absorbed decreases as the concentration of the organic solvent increases. In other words, the amount of water decreases with a reduction in the degree of swelling. The amount of organic solvent first increases and then decreases. As a result, when the degree of swelling is reduced, the concentration of the organic solvent in the gel is smaller than that in the solution and water is selectively absorbed. The extent of absorption depends on the polarity of the organic solvent. The lower the polarity the more selective the water absorption.

When amphoteric dimethylacrylamide gel is used, a similar phenomenon is observed with a polar/nonpolar mixed solvent system. Figure 5 illustrates, as an example, the degree of swelling in ethanol/benzene- and ethanol/cyclohexane-mixed solvent systems. Unlike in the case of the aforementioned acrylamide gel, reduction in degree of swelling is not large even when the amount of solvents that have almost no compatibility with the gel (such as benzene and cyclohexane) is increased. In particular, there is no reduction in the degree of swelling until a high concentration is reached in the case of benzene. Judging from the amount of solvent absorbed into the gel, as shown in Fig. 6, there is almost no difference in the amount of ethanol absorption. However, in the case of benzene and cyclohexane, benzene is absorbed more than cyclohexane. Consequently, the selective absorption of ethanol is markedly different in both mixed solvents. This implies that not only compatibility between the gel and the solvent but also compatibility between the solvents themselves contribute significantly to selective absorption of the solvent.

Judging from the swelling characteristics in various solvents as shown in Fig. 2, methacrylate derivative gels are expected to show selective absorption of aromatic solvents from the aromatic solvent/paraffin mixed solvent. Figures 7 and 8 show the degree of swelling and selective absorption of benzene into the gel for benzene/cyclohexane-mixed solvent solutions where the size of the molecules is similar. Even if the degree of swelling decreases, selective absorption does not change noticeably. Thus it can be said that selective absorption is also influenced by the molecular size of the solvent.

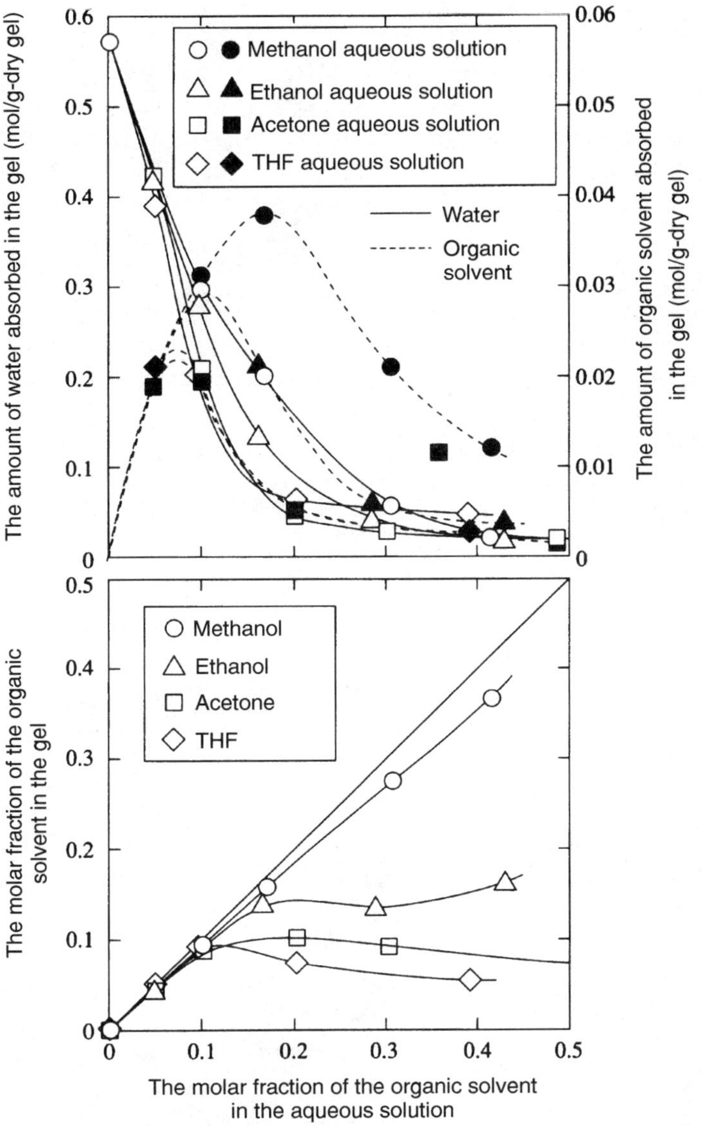

Fig. 4 The amount of water and organic solvents absorbed and the molar fraction of the organic solvent in an acrylamide gel in various organic solvent aqueous solutions.

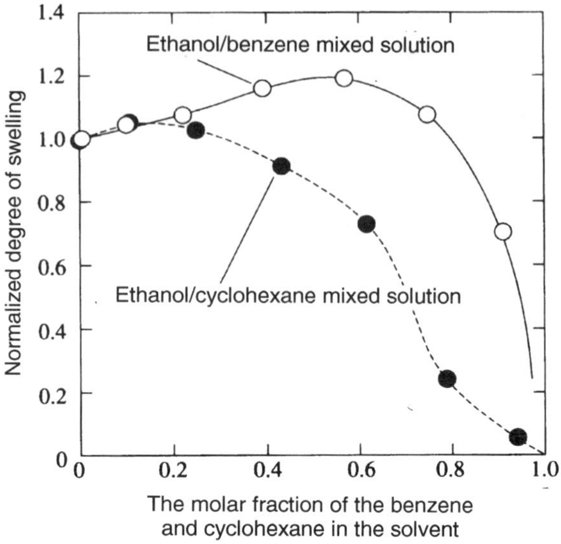

Fig. 5 The degree of swelling of dimethylacrylamide gel in ethanol/benzene- and ethanol/cyclohexane-mixed solvents.

3.1.3.3 The effect of crosslink structure on the selective absorption of solvents

It is important to control the crosslink structure, that is the effective pore size of the network, and the chemical structure in order to improve the selective absorptivity of a solvent in a gel [8, 11]. For this, crosslink density or the length of the crosslinking agent can be changed, resulting sometimes in significant changes in selective absorption of the solvent.

In the case of hydrophilic gels, as shown in Fig. 9, the gel becomes turbid if the concentration of the crosslinking agent is increased. Turbidity is enhanced if the primary monomer concentration is decreased or the gelation temperature is reduced. Such turbidity is closely related to the homogeneity of the gel network structure. The network of a transparent gel is regarded as macroscopically homogeneous. On the other hand, the turbid gel is heterogeneous, consisting of dense and rare areas [12]. Such network heterogeneity influences the degree of swelling and selective absorptivity of the gel. Figure 10 shows an example in which the degree of swelling and the molar fraction of ethanol in the gel are plotted for acrylamide gels synthesized with three different crosslinking agent concentrations. As the concentration of the crosslinking agent increases,

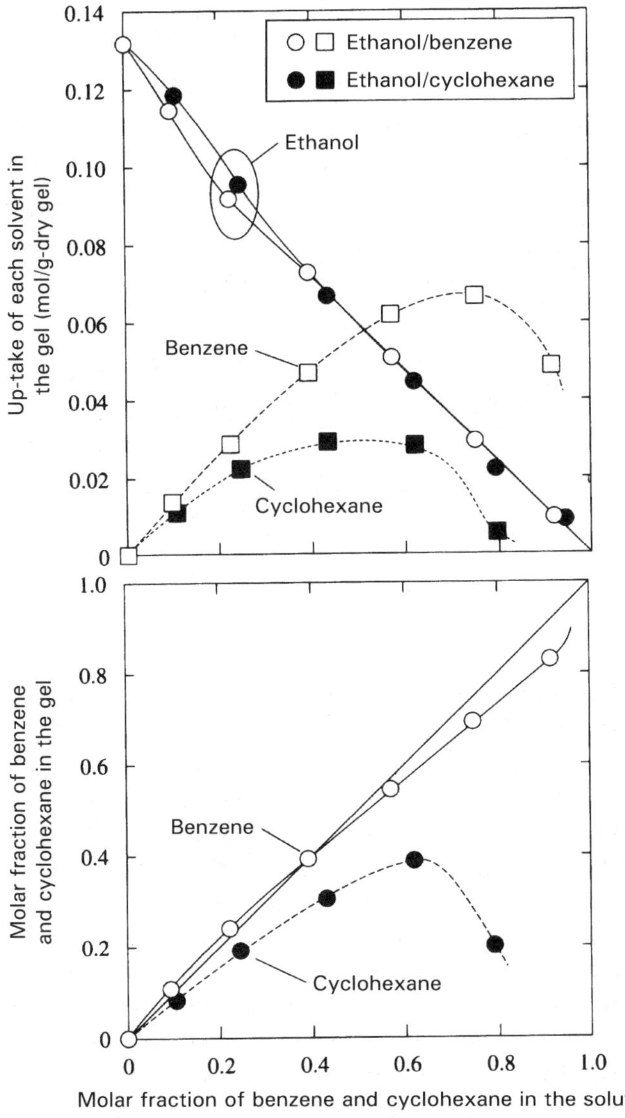

Fig. 6 The amount of ethanol, benzene and cyclohexane absorbed in dimethylacrylamide gel and molar fraction of benzene and cyclohexane in the gel.

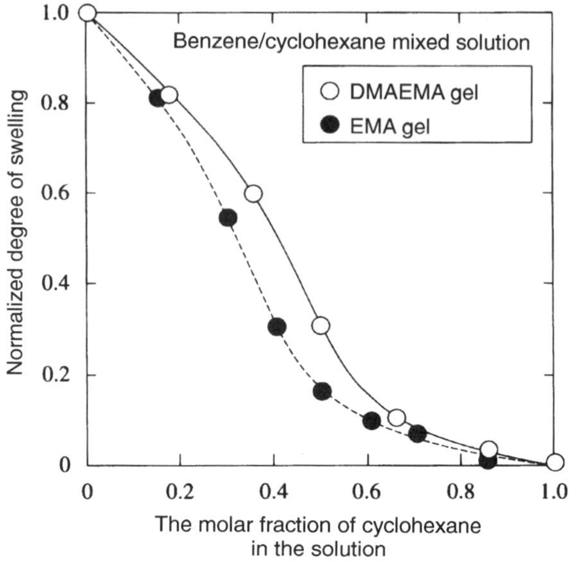

Fig. 7 The degree of swelling of DMAEMA gel and EMA gel in benzene/cyclohexane-mixed solvent.

the degree of swelling in pure water decreases. However, in the ethanol aqueous solution, the reversion can be observed around the area indicated by a circle, where the turbid gel with a high crosslinking agent concentration swells more. The ethanol concentration is higher with a higher concentration of crosslinking agents. This is probably because the ethanol is absorbed into the rare region as a result of the heterogeneity formation. Accordingly, an increase in crosslinking agent concentration of this type of gels results in a decrease of the selective absorption of water.

Various divinyl monomers can be used as the crosslinking agents for methacrylate derivatives. Figure 11 illustrates the swelling characteristics and molar fraction of cyclohexane in the DMAEMA gels, which are prepared by using various length mono-, di-, and triethylene glycol dimethacrylates (EGDMA, DEGDMA, and TEGDMA). The degree of swelling with benzene increases in the order EGDMA < TEGDMA < DEGDMA. For high cyclohexane concentration, DEGDMA shows the lowest degree of swelling and the cyclohexane concentration is the lowest.

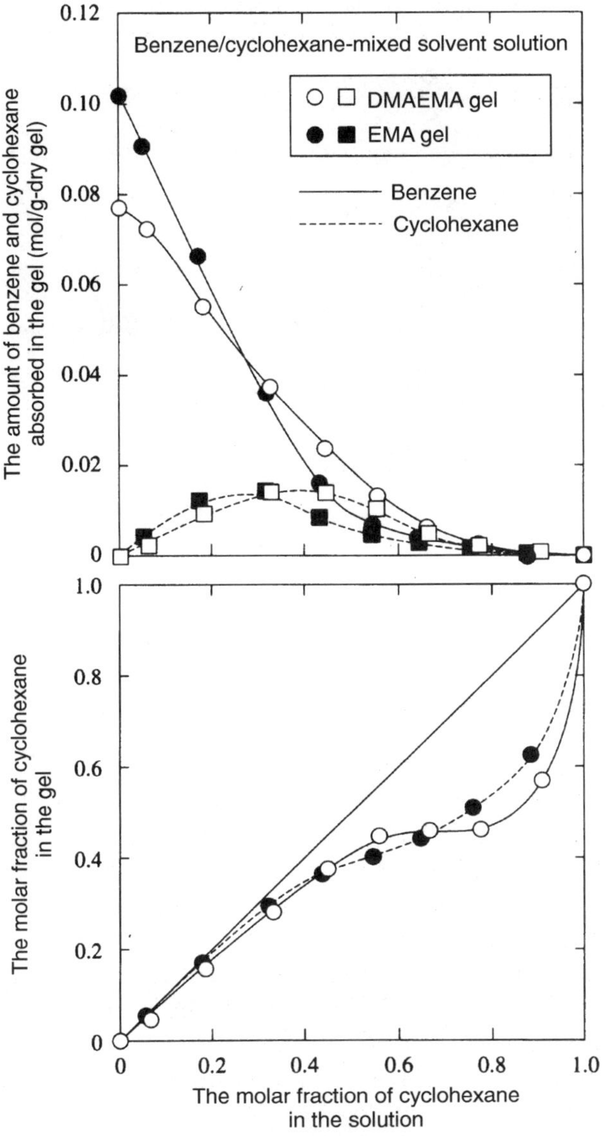

Fig. 8 The amount of benzene and cyclohexane absorbed in DMAEMA gel and EMA gel and the molar fraction of the cyclohexane in the gel.

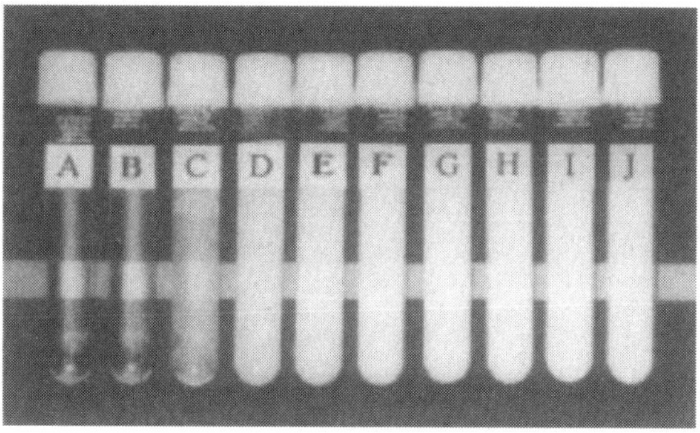

Gelation temperature: 50°C
Acrylamide concentration: 1000 mol/m^3

Crosslinking agent (methylenebisacrylamide) concentration (mol/m^3)

A	B	C	D	E	F	G	H	I	J
30	40	50	60	70	80	90	100	110	120

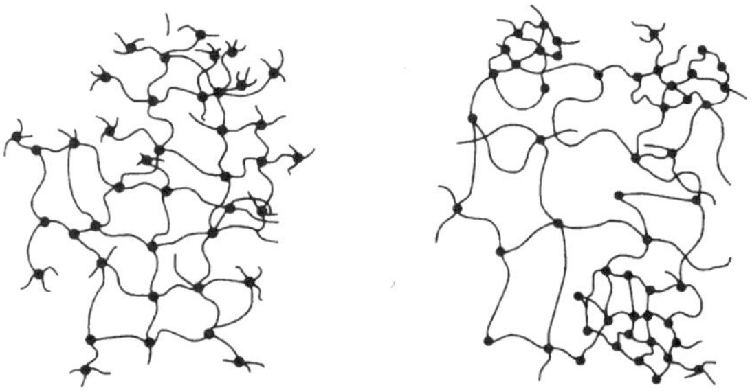

Homogeneous structure (transparent) → Heterogeneous (rare/dense) structure (turbid)

Increased crosslinking agent concentration
Reduction of the primary monomer concentration
Reduction of the gelation temperature

Fig. 9 The process of developing turbidity of acrylamide gels as a function of the crosslinking agent concentration and the conceptual diagram of the homogeneity of the network structure.

Accordingly, the control of the crosslink structure in addition to the chemical structure of a gel is essential for selective absorption. In order to improve selective absorptivity, it is necessary to properly select the concentration and length of the crosslinking agent.

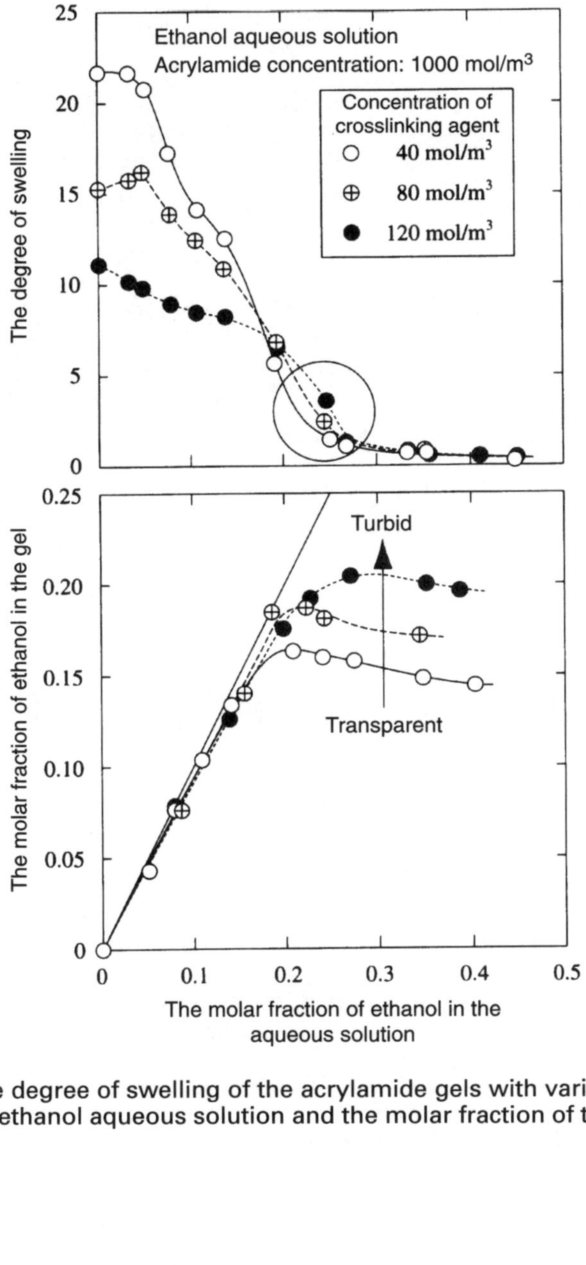

Fig. 10 The degree of swelling of the acrylamide gels with various crosslink densities in ethanol aqueous solution and the molar fraction of the ethanol in the gel.

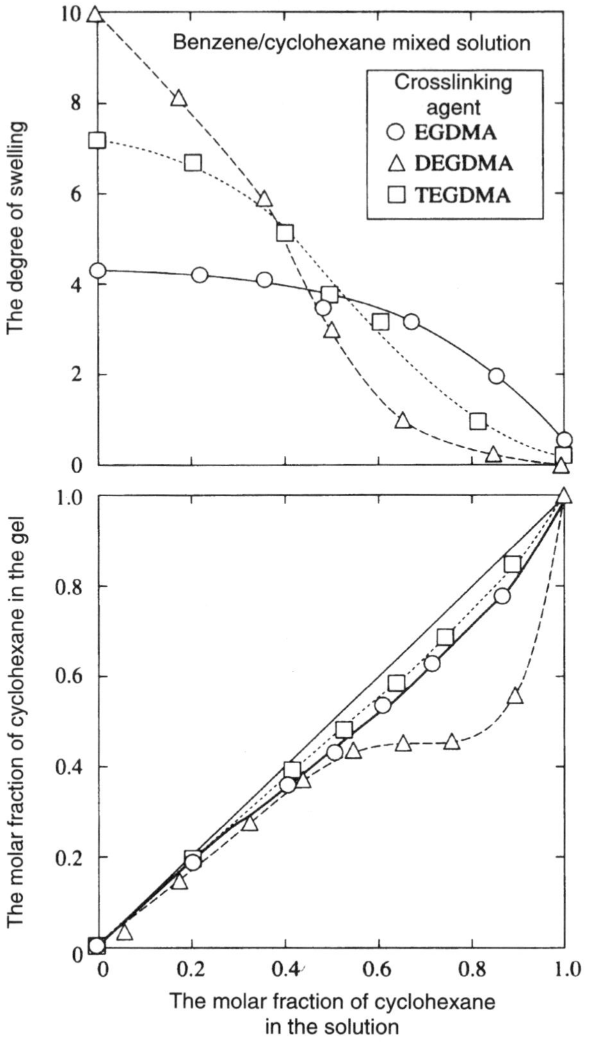

Fig. 11 The degree of swelling of DMAEMA gels synthesized with cross-linking agents and various lengths in benezene/cyclohexane mixed solution and the molar fraction of cyclohexane in the gel.

3.1.4 Separation of Mixed Solvents by Gel Membranes

Thus far, we have seen that the proper selection of the chemical and crosslink structures allows for selective absorption of the solvent of interest into the gel. However, selective absorption takes place at a low degree of swelling and thus it is impractical to directly use this phenomenon for concentration and separation processes. It is important to consider how such a function actually can be used. As an approach, it is possible to prepare a gel membrane and use it for concentration and separation processes.

3.1.4.1 Preparation method of thin gel membranes and separation experiments

When a gel is used as a membrane, the most difficult problem is how to prepare a thin membrane film due to the fact that it has low mechanical strength [13]. The use of a substrate can be a natural solution. If a gel is supported in the small pores of a porous polymeric membrane or an inorganic membrane, a mechanically stable thin gel membrane can be prepared. The authors used a thin silica-alumina porous membrane as shown in Fig. 12 in order to avoid the swelling of the substrate itself. This substrate was prepared by silica treatment on the surface of the thin alumina membrane, which was prepared by the sol-gel technique on the outer surface of a porous α-alumina tube. The thickness of the membrane was approximately several μm. The pore size can be somewhat controlled by the particle size of the alumina sol and its number of coating operations. The control of the pore size of the thin membrane is extremely important for proper use of the gel characteristics. There seems to be an optimum micropore size [14]. The majority of pore size used in this experiment was approximately several tens of nm. As pore size was large, there was no selective absorption of solvent by the substrate itself.

Gel support was accomplished by absorbing a polymerization initiator in the micropores and the tube was immersed into the mixed solution of the primary monomer and crosslinking agent. The separation experiment is performed by the pervaporation method.

3.1.4.2 Control of the effective micropore size of gel networks and separation characteristics

In order to achieve sifting on the molecular level, it is necessary to make the effective micropore size sufficiently small [15]. As stated earlier, if the crosslink density is incresed to achieve this, heterogeneous networks will

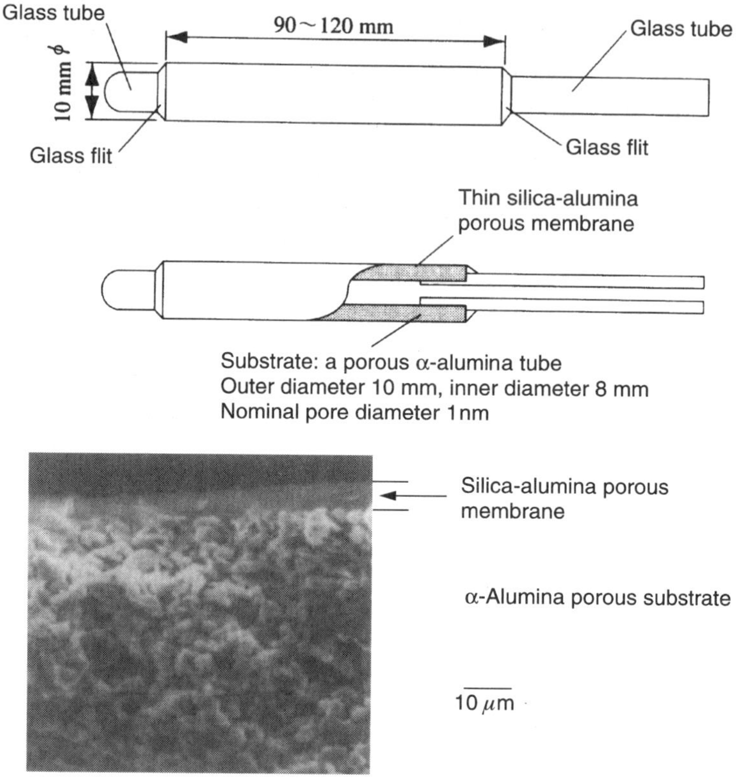

Fig. 12 The appearance of a thin membrane module of porous silica-alumina as the support for the gel and SEM photomicrograph of the membrane cross-section.

be created in the case of hydrophilic gels and selectivity is reduced. Therefore, a transparent gel, that is a gel with approximately homogeneous networks, is used to reduce the effective micropore size by substrate support. This is nothing but formation of interpenetrating polymer networks [16].

Figure 13 shows the effect of repeated use of supported gel membranes on the separated ethanol aqueous solution using as an example an acrylamide-supported gel membrane. The composition of the supported gel membrane is listed in Table 3 and the gel formed with this composition is approximately transparent as shown in the photograph in Fig. 9. Even if a supported gel membrane is used, the permeation flux of pure water is large, indicating that a sufficiently thin membrane has been properly prepared. As the ethanol concentration increases in the upper stream, the permeation flux of water monotonously decreases. In contrast, the permeation flux of the ethanol first increases, then decreases, and again increases. This indicates that, as expected, the gel membrane exhibited selective separation at high concentrations. However, the permeation flux of the ethanol with a single application of the gel is fairly high and the ethanol fraction of the upper stream is about that expected by Fig. 3. It is thus difficult to claim that a sufficient separation took place. If the gel is further added to the substrate, the permeation flux of the ethanol is drastically reduced. On the other hand, the permeation flux of the water reduces in a convex manner as the ethanol concentration increases in the upper stream. The permeation flux increases when the concentration of the ethanol in the upper stream is more than 20 mol% for the gel with triple applications. Such a phenomenon appears when the degree of swelling suddenly decreases as the selectively absorbing solvent (in this case ethanol) increases slightly. As a result, the ethanol concentration in the down stream reduces drastically, indicating successful preparation of a highly selective separation membrane.

Accordingly, control of the effective micropore size of the networks by forming a gel membrane can be achieved relatively easily, allowing for concentration and separation functions in the gel.

3.1.4.3 Separation of various organic solvent aqueous solutions by a supported acrylamide membrane

Figure 14 shows the results of the separation experiments of various organic solvent aqueous solutions using a repeatedly applied acrylamide membrane [10]. The gel was applied 6 times. The permeation flux of water differs significantly depending on the solvent. However, the permeation flux of water in any aqueous solution is quite high. In addition, in some solvents flux reduction appears in a convex manner. This is closely related to the gas-liquid equilibrium of organic solvent aqueous solutions. The permeation flux of organic solvents is quite low except for methanol and ethanol, which have strong compatibility with water. As a result, very high separation efficiency is obtained. The order of the efficiency is the same as in Fig. 3, indicating that the selective absorption of these solvents is recreated by the gel membrane.

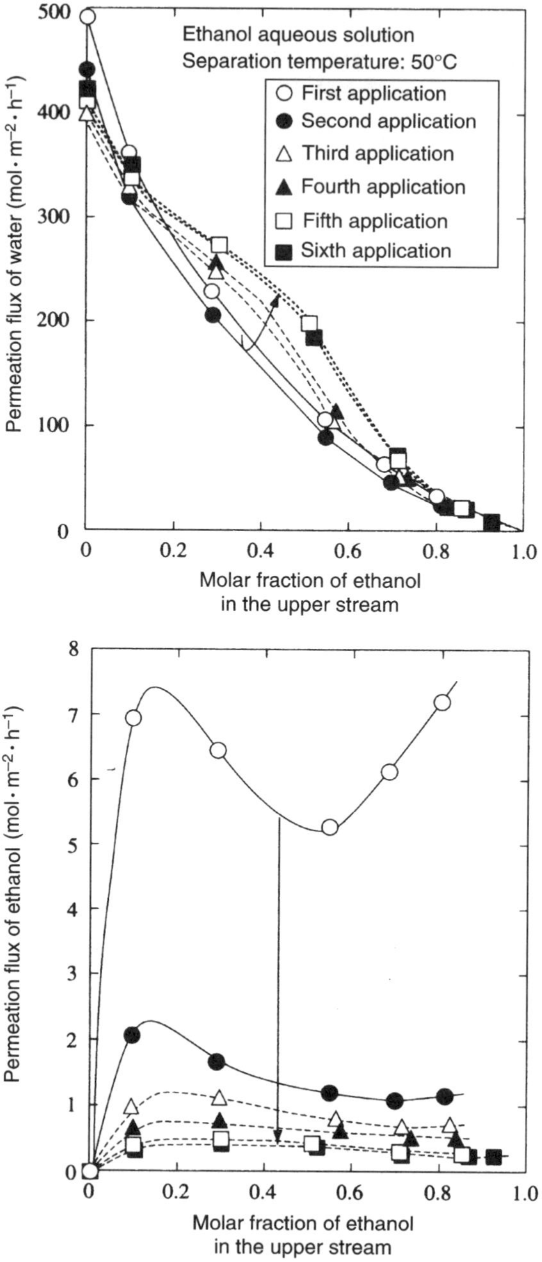

Fig. 13 Effect of repeated application of the gel in the separation of ethanol aqueous solution using an acrylamide-supported gel membrane.

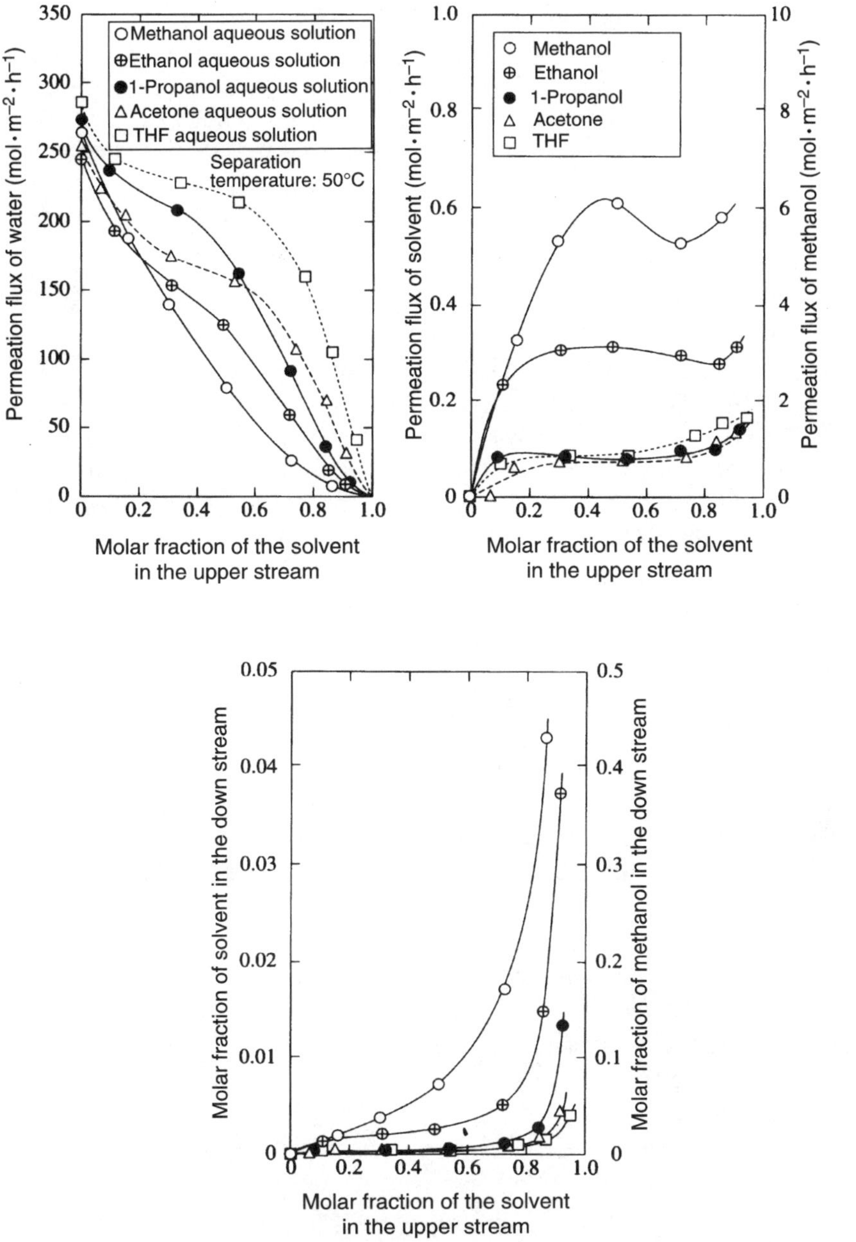

Fig. 14 Results of the separation experiments of various organic solvent aqueous solutions using a supported acrylamide membrane.

If the separation efficiency is evaluated by the separation coefficiency as defined by the following equation, it is on the order of several thousands except in the case of methanol and ethanol,

$$\alpha = \frac{x}{1-x} \cdot \frac{1-y}{y} \tag{1}$$

where x and y are the molar fraction of organic solvents in the up and down streams, respectively.

This gel membrane was used for several months by changing the organic solvent aqueous solutions. Every time the organic solvent aqueous solution was changed, the permeation flux of water increased. Thus, repeat experiments were attempted for the same organic solvent aqueous solution and separation efficiency was evaluated. In this case, the permeation flux of water increases whereas that of the organic solvent stayed almost the same or rather increased, indicating improvement in separation efficiency. This was probably because the amide group was hydrolyzed into a more hydrophilic carboxyl group. Although this phenomenon is a problem from the point of view of gel stability, it is an interesting phenomenon if one is interested in designing and synthesizing gels with higher separation efficiency.

3.1.4.4 Separation of a polar/nonpolar mixed solvent by a supported dimethylacrylamide membrane

As already described, a polar solvent can be selectively separated from a polar/nonpolar mixed solvent using a supported amphoteric dimethylacrylamide gel membrane [1]. Figure 15 shows examples of the separation from ethanol/benzene or ethanol/cyclohexane mixed solution. The membrane used was prepared by repeating the gel application four times. Ethanol, which has better compatibility with the gel, selectively permeates. However, similar to Fig. 6, benzene leakage is significant while the permeation flux of cyclohexane is very small. Expectedly, very high separation efficiency has been obtained for the ethanol/cyclohexane mixed solution.

An interesting relationship can be found by comparing the compatibility of the gel and these solvents with the permeation characteristics of these solvents through the gel membranes. Figure 16 illustrates the apparent permeability of this gel membrane in polar solvents like methanol and ethanol and nonpolar solvents like benzene and cyclohexane at various temperatures. Of particular interest is that the permeability of

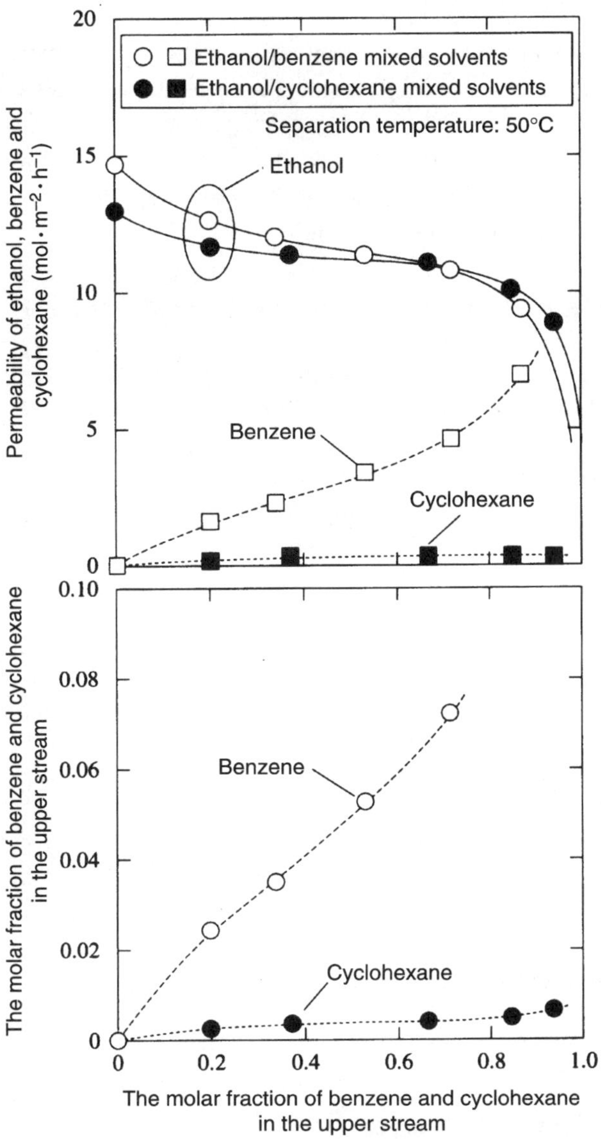

Fig. 15 Results on the separation of ethanol/benzene and ethanol/cyclohexane mixed solvents by a supported dimethylacrylamide membrane.

benzene is greater than that of ethanol. Moreover, the relation of temperature dependence to benzene permeability is different than for other solvents. The order of permeability does not necessarily agree with the order of the compatibility of the gel with the solvents. This result may show that in the separation of these mixed solvents, ethanol, which has great compatibility with the gel, absorbs strongly with the networks and interferes with benzene permeation.

3.1.4.5 Separation of benzene/cyclohexane mixed solvents by a supported DMAEMA gel membrane

Membrane separation is especially beneficial for separating mixed solvents with similar physical properties because they are difficult to separate by ordinary methods [8]. Benzene/cyclohexane is the model solution used for this purpose and various membranes have been evaluated for this solution. Unfortunately, sufficient separation has not been achieved at this point.

As shown in Figs. 8 and 11, DMAEMA gel shows selective absorption of benzene. Thus, it is possible to separate it using a supported membrane as shown in Fig. 17. This figure compares the aforementioned

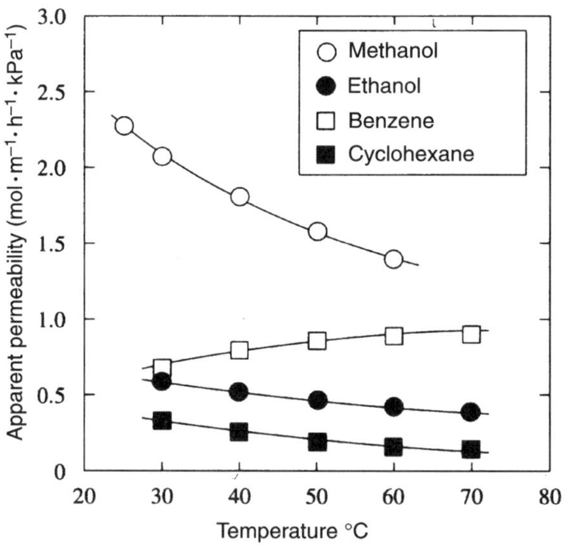

Fig. 16 Apparent permeability of a supported dimethylacrylamide membrane in various solvents.

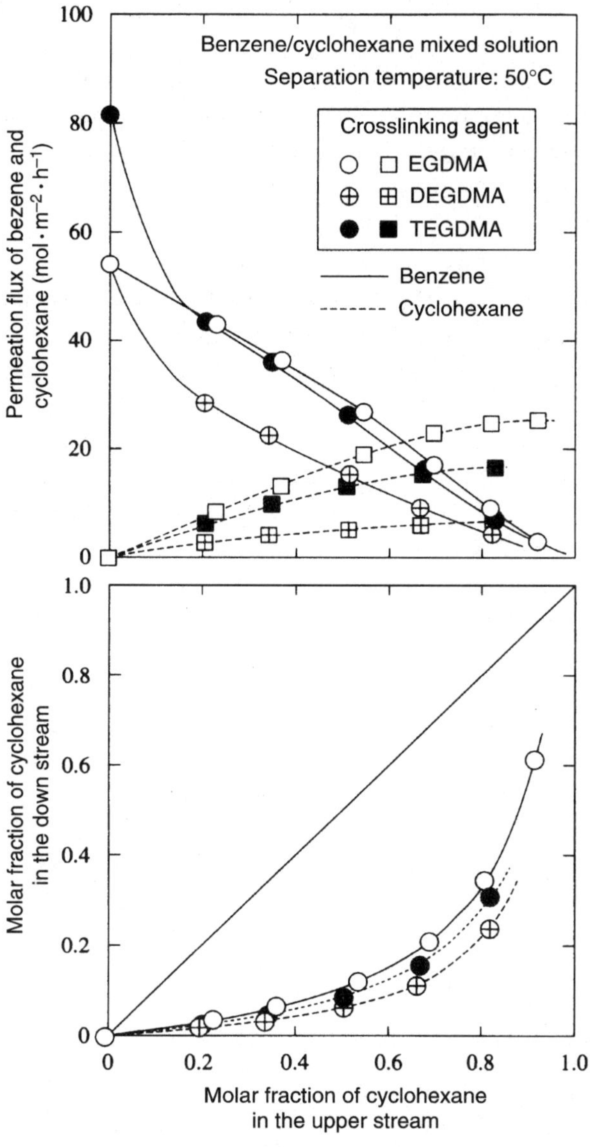

Fig. 17 The influence of the length of the crosslinking agent on the separation of benezene/cyclohexane mixed solutions using a supported DMAEMA gel membrane.

three kinds of gels whose crosslinking agent lengths differ. The different permeation flux for each membrane is probably due to differences in thickness. However, as expected, benzene selectively permeates. Of those three crosslinking agents, DEGDMA shows the highest selective absorption, as shown in Fig. 8. This membrane was not prepared by repeat polymerization and, thus, selective separation is not sufficient. However, selective separation of the gel membrane is considerably higher than the gel itself, which is shown in Fig. 11. It is expected that further improvement can be achieved by evaluating the micropore structure and supporting method of the inorganic membrane, which is used as support for the gel.

3.1.5 Conclusions

The selective absorption of solvents in a gel is fundamentally determined by the chemical structure of the primary monomer. However, to improve selectivity, it is extremely important to control the crosslinking structure. In this section, gel membrane application is introduced as an example of practical use. However, further usefulness is expected in the area of membrane separation for organic solvent mixtures where many azeotropic solvent mixtures exist. For practical application, it is necessary to study the relationship between the chemical or crosslinking structure that has selective absorption, and the mechanism of the solvent, which permeates through the membrane.

3.2 ADSORPTION

JIAN-PING GONG

3.2.1 Introduction

Gels that have fixed ions on the network (i.e., polyelectrolyte gels) exhibit anomalous phenomena such as volumetric phase transition [17], electric shrinkage [18] and existence of nonfrozen water [19]. Authors calculated the static potential of ionic polymer networks with the 3D numerical method using the approximation from the two-dimensional (2D) stacking model and further using the Gauss-Zeidel method [20]. In such a rigid network model, the mobility of the networks and the condensation effect of the counter ions [21] are not considered. However, it is found by this calculation that a deep potential valley near the chain and a deep potential well exist at every joint where there are strong electric fields. High water absorption, metallic ion absorption, ion exchange ability, and the existence of a large amount of nonfrozen water of polyelectrolyte gels can be related to this strong electric field.

In addition to static forces, the driving forces for absorption of polymer gels are hydrophobic interactions between polymer networks and adsorbate, hydrogen bonding, and the van der Waals force. In this section, the goal is to avoid a discussion on ordinary adsorption. Instead, interactions between surface active agents and polymer gels, which use static and hydrophobic interactions as driving forces, will be described. For ion exchange resins, readers are referred to several monographs [22–25].

When a highly swollen polyelectrolyte gel is immersed in an aqueous solution of a surfactant, the volume of the gel suddenly shrinks. Further, if a hydrophobic gel that does not swell in water is placed in the same solution, the gel swells. These phenomena are caused by the adsorption of the surfactant onto the gel. The adsorption of the surfactant onto the gel is caused by static interaction (when it is a polyelectrolyte gel) or hydrophobic interaction. Gels can change from hydrophilic to hydrophobic and vice versa by interacting with surfactant molecules. As a result, the properties of the gel change drastically. There are many reports on the interaction between surfactants and linear polymers [26–29]. A Russian group reported on work on three-dimensionally crosslinked polymer gels [30–32]. Later, other authors developed a worm-like device that moves around by stretching and shrinking its body using this interaction [33–35]. Such a device attracted much attention.

3.2.2 Interaction Between Nonelectrolyte Gels and Surface Active Agents

It is well known that amides possessing a strong hydrophobic group, such as N-isopropylacylamide (NIPAAm), exhibit a volumetric phase transition at high temperatures in water [17]. Gels that possess such a hydrophobic group interact with a surfactant. As a result, both phase transition temperature and width of the transition significantly increase. Figure 1 shows the phase transition of a NIPAAm gel in an anionic surfactant (sodium dodecyl sulfate; SDS), a cationic surface active agent (dodecyl trimethyl ammonium chloride; DTAC), and a neutral surface active agent (nona-oxy-ethylene dodecyl ether; NODE) [36]. These three agents possess an alkyl group with the same length. Nonetheless, the effect on phase transition is different as shown in Fig. 1. This is because the increase in the phase transition temperature is due to the adsorption of the surfactant onto the gel by hydrophobic interaction. Thus, the gel turns into an ionic gel. Figure 2 depicts the adsorption of the surfactant in a SDS

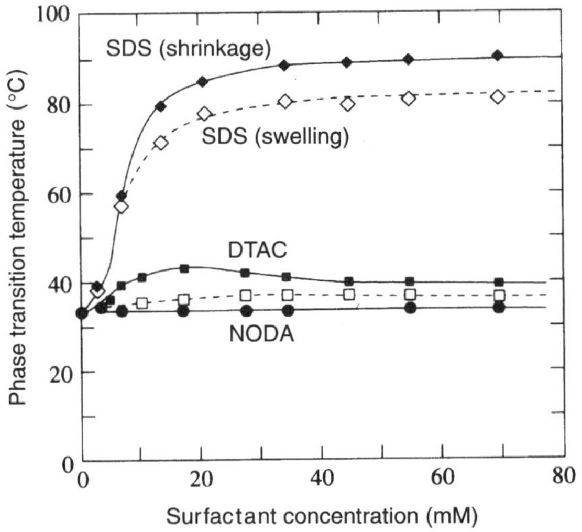

Fig. 1 Surfactant concentration dependence on volumetric phase transition temperature of a NIPAAm gel in SDS, DTAC and NODE solutions [36].

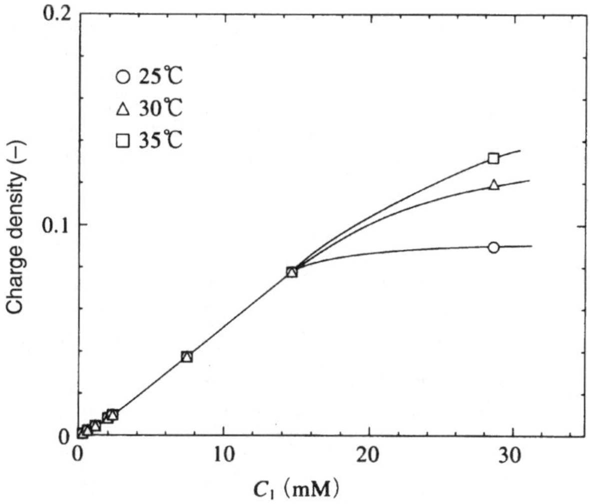

The initial concentration of SDS is shown on the abscissa; the ratio of the absorbed SDS with respect to the functional group of NIPAAm is shown on the ordinate.

Fig. 2 The adsorption ratio of SDS at various concentrations on a NIPAAm gel [37].

solution onto the NIPAAm gel [37]. The abscissa of Fig. 2 gives the SDS concentration and the ordinate gives the ratio of the SDS that adsorbed onto the functional group of NIPAAm. Adsorption reached equilibrium at $\approx$10% of the ratio. Static repulsion apparently increased by adsorption and the adsorption ceased when the hydrophobic interaction and the static repulsive forces were balanced.

3.2.3 Interaction Between Polyelectrolyte Gels and Surfactants

Polyelectrolyte gels with strong static potential also adsorb ionic surfactant with opposite charge. Figure 3 shows the adsorption isotherms of the cationic surfactant, N-dodecylpyridinium chloride (C_{12}PyCl), on the anionic polymer gel, poly (2-acrylamide-2-methylpropanesulfonic acid) (PAMPS). This figure is compared with the corresponding linear polymer solution [38].

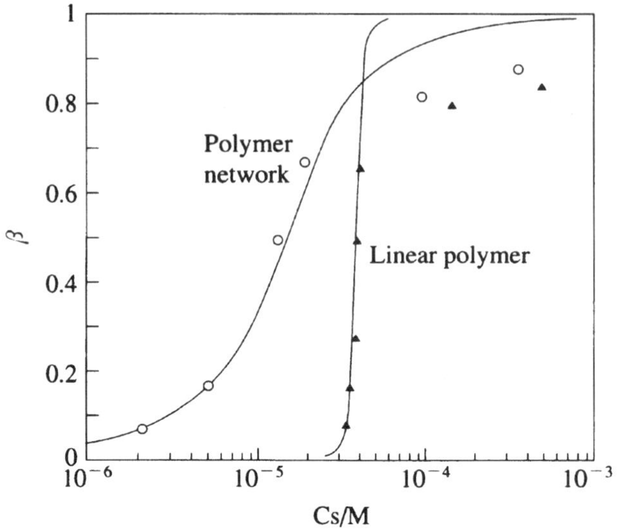

The curve is the theoretical value and the points are experimental values.

Fig. 3 Adsorption isotherms for the complex formation of $C_{12}PyCl$ and linear PAMPS or PAMPS gel [38, 41].

In solution, adsorption suddenly begins at a certain concentration of surfactant. The slope is steep and the reactivity exhibits a high cooperative nature. On the other hand, the surfactant adsorbs at a much lower concentration. Moreover, the slope is shallow and the reaction shows little cooperative tendency. However, when a salt is added to this gel, the adsorption isotherm of the gel shifts to a higher temperature and the slope becomes steep. In this case, the contribution of hydrophobic bonding becomes important [40] (see Fig. 4).

This adsorption can be divided into: (1) the "initiation process" where the first surfactant bonds to polymeric ions; and (2) the "growth process" (cooperative process) where the surfactant molecules interact by hydrophobic interaction with each other. Judging from the salt effect, as can be seen from Fig. 5, the initiation process is controlled by static interaction.

On the contrary, when the surfactants adsorb side by side, hydrophobic interaction between the side chains acts in addition to static interaction, resulting in a gain of excess energy. This is the main reason why cooperativity appears. The gel loses its static potential by adsorption

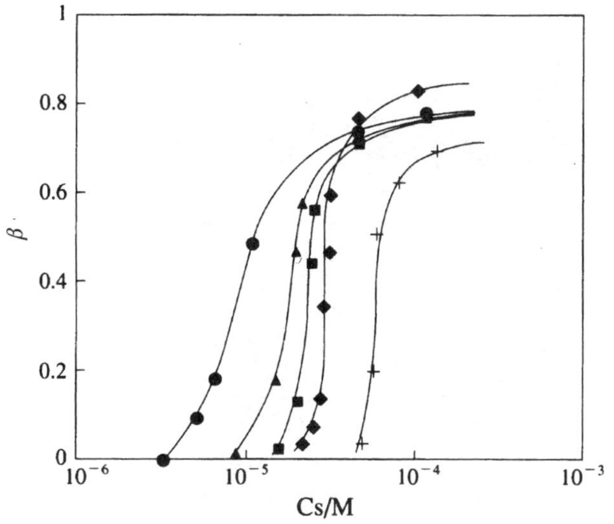

Concentration of sodium sulfate: (●) No salt, (▲) 1×10^{-3}M, (■) 3×10^{-3}M, (◆) 5×10^{-3}M, (+) 1×10^{-2}M

Equilibrium time: 14 days, 25°C

Fig. 4 The adsorption isotherms at the complex formation of C_{12}PyCl and PAMPS gel in sodium sulfate at various concentrations [39].

and thus exhibits volumetric shrinkage. At that time, the gel develops a different counter ion distribution in and out of the gel. This leads to generation of osmotic pressure, which then prevents the gel from shrinking easily.

On the other hand, the isolated linear polymer can deform freely. This is why cooperativity is lost in a gel. Cooperativity appears upon the addition of a large quantity of salt because the networks can freely deform due to the reduction in osmotic pressure (see Fig. 5).

3.2.4 Thermodynamic Models for Adsorption

The adsorption as described here is caused by: (1) the "initiation process" where the first surface active agent bonds to polymeric ions; and (2) the "growth process" (cooperative process) where the surfactant molecules interact with each other hydrophobically. When the surfactant molecules form a complex with the polymer chain, hydrophobic interaction between the neighboring surfactant molecules is formed and provides more energy

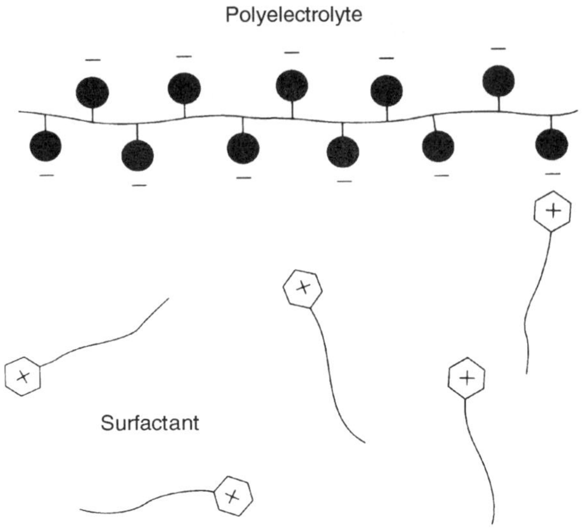

(a) Initiation process
(mainly static interaction drives the process)

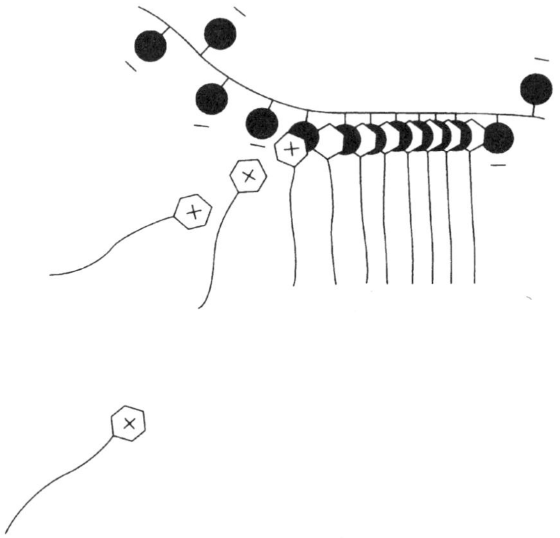

(b) Cooperative process
(hydrophobic interaction cooperatively drives the process

Fig. 5 The adsorption process of a cationic surfactant on an anionic polymer.

than random adsorption can. Hence, adsorption takes place under certain sequences rather than randomly. The strength of the hydrophobic interaction between the surfactant molecules depends on the hydrophobicity of the surfactant, the steric structure, and the distance between the adsorption sites on the polymer chain. The stronger the hydrophobic interaction between the surfactant molecules, the more the tendency to adsorb sequentially. Before discussing the adsorption theory of 3D networks, we will discuss the adsorption of surfactants in a linear polyelectrolyte solution.

3.2.4.1 Interaction with linear polyelectrolytes

Let us consider N mol of polyanion and M mol of surfactant in volume V [41]. The free energy of this solution F can be approximated by the sum of the following three terms:

$$F = F_{\text{int}} + F_{\text{mobile}} + F_{\text{comp}} \tag{1}$$

where F_{int} is the free energy of mixing of the polymer and solution, F_{mobile} the translational kinetic energy of low molecular weight ions, and F_{comp} the free energy of the complex formation between the surfactant and polyelectrolyte. According to the lattice model of Flory and Huggins [42], the following equation is given,

$$F_{\text{int}} = RT\frac{V}{v_c}\left[\frac{\phi}{m}\ln\phi + (1-\phi)\ln(1-\phi) + \chi\phi(1-\phi)\right] \tag{2}$$

where V is the volume of the solution, ϕ the volumetric fraction of the polymer, χ the Flory-Huggins interaction parameter, m the degree of polymerization of the polymer chain, and v_c the average molar volume of the chemical repeat unit of the polymer and solvent. Here R and T are the gas constant and absolute temperature, respectively.

If the ratio of the number of moles of the adsorbed surfactant and the chemical repeat units of the polymer is expressed by the degree of bonding β, the concentration of the surfactant in the solution at equilibrium can be expressed as $C_{\text{s}}^{\text{p}} = (M - N\beta)/V$. The superscript p and subscript s indicate a linear polymer and polymer gel, respectively. The kinetic energy of the low molecular weight material, F_{mobile}, can be expressed by the following equation:

$$F_{\text{mobile}} = RT\left[(M - N\beta)\ln\frac{(M - N\beta)v_c}{V} + M\ln\frac{Mv_c}{V} + N\ln\frac{Nv_c}{V}\right] \tag{3}$$

Next, let us consider the free energy F_{comp} upon complex formation between a surfactant and a polyelectrolyte. As described at the beginning of this section, the formation of a complex between a surfactant and polyelectrolyte is due to the static interaction between the surfactant and polyelectrolyte and hydrophobic interaction between the surfactant molecules. Let us express the static energy change ΔF_e upon formation of the complex between the surfactant and polyelectrolyte. When the chain length of the polymer is sufficiently long and terminal effects can be ignored, ΔF_e corresponds to the depth of the static energy valley on the polymer chain surface. For simplicity, let us assume that ΔF_e is constant upon complex formation. Only the nearest neighbor interaction will be considered for the hydrophobic interactions among surfactant molecules. Thus, when the surfactant forms a complex with the polymer chain, the surfactant molecules develop hydrophobic interaction with the energy gain of ΔF_h. When only nearest neighbor interaction is considered, F_{comp} can be expressed by expanding the Ising model [43]. Also see Reference [44].

$$\frac{\partial F_{comp}}{\partial \beta} = N_A N(\Delta F_e + \Delta F_h) + RTN \times \ln \frac{\sqrt{4\beta(1-\beta)\left[\exp\left(-\frac{\Delta F_h}{kT}\right) - 1\right] + 1} + 2\beta - 1}{\sqrt{4\beta(1-\beta)\left[\exp\left(-\frac{\Delta F_h}{kT}\right) - 1\right] + 1} - 2\beta + 1} \tag{4}$$

where N_A is the Avogadro number. The partial differentiation of the free energy F of the system with respect to β is zero for the interaction of the surfactant and polyelectrolyte at equilibrium. From this condition, the adsorption isotherm equation can be obtained as follows:

$$\ln C_s^p v_c = \frac{\Delta F_e + \Delta F_h}{kT} - 1 + \ln \frac{\sqrt{4\beta(1-\beta)\left[\exp\left(-\frac{\Delta F_h}{kT}\right) - 1\right] + 1} + 2\beta - 1}{\sqrt{4\beta(1-\beta)\left[\exp\left(-\frac{\Delta F_h}{kT}\right) - 1\right] + 1} - 2\beta + 1} \tag{5}$$

The first term on the right-hand side of Eq. (5) is based on the strength of the static and hydrophobic interactions and is independent of the degree of adsorption β. The second term is a step function of the degree of adsorption β. Its slope becomes steep as ΔF_{h} increases. Accordingly, the adsorption isotherm curves of linear polymers can be characterized by the critical adsorption concentration in the first term (initiation process) and cooperativity of adsorption in the second term. The critical adsorption concentration depends not only on static interaction but also on the strength of the hydrophobic interaction. In contrast, the cooperativity of adsorption depends only on hydrophobic interaction. If the cooperative interaction of the surfactant and polyelectrolyte is expressed by the inverse slope of the adsorption isotherm curve at $\beta = 0.5$, the following relationship is obtained for a linear polymer,

$$\left(\frac{d \ln C_{\mathrm{s}}^{\mathrm{p}}}{d\beta}\right)_{\beta=0.5} = 4 \exp\left(\frac{\Delta F_{\mathrm{h}}}{2kT}\right) \tag{6}$$

Therefore, the slope of the curve reflects the strength of the hydrophobic interaction between the nearest neighbor surfactant molecules. When $\Delta F_{\mathrm{h}}/kT \ll 0$, the right-hand side of the preceding equation approaches zero. At the same time, the slope of the adsorption isotherm curve becomes very large and strong cooperativity can be observed on the interaction between the surfactant and the polyelectrolyte. Here we will define the cooperativity parameter u and stability constant of the initiation process, K_0, as follows:

$$\left(\frac{d \ln C_{\mathrm{s}}^{\mathrm{p}}}{d\beta}\right)_{\beta=0.5} = \frac{4}{\sqrt{u}} \tag{7}$$

$$\left(\frac{1}{C_{\mathrm{s}}^{\mathrm{p}}}\right)_{\beta=0.5} = K_0 u \tag{8}$$

Inserting these into Eq. (5), we obtain,

$$u = \exp\left(-\frac{\Delta F_{\mathrm{h}}}{kT}\right) \tag{9}$$

$$K_0 = e v_{\mathrm{c}} \exp\left(-\frac{\Delta F_{\mathrm{c}}}{kT}\right) \tag{10}$$

Apparently, K_0 expresses only the static interaction between the surface active agent and polymer ions and corresponds to the adsorption constant of the isolated surfactant molecules. The ability to associate neighboring surfactant molecules via hydrophobic interaction is expressed by u. When $\Delta F_h < 0$, the former adsorption accelerates the next adsorption and, thus, it exhibits cooperativity. When $\Delta F_h > 0$, the former adsorption retards the next absorption. When $\Delta F_h = 0$, the adsorptions are independent of each other and become random.

3.2.4.2 Interaction with polyelectrolyte gels

For linear polymers, let us consider the system with total volume V, gel volume V_g, the number of polyanion units N, and the concentration of the cationic surface active agent M mole. The free energy of the system F is the sum of the F_s of the solution and F_g of the gel phase. The unit F_g includes the elastic energy of the gel F_{el}, in addition to F_{init}, F_{mobile}, and F_{comp}:

$$F_g = F_{int} + F_{mobile} + F_{comp} + F_{el} \tag{11}$$

If we assume the surfactant, its counter ion, and the counter ion of the polymer ion in the external solution ($i = s$) are S_i^+, S_i^-, and P_i^+, respectively, from the electroneutrality principle, we obtain,

$$S_s^+ + P_s^+ = S_s^- \tag{12}$$

$$S_g^+ + P_g^+ = S_g^- + N(1-\beta) \tag{13}$$

Furthermore, from the mass conservation law,

$$M = S_s^+ + S_g^+ + N\beta \tag{14}$$

$$M = S_s^- + S_g^- \tag{15}$$

$$N = P_s^- + P_g^- \tag{16}$$

If we write $\alpha = S_g^+/N$, $\gamma = S_g^-/N$, then $S_s^+ = M - N(\alpha+\beta)$, $S_s^- = M - N\gamma$, $P_g^+ = N[1+\gamma-(\alpha+\beta)]$, and $P_s^+ = N[(\alpha+\beta)-\gamma]$. Since the free energy of the external solution phase is due to the translational movement of low molecular weight ions, the following relationship is obtained,

$$F_s = RT\left\{[M - N(\alpha+\beta)]\ln\frac{[M - N(\alpha+\beta)]v_c}{V - V_g}\right\} + (M - N\gamma)\ln\frac{(M - N\gamma)v_c}{V - V_g} + N(\alpha+\beta-\gamma) \times \ln\frac{N(\alpha+\beta-\gamma)v_c}{V - V_g} \tag{17}$$

where V_g is the equilibrium volume of the gel in the surfactant solution. If the degree of polymerization m is replaced by infinity and the volume is V_g in the linear polymer solution equation, F_{int} of the gel is obtained,

$$F_{int} = RT\frac{V_g}{v_c}[(1-\phi)\ln(1-\phi) + \chi\phi(1-\phi)] \tag{18}$$

The translational kinetic energy of the low molecular weight ions in the gel is

$$F_{mobile} = RT\left[N\alpha\ln\frac{N\alpha v_c}{V_g} + N\gamma\ln\frac{N\gamma v_c}{V_g} + N(1+\gamma-\alpha-\beta)\ln\frac{N(1+\gamma-\alpha-\beta)v_c}{V_g}\right] \tag{19}$$

The elastic energy of the gel is

$$F_{el} = \frac{3}{2}RTv_e\left[\left(\frac{\phi_0}{\phi}\right)^{2/3} - 1 - \ln\left(\frac{\phi_0}{\phi}\right)^{1/3}\right] \tag{20}$$

where ϕ_0 is the volume fraction of the gel in the reference condition. The free energy F_{comp} upon the complex formation between the surface active agent and polyelectrolyte gel is the same as that of the linear polymer. Here, we assume that the change of static energy ΔF_e upon the complex formation between the surface active agent and polymer chains is constant along the polymer chain in the same manner as for the linear chain.

Similar to the linear polymer case, because the partial differential of the free energy of the system with respect to β is zero, the adsorption isotherm equation of the gel is obtained as follows:

$$\begin{aligned}\ln C_s^g v_c = {} & \frac{\Delta F_e + \Delta F_h}{kT} - 1 \\ & + \ln\frac{\sqrt{4\beta(1-\beta)\left[\exp\left(-\frac{\Delta F_h}{kt}\right) - 1\right] + 1} + 2\beta - 1}{\sqrt{4\beta(1-\beta)\left[\exp\left(-\frac{\Delta F_h}{kt}\right) - 1\right] + 1} - 2\beta + 1} \\ & + \ln\frac{(\alpha+\beta-\gamma)V_g}{[1-(\alpha+\beta-\gamma)](V-V_g)}\end{aligned} \tag{21}$$

where V_g is the volume of the gel and V is the volumetric sum of the gel and the external solution.

As can be seen by comparison with the adsorption isotherm equation [Eq. (5)] of linear polymers, the last term of Eq. (20) is due to the crosslinking of the gel. Since the term can be rewritten as follows, it can be seen that this is due to the osmotic pressure caused by the difference in macrocounter ions in and out of the gel:

$$\ln \frac{(\alpha+\beta-\gamma)V_g}{[1-(\alpha+\beta-\gamma)](V-V_g)} = \ln \frac{[P_s^+]}{[P_g^+]} \tag{22}$$

where $[P_g^+]$ and $[P_s^+]$ are the macrocounter ion concentrations inside and outside of the gel, respectively. In the case of polyelectrolyte gels, the inverse slope of the adsorption isotherm curve is

$$\left(\frac{d \ln C_s^g}{d\beta}\right)_{\beta=0.5} = 4\exp\left(\frac{\Delta F_h}{2kt}\right) + \frac{4}{1-4(\alpha-\gamma)^2} + \frac{1}{V_g}\frac{dV_g}{d\beta} \tag{23}$$

and the right-hand side of Eq. (23) will not approach zero even if $\Delta F_h/kT \ll 0$. Hence, the slope of the curve is insensitive to the hydrophobicity of the surfactant and is almost constant. It is seen that the cooperativity of the hydrophobic interaction is controlled by the network effect.

Similar to the linear polymer case, if the following definitions are adopted,

$$\left(\frac{d \ln C_s^g}{d\beta}\right)_{\beta=0.5} = \frac{4}{\sqrt{u^g}} \tag{24}$$

$$\left(\frac{1}{C_s^g}\right)_{\beta=0.5} = K_0^g u^g \tag{25}$$

we obtain,

$$u^g \approx 1 \tag{26}$$

This equation means that, in polymer gels, hydrophobic interaction is inhibited and adsorption usually takes place noncooperatively. In order for the neighboring surfactant molecules to interact hydrophobically, the polymer chain must be flexible and little energy consumption should accompany the deformation of the polymer chains upon association of surfactant molecules.

Cooperativity is inhibited in three-dimensionally crosslinked gels because the conformational changes of the main chains accompanying the aggregate formation are inhibited by the high osmotic pressure of the counter ions (see Fig. 6). Thus, if a large amount of low molecular weight salt exists, the osmotic pressure of the counter ions decreases. Accordingly, the conformation of the main chain can be changed freely and adsorption starts showing cooperativity. However, the static potential of the polymer chain is shielded by the salt and the initiation process of the adsorption weakens. Hence, the adsorption isotherm curve shifts to higher concentration.

Figure 3 shows the theoretical adsorption isotherm curves of a linear polyelectrolyte and polyelectrolyte gel using theoretical equations (5) and (21). Further, Table 1 lists the cooperativity parameter u, and experimental and theoretical stabilization constants K_0 of the initiation process when PAMPS gels with various degrees of swelling interact with CnPyCl. For comparison, the values of the linear polymers are also shown. According to Fig. 3, the theoretical values agree well with the experimental values

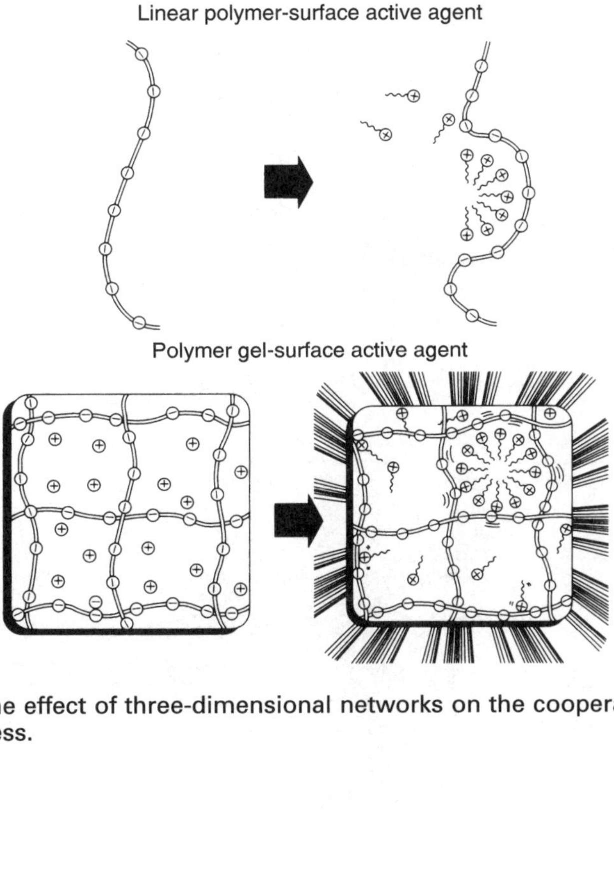

Fig. 6 The effect of three-dimensional networks on the cooperative adsorption process.

with $\beta = 0.7$. Thus, this theoretical model correctly depicts the interaction between the surfactant and the polyelectrolyte. However, when b increases, ΔF_e, which accompanies the adsorption, cannot be ignored. As the theoretical curve deviates from the experimental curve, the theory needs to be modified.

In addition to the thermodynamic models, Khokhlov *et al.* proposed a model for the hydrophobic interaction of an ionic surface active agent and electrolyte gel where the volume changes discontinuously in a gel due to the micelle formation [45]. For actual descriptions, the readers are referred to the original monograph.

3.2.5 Cooperativity of Adsorption and Steric Structure of Molecules

As discussed in the previous section, polymer chain flexibility is important for cooperative adsorption. Both a single polymer chain and 3D networks in a salt solution exhibit flexibility. In the case of the adsorption of N-pyridinium chloride (CnPyCl, $n = 4$, 8, 10, 12, 16, and 18) onto such flexible polymers, the cooperativity increases in proportion to the hydrophobicity (the length of the aliphatic chain) [40, 41]. In this section, the character of the adsorbed surfactant will be discussed.

Table 2 lists the cooperativity parameter u and total stability constant K, which are calculated from the adsorption isotherm curves of a bulky surface active agent, tetraphenylphosphonium chloride (TPPC) and PAMPS gel [46]. For comparison, the data for the interaction between $C_{12}PyCl$ and PAMPS gel are also shown. When a low molecular weight salt is lacking, both cases show a small cooperativity parameter u and the effect of polymer networks. Since the polymer network effect diminishes under the presence of a large amount of sodium sulfate (1×10^{-2} M), the

Table 1 Initiation stability constant K_0 and cooperativity parameter u for the interaction between surfactant, $C_{12}PyCl$, and PAMPS gel.

q^a	50	230	350	490	790	1700[b]
K_0 (l/mol) (calculated)	5.0×10^4	4.9×10^4	4.9×10^4	5.0×10^4	5.4×10^4	52
u (calculated)	1.3	1.6	1.7	1.8	1.8	490
K_0 (l/mol) (observed)	4.4×10^4	2.4×10^4	2.1×10^4	1.4×10^4	0.8×10^4	44
u (observed)	1.3	1.9	1.7	2.3	1.8	630

[a] The degree of swelling of the gel
[b] Linear polymer solution

Table 2 The interaction parameter of surface active agents and a PAMPS gel.

		TPPC	C_{12}PyCl
K/M^{-1}	No salt	3.2×10^3	3.2×10^4
	With salt[a]	3.1×10^2	1.7×10^4
u	No salt	0.8	5.5
	With salt[a]	6.6	710

[a] 1×10^{-2} M sodium sulfate

u of C_{12}PyCl dramatically increases and shows cooperativity. However, in the case of TPPC, the concentration at which the adsorption starts shifts to a higher value despite the introduction of a salt and u increases slightly. In other words, in spite of the bulky hydrophobic groups in TPPC, the neighboring hydrophobic interaction is extremely weak. This is possibly due to the effect of the 3D regular tetrahedron of TPPC, which exceeds the effect of molecular packing.

Accordingly, the hydrophobic interaction among surfactant molecules is not only determined by the hydrophobicity but also by the unique packing of the interacting molecules. Even though the molecules are highly hydrophobic, if the steric structure is not appropriate for suitable packing, hydrophobic interaction is weak and adsorption lacks cooperativity.

3.3 INTERACTION WITH NATURAL MATERIALS

AKIHIKO KIKUCHI AND MITSUO OKANA

3.3.1 Introduction

A hydrogel is a soft material that is swollen by water. Due to its softness and high water content, the hydrogel exhibits extremely unique properties when it is in contact with natural materials. First of all, due to its softness, it does not harm natural systems. In addition, the high water content enhances material permeability. Because the outermost layer of cells is covered by a polysaccharide of high water content, hydrogels may exhibit gentle interaction with natural cellular components [47]. In this section, the interaction between hydrogels and proteins or cells will be investigated. Furthermore, the application of a hydrogel for the separation of proteins or cells also will be discussed.

3.3.2 Interaction between Hydrophilic Surfaces and Natural Components

3.3.2.1 Interaction between hydrophilic polymer membranes and cells

In general, the interaction between a modified polymer surface and cells or proteins is considered to be an adhesion (adsorption) process and it is studied from an interfacial thermodynamic point of view. Folkman and Moscona [48] coated a solid surface with a relatively hydrophilic poly(2-hydroxyethyl methacrylate) (PHEMA) of various thicknesses and studied the adhesion and growth of a cell on this surface. They reported that, when membrane thickness was increased, the growth rate of the cell decreased. This may be due to enhanced hydrophilicity as thickness is increased. Andrade *et al.* [49] reported that the polymer surface, which minimizes interfacial free energy with blood components, minimizes the adhesion of blood cells (such as platelets) because it is not recognized as a foreign material. Therefore, such materials are useful as anticoagulants. Lydon *et al.* [50] evaluated a synthetic polymer surface and cell for interaction of the physicochemical properties of the interface. They used PHEMA as a hydrophilic material. By copolymerizing ethyl methacrylate (EMA) or styrene to PHEMA, they prepared the surface with controlled hydrophilicity or hydrophobicity and studied systematically the adhesion of the cell onto these surfaces. As a result, they found that the cells could not adhere onto the hydrophilic PHEMA surface. On the other hand, copolymerization of styrene or EMA contributed to the adhesion of the cells by decreasing the equilibrium degree of swelling. They also indicated that not only the water content but also the surface functional groups and the mechanical properties of the polymer surface influence the cell adhesion. Similar to the work by Lydon *et al.*, Horbett *et al.* [51] studied the adhesion and desorption of cells on the HEMA-EMA copolymer surfaces. They reported that the increase in the hydrophilic HEMA component reduces the adsorption of proteins and weakens cell adhesion. Ueda-Yuasa and Matsudo [52] prepared discretely changing surfaces from hydrophilic to hydrophobic on a single surface by hydrolyzing poly(vinylene carbonate) into poly(hydroxymethylene). When the concentration of hydroxyl groups increased by hydrolysis and subsequently increased hydrophilicity, adhesion and growth of cells from the internal blood vessel walls was reduced. It has been confirmed that the growth of adhered cells can be readily controlled. Accordingly, on hydrophilic surfaces, the adhesion of natural components is generally inhibited. The following studies have been reported.

3.3.2.2 Formation of highly hydrated surface layers using hydrophilic polymers

There have been attempts to inhibit the adhesion (adsorption) of natural materials such as cells by grafting a hydrophilic, flexible polymer, poly(ethylene glycol), onto a substrate surface [53–64]. At the surface, which is formed by grafting hydrophilic PEO chains, the interfacial free energy decreases in water as a result of sufficient hydration of the polymer chains. By the formation of highly hydrated surface layers, cells, or proteins such as albumin (alpha, beta, gamma, fibrinogen)-globulins. Compression of the surface-hydrated polymers by the adsorption (adhesion) of proteins is thermodynamically undesirable. There appears to be elastic repulsion between the PEO chains and adsorbates caused by an osmotic pressure mechanism. Namely, this inhibition is considered to be due mainly to the excluded volume effect (steric hindrance) of the polymer chains at the interface where formation of a highly hydrated surface with high mobility occurs. Upon evaluation of a PEO-grafted surface implanted in a stomach cavity, no inflammation or fibrilar encapsulation was observed. On the contrary, the untreated surface showed signs of inflammation followed by excessive fibrilar encapsulation from the very beginning of the test [65]. This indicates that the formation of a highly hydrated surface layer is useful in reducing the rejection reaction of the body. On such a surface, significant reduction not only of cells and proteins but also bacteria and viruses can be seen. Hence, it might be useful as a surface treatment for catheters [66].

In addition to inhibition of adsorption (adhesion) of natural body components due to PEO, an attempt was made to enhance blood type compatibility to the material surface using anticoagulant molecules. Heparin is an anticoagulant manufactured by the body. Many researchers [67–75] have reported that adhesion and aggregation of platelets can be significantly inhibited by bonding heparin to PEO molecules because of the hydrophilicity of PEO and the antithrombic activity of heparin. Heparin's antithrombic activity is caused by its role in preventing prothrombin from being converted to thrombin. Thus, Han *et al.* evaluated the anticoagulation ability on a surface where a sulfonate group (heparin is a polysulfuric acid ester in the mucoitin family) was added to the proximal end of a surface-grafted PEO as shown in Fig. 1. They found that this system exhibited anticoagulant ability similar to that of fixed heparin [76, 77].

Although improvements in blood compatibility of highly hydrated PEO surfaces yielded favorable results in *in vitro* or *ex vivo* evaluation, *in vivo* evaluation did not necessarily yield favorable results even when highly hydrated PEO surface layers formed but the surfaces were left for lengthy periods of time in slowed or decreased blood flow areas (e.g., extremities); both adsorption of proteins and adhesion of blood components were observed [58, 61, 75]. There seems to be an optimum graft density and graft length when protein adsorption and anticoagulation are the issues. For details, readers are referred to several reviews [78–80].

Grafted enzymes or proteins are applied at the free end of the PEO graft on the surface when the enzyme-linked immunosorbent assay (ELISA) method is used or when separation of specific cells is the goal (see Fig. 2). In such systems, the reduction in activity of the fixed enzyme (as described in a later section) can be avoided by the high molecular mobility of the grafted PEO chains on the surface. Moreover, nonspecific adsorption of proteins or cells can be inhibited. Therefore, a highly sensitive, highly efficient ELISA method is expected [81].

Using the inhibition of protein adsorption and nonadhering cells by highly hydrophilic PEO, and further preparing *in situ* a PEO

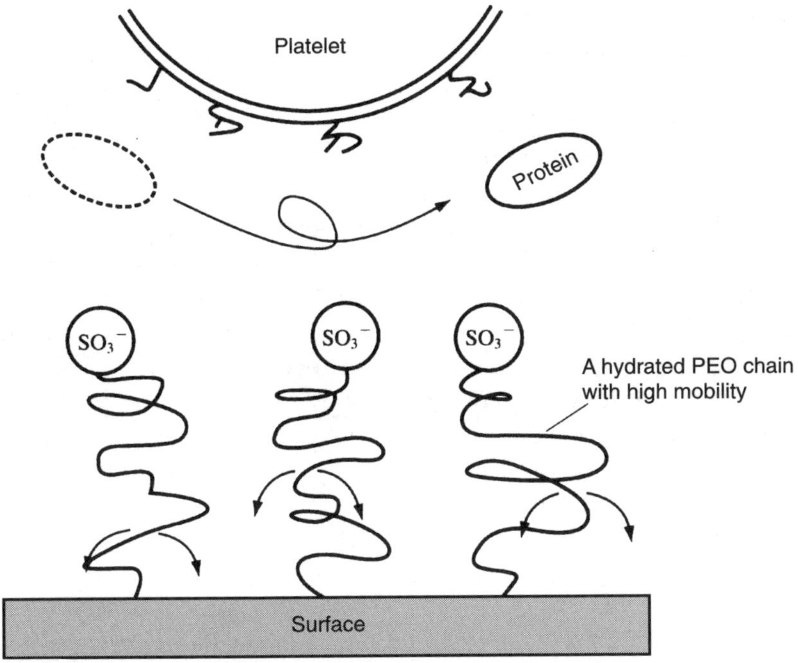

Fig. 1 The inhibition model of the interaction between grafted poly(ethylene oxide) (PEO) which has a terminal negative charge and proteins or platelets [76].

hydrogel endcapped at both ends with an acrylate group that uses PEO photopolymerization, attempts have been made to use these materials as antibody separation membranes [82–84]. Other methods that attempt to prevent adhesion during wound healing have used the PEO gel and a biodegradable fixing group [85–87]. At this time, if gelatin is mixed with the gel, a flexible, blood clotting adhesive can be prepared utilizing the adhesive nature of gelatin to natural systems or cells [88]. It is also possible to design an antisticking material, which gels quickly *in situ* by photoirradiation to a highly viscous solution with the photocrosslinking ability by introducing cinnamate to natural muco-poly-saccharides like hyaluronic acid and chondroitin sulfate. These polysaccharide gels do not stick to cells or organs. They are also interesting materials because they have low-inflammation, nonantibody, biodegradable, and bioabsorptivity characteristics simultaneously [89, 90].

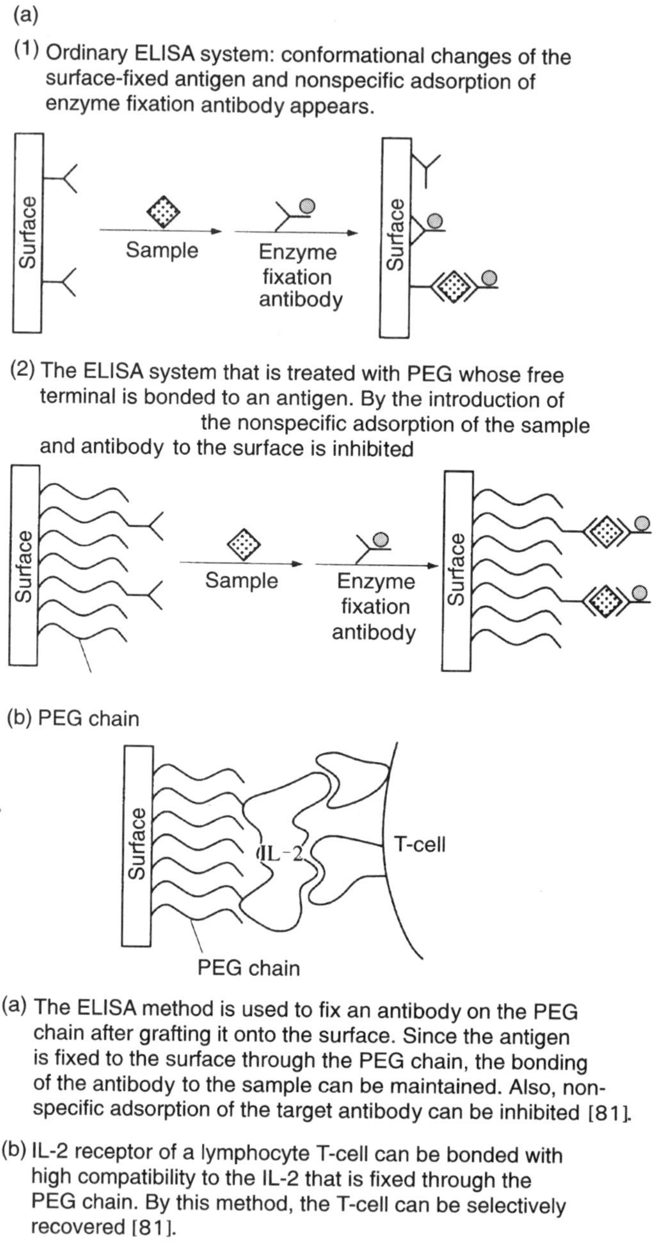

Fig. 2 Interaction between (a) ELISA system and (b) IL-2 fixed surface and lymphocyte.

For formation of highly hydrated surface layers through surface grafting of a highly hydrophilic polymer and concomitant inhibition of adhesion of proteins and cells, not only PEO but also neutral polymers such as poly(vinyl alcohol) (PVA) and poly(acrylamide) (PAAm) have been evaluated. They have been studied from various points of view [91–95]. The adsorption of serum albumin and fibrinogen onto such surfaces is reduced with increases in hydrophilicity and water content. At this time, reduced platelet adhesion to the surface has also been found. Hydrophilic materials with negative charges such as poly(acrylic acid) absorb large amounts of proteins. As a result, platelet interactions increase and adsorption tends to increase [92]. Thus, on highly hydrated neutral polymer surfaces, adhesion and activation of platelets are thought to be inhibited, which leads to anticoagulation. On these surfaces, substrate dependant cells like fibroblast cannot adhere. Furthermore, growth is also inhibited [93]. The fibroblast plays a key role in the rejection reaction when a foreign material enters the body. Thus, materials that can inhibit adhesion and growth of such cells require good biocompatibility. As neutral and highly hydrated polymers, PEO, PVA, and PAAm, dextrin and poly(N-vinyl pyrrolidone) are also being evaluated.

3.3.3 Stimuli Responsive Solid Surfaces

3.3.3.1 Design of surfaces that respond to external stimuli

By introducing stimuli responsive polymers to the surfaces of polymer, metal, glass and other materials it is possible to prepare high-performance surfaces. The structure of these surfaces can be changed by external signals like pH, temperature, and light. From this point of view, these surfaces can be called high-performance functional surfaces rather than the traditional functional surfaces. Recognizing that surface functions can be controlled by changes in surface structures using external stimuli, new materials separation and purification have been proposed (see Fig. 3) [96, 97]. In particular, separation and purification that control materials structurally through external stimuli are a possibility. This is done by a stimuli-responsive polymer introduced to the surface of a porous or swollen membrane in a high-to-low density range and/or a hydrophilic-to-hydrophobic range. Nagasaki *et al.* prepared poly(silamine), which has silyl and amine groups in the molecular chain and changes its molecular conformation in response

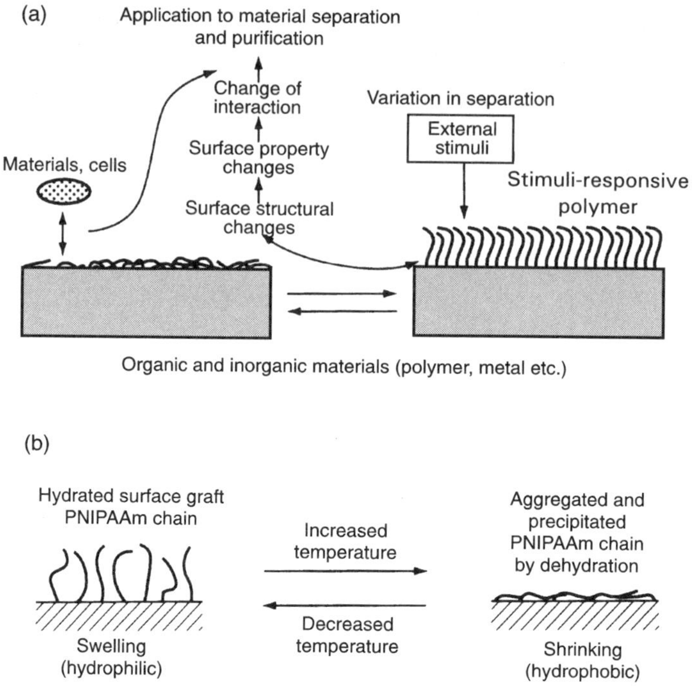

Fig. 3 (a) Application of a high performance surface treated with a stimuli-responsive polymer for separation and purification of materials. (b) Schematic diagram of the hydrophilicity and hydrophobicity changes of a thermoresponsive PNIPAAm graft surface upon temperature changes.

to pH and temperature [98, 99], using this polymer for surface treatment of a solid [100, 101]. The amine group in poly(silamine) changes its degree of protonation with changing pH. The degree of protonation is also influenced by temperature. It has been confirmed by an atomic force microscope that the change in the degree of protonation significantly influences molecules at the surface.

Furthermore, this type of molecule possesses reactive functional groups on both ends; one side can be used to fix the molecules on the solid surface and the other end can be used to fix a protein like an enzyme. Thus, it is possible to use it as a functional surface whose enzyme-substrate reaction or adhesion of cells can be controlled by responding to temperature and pH. It is hoped that progress will occur in this field in the future.

3.3.3.2 Changes in hydrophilicity and hydrophobicity on temperature changes and interaction with natural components

Takei *et al.* [102] attempted to add temperature responsiveness to a solid surface by grafting a temperature-responsive poly(N-isopropylacrylamide) (PNIPAAm) to an aminated glass surface [104]. The PNIPAAm dissolved in water <32°C, while at higher temperatures it became rapidly insoluble due to dehydration. The surface with this PNIPAAm graft exhibited hydrophilicity due to hydration of the PNIPAAm chains at low temperatures. This surface aggregates upon heating by dehydrating the water molecules and becoming hydrophobic (see Fig. 3(b)). As a result, the contact angle of the surface is 51° at 20°C in contrast to 87° at 26°C. Whereas drastic changes in hydrophilicity and hydrophobicity can be observed even with such a small temperature change, it is possible to control the interaction of this surface with natural materials and cells by changing temperature.

3.3.3.3 Interaction between thermoresponsive surface and proteins

Kawaguchi *et al.* prepared microparticles that had a PNIPAAm shell using seed polymerization and evaluated the adsorption and desorption of proteins onto this surface upon temperature changes. In this case, a shell structure can be modeled by the PNIPAAm gel covering the microparticle. The adsorption of proteins (lactalbumin, mioglobin, lysozyme, ribonuclease, peroxidase) onto the PNIPAAm microparticles is found to be less than that of polystyrene microparticles at 40°C. When the temperature is changed to 25°C, proteins that adsorbed onto the microparticles desorbed partially due to the change of the microparticle surface into a hydrophilic one by hydration of PNIPAAm (see Fig. 4). A more hydrophilic and flexible mioglobin desorbed from the PNIPAAm surface. When the proteins adsorbed at 40°C, they underwent denaturization and likely resulted in irreversible adsorption [103–105]. Interaction between such a surface and white blood cells was evaluated using oxygen consumption as the measure. When microparticles are in contact with the cells oxygen consumption of the cells increased at 37°C. By contrast, when the microparticle surface is hydrophilic at 25°C, oxygen consumption was not observed.

Yoshioka *et al.* [106] fixed the PNIPAAm so that an amine group was introduced at one end of the molecule—glutaldehyde onto an amine-based silica gel. They studied the adsorption and desorption of albumin

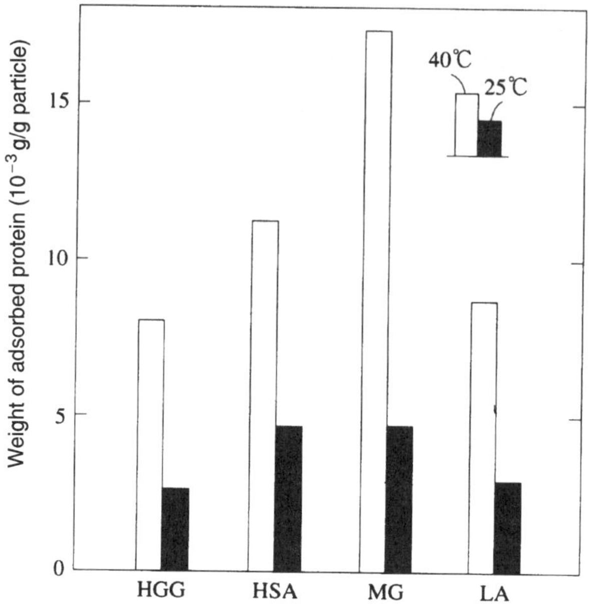

PNIPAAm microsphere 9.90 mg/ml (0.015 M KCl), protein concentration 0.37 mg/ml, human γ-globulin (HGG), human serum albumin (HSA), mioglobin (MG), lactalbumin (LA).

Fig. 4 Adsorption of proteins on PNIPAAm microspheres at 40 and 25°C [104].

and IgG as a function of temperature. Albumin is a hydrophilic protein and so it did not adsorb onto the treated silica surface at either 24 or 37°C, whereas the relatively hydrophobic IgG adsorbed onto the treated silica at 37°C. Of the adsorbed IgG, only 60% desorbed upon treatment at a low temperature. This is probably because the adsorbed protein molecules denatured and irreversible adsorption occurred. In the preceding two examples, there is a difference between the adsorption and desorption behaviors. This is probably due to the influence of the density and structure of the PNIPAAm molecules on the surface.

Wang *et al.* [107], using the swelling-shrinking paradigm of a PNIPAAm gel in response to temperature variation, studied soy protein condensation.

This system concept is depicted in Fig. 5. A gel is used as a molecular sieve. When a shrunken gel is immersed in a protein solution, the gel absorbs both water and low molecular weight materials. Conver-

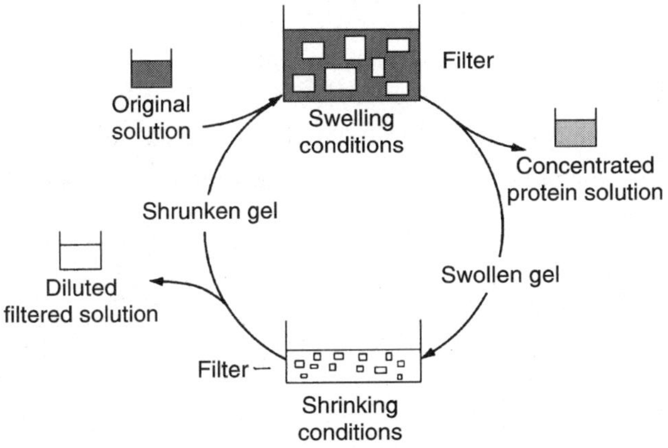

Fig. 5 Separation system using a gel [101].

sely, a high molecular weight protein will be excluded. Upon removal of the swollen gel, a concentrated solution is obtained. Forty five kg of products of over 90% protein purity and +80% yield from 100 kg of soy flakes were produced by this method. Among the advantages, the work can be performed at relatively low temperatures, denaturization can be inhibited by the physical process, and a high-purity protein can be recovered. However, adsorption of the protein onto the gel has been observed. Thus, the adsorbed protein needs to be recovered by a washing-out procedure.

3.3.3.4 Control of the desorption of cells from a thermoresponsive surface

Takei *et al.* prepared a microsphere that changed its surface hydrophilicity and hydrophobicity in response to temperature changes. They attempted to control interaction between the surface and cells using temperature variation [108]. The PNIPAAm grafted microparticles reversibly suspend below the transition temperature. They aggregate among hydrophobic particles and precipitate above this temperature. Using these thermoresponsive microparticles, interaction of the cells with particles was studied at various temperatures. Interaction between the PNIAAm-grafted microparticles and platelets was investigated using calcium concentration changes within the cell wall as seen with a fluorescent die, Fura2, which intensity changes in response to calcium concentrations. At low temperatures, calcium concentrations changed insignificantly within the cell wall. However, at higher temperatures the concentration suddenly increased. This indicates that below the transition temperature the PNIPAAm-grafted surface is covered with highly hydrated material and thus there is little interaction with the platelet. The temperature at which calcium concentration begins to climb corresponds to the temperature at which the PNIPAAm-grafted surface changes from hydrophilic to hydrophobic. Accordingly, interaction between the cells and a PNIPAAm-grafted surface corresponds to changes in hydrophilicity and hydrophobicity at the surface. It has become apparent that strong interaction appears only when the surface becomes hydrophobic. When interaction was evaluated using a lymphocyte, the microspheres and lymphocytes appeared as a homogeneous suspension at low temperature, whereas above the PNIPAAm transition temperature they formed aggregates and precipitated. Using these systems, it is possible to separate a specific cell from a cell suspension at a temperature at which interaction occurs only with the specific cell. It is also possible to on-off control the adhesion of both microparticles and cells. The microparticles, which carry drugs, can be regarded as a basic technology for physical targeting of a drug delivery system (DDS).

The cell culture that has been used to study cells in the live body or functions of organs has been increasing in importance due to recent developments in biotechnology that involve production of useful materials and screening for new drugs. Cells that can serve as culture substrates are numerous, but most exhibit their functions by adhering to a substrate. To desorb and recover such substrate dependent cells, both physical desorption and chemical desorption methods that use chelating agents or protein digestive enzymes are used. A chelating agent for calcium, EDTA relates to cell adhesion. The use of EDTA in cell recovery is somewhat limited and thus it is often necessary to use proteolytic enzymes like trypsin and collagenase. Unfortunately, these enzymes also will digest the connective membrane on cellular surfaces. This then causes cell functions to be reduced. In practice, collagenase has been used to collect lymphocytes near the intestine. The use of these protein digestion enzymes will not cause morphology of the cell.

However, it has been reported that the marker protein, which exists on the surface of the cell membrane, reduced or diminished [109].

Many authors [110–118] have attempted desorption of cultured cells using changes in hydrophilicity and hydrophobicity of a PNIPAAm-grafted surface in response to temperature changes. If an electron beam is irradiated onto the surface of a polystyrene dish to which a coating of isopropyl alcohol solution of NIPAAm is applied, the NIPAAm polymerizes while covalently attached to the surface. The amount of polymer fixed to the surface can be quantitatively determined by attenuated total reflection infrared spectroscopy (ATR-IR). It is possible to control the amount of the grafted polymer by controlling the concentration of the monomer used for polymerization [111]. As the amount of grafted PNIPAAm increases, the contact angle of this surface to water will gradually increase. The contact angle at 15°C is smaller than that at 37°C because the grafted PNIPAAm hydrated at the low temperature and its contact angle with water was reduced. By contrast, dehydration took place at 37°C and the surface became hydrophobic, which resulted in an increased contact angle. In comparison to these phenomena, a commercial polystyrene dish used for cell culture did not show any change in contact angle despite temperature change. When a rat liver cell is cultured on a thermoresponsive surface where both hydrophilicity and hydrophobicity change as a function of temperature, the adhesion ratio after 24 h is the same as that of the commercial polystyrene dish at 70%. When the desorption rate is investigated using temperature changes, no desorption of cells took place from the polystyrene dish but the PNIPAAm-grafted dish showed desorption simply by keeping it at 4°C for 60 min [110]. Further, when cultured at 10°C, the liver cells did not adhere to the PNIPAAm-grafted surface at all, showing inhibition of adhesion onto the hydrophilic surface. When the liver cells that had been desorbed by low-temperature treatment were transferred to a new culture dish, the cells exhibited a high adhesion ratio that was comparable to first generation culture. Liver cells recovered by ordinary means using trypsin for desorption and cell recovery had <20% adhesion ratio, suggesting reduction in cellular function caused by the disruption of cell membrane surface proteins. Liver cell function was evaluated using albumin production as a measure. The cells desorbed and recovered with trypsin showed a productivity of $52\,ng/h/10^5$ cells, which is <20% of the first generation cells ($260\,ng/h/10^5$). However, the next generation liver cells recovered using the low-temperature treatment maintained their productivity at 250 ng/h/10^5 cells, which was nearly the same as that of the first generation cells. Similar results with desorbed and recovered cells has also been observed using cells from the internal wall of a bovine blood vessel in the thoracic cavity. This observation confirmed the characteristics of the PNIPAAm-grafted dish regarding cellular desorption and recovery in a low-temperature environment [114].

The preceding results indicate that the PNIPAAm-grafted surface when subjected to easily created temperature differences is able to change its nature from hydrophobic to hydrophilic. This new system will be able to desorb and recover cells without harm to cellular functions [112]. Cells that adhered at 37°C grew and increased in number; this rate of growth was the same as that on the commercial dish. Figure 6 shows the desorption ratio of the cultured internal wall cells and liver cells on the

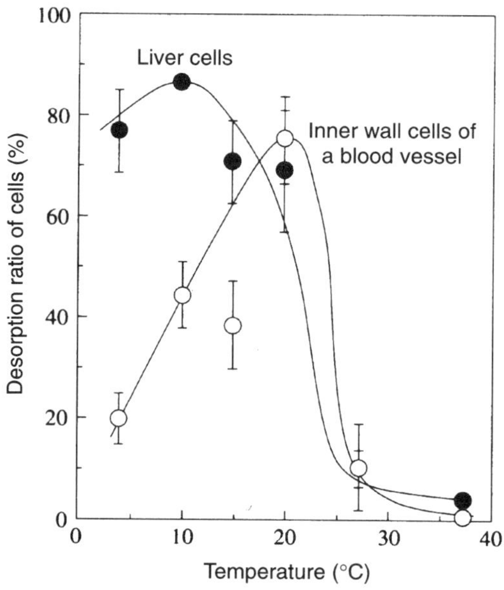

The graph indicates the desorption ratio after 30 min incubation at the specified temperature followed by an additional incubation at 25°C for 5 min. It can be seen that the desorption temperature varies depending on the type of cells.

Fig. 6 Changes in desorption ratio of the cultured cells from the PNIPAAm-grafted surface in response to temperature change.

PNIPAAm-grafted surface. The time required for low-temperature treatment was 30 min. As is clear from the figure, if the type of cell differs, desorption temperature varies. Traditionally, differences in certain types of molecules as relates to adhesion are used when different types of cells are desorbed. However, this figure indicates the potential for a new separation material that will be able to deal with the different desorption characteristics of different adherent cells [113].

The desorption behavior of the liver cells on the PNIPAAm-grafted dish has been examined in detail. It was found that incubation at 10°C for 30 min followed by treatment at 25°C for 5 min provides a better recovery rate than treatment at only 10°C. Investigation of cell morphology revealed that the cells showed a flat shape when they were incubated at 10°C for 30 min, whereas when they were subjected to 25°C, they became spherical and desorbed. If the cultured cells are treated by a metabolism inhibitor, sodium azide, the desorption ratio decreased. With low-temperature treatment, the grafted PNIPAAm molecule started hydrating. The hydration process initiates desorption of the substrate dependent cells, with the result of this sodium azide treatment suggesting that both cellular metabolism and desorption are altered. It is also suspected that the aforementioned variations in desorption temperatures might vary from cell type to cell type as a result of temperature-sensitive metabolism processes of different cells.

Further, the authors have succeeded in recovering a cultured cell sheet that had been cultured as a single confluent layer on a PNIPAAm-grafted dish [116–118]. This was done by culturing the internal wall cells of blood vessels or liver cells until confluent and then applying a low-temperature treatment. To prevent the desorbed sheet from aggregated bulk, a chitinous membrane or cell culture insert was placed atop the cell sheet as a substrate. When a chitinous membrane was used, the cells were desorbed completely from the PNIPAAm dish. Furthermore, this cell sheet transferred to another culture dish in sheet form with a high survival rate. Even when a culture insert is used, when collagen is used to coat the bottom of the sheet prior to use, the cell sheet transfers to the bottom and additional cultures continue to have a very high survival rate. A substrate-like pattern was seen at the bottom of the recovered cells. When the cell sheet desorbed from the PNIPAAm-grafted dish, desorption took place from the protein component, which adsorbed onto the PNIPAAm or was produced by the cells themselves. Observation of the recovered cells by electron microscopy showed that the sample maintained tight cellular junctions. Physiological functions (e.g., for liver cells, albumin production, and the production of plasminogenic activators) are maintained or enhanced [116–118].

Fukushima *et al.* [119] desorbed a single sheet of inner blood vessel wall cells using a PNIPAAm-grafted dish and low-temperature treatment. They transplanted this sheet between a wound that lacks skin and a grafted skin. Several days later the autopsy was performed. The blood vein in the grafted skin and the adhesive layer clearly grew, demonstrating that the transplantation of cultured sheet-like cells of inner blood vein wall regenerated blood vein growth within the grafted skin. This indicates accelerating cell recovery is possible by using a thermoresponsive culture dish.

Takezawa *et al.* coated PNIPAAm onto a dish and attempted to recover cultured cells using low temperature PNIPAAm dissolution [120–122]. On the surface where a homogeneous solution of I-type collagen and PNIAAm is coated, fibroblast can be cultured until they become a single sheet. By lowering the temperature, the cell sheet desorbs while the connection among the cells is maintained and forms aggregates of multiple cell colonies. If these aggregates are transferred to an organ culture dish and culture is continued, aggregation can be maintained. Thus, it is possible to consider applications in the area of the drug evaluation system for replacement of transplants or animals [120]. Rat liver cells are co-cultured on top of the confluent cells. This co-cultured system was then treated at low temperature and desorbed to form an aggregate. From this aggregate, albumin was produced for three weeks. Furthermore, it is interesting to notice that a primitive blood vessel-like system was also produced [122]. Unfortunately, their system created the possibility that the desorbed cells could be contaminated because the surface-coated PNIPAAm dissolves during low-temperature treatment.

In addition to the aforementioned examples of a cultured system that uses a thermoresponsive PNIPAAm, Morra and Cassinelli [123] reported coating a polystyrene surface via covalent bonds using UV radiation. Rollason *et al.* [124] reported on work that explored increasing hydrophobicity and handling a culture by copolymerizing PNIPAAm with N-t-butylacrylamide and by lowering LCST. Morra and Cassinelli reexamined and confirmed the work done by Okano *et al.* [112, 114]. They reported that the thickness of the surface gel layer and its hydration at low temperature strongly influence desorption.

Kubota *et al.* [125] evaluated a system that made removing a wound dressing gauze less traumatic by coating it with thermoresponsive PNIPAAm and exposing the covered area to low temperature. Ordinary gauze adheres to a completely exposed skin wound in only four days. The newly grown skin adheres to the gauze and when the gauze is peeled from the wound, the newly grown skin was damaged. However, with PNIPAAm coating and low-temperature treatment, peeling was easily done even after 7 days.

No adhered materials were observed using an optical microscope to examine the wound, and the wound surface was also smooth. Because white blood cells and giant cells were not observed, there was obviously no inflammation in the PNIPAAm samples. These results indicate that PNIPAAm does not interact strongly with natural components. It is therefore possible to apply PNIPAAm to implanted drug delivery devices.

3.3.3.5 Application of stimuli-responsive surface as molecular valves

It is possible to apply onto a porous membrane a polymer that changes its spread or mobility in response to physical or chemical stimuli as a way of preparing molecular valves that open and close.

Itoh *et al.* attempted to control molecular permeation using the system shown in Fig. 7(a) [126]. Acrylic acid was plasma polymerized

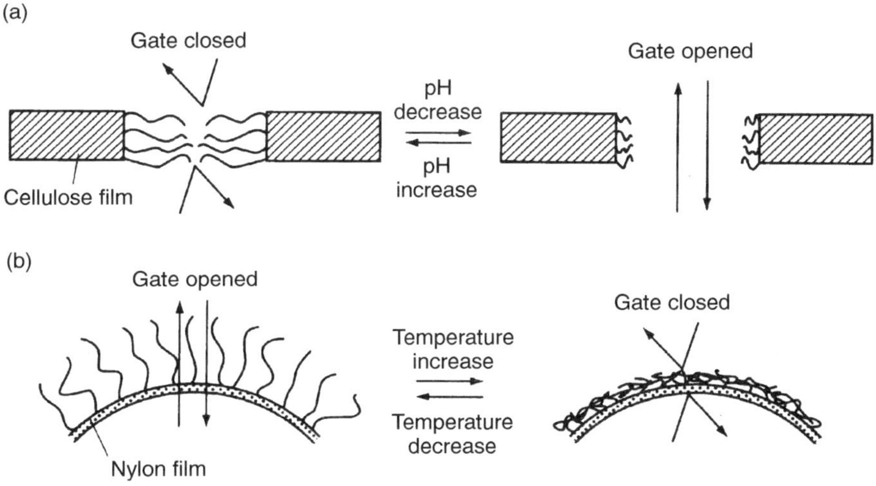

Fig. 7 Conceptual diagram of stimuli-responsive molecular valve.

onto a regenerated cellulose film surface. Then, glucosidase (GOD) was introduced in the poly(acrylic acid) using carbodiimide. When no glucose is present, poly(acrylic acid) is in the dissociated state and the polymer chain is widely spread. On the other hand, when glucose is present, hydrogen peroxide is produced by GOD and the pH falls. As a result, the poly(acrylic acid) changes into a nondissociated form and shrinks. Such a change in polymer chains was used as a molecular valve and this system was used to construct a controlled insulin delivery system to control diabetes.

Okahata *et al.* [127] grafted PNIPAAm onto a porous nylon capsule and investigated the release of sodium naphthalene disulfonic acid. Whereas PNIPAAm exhibits its cloud point at around 35°C, the PNIPAAm shrank above the cloud point (Fig. 7(b)) and the release of ions decreased. In contrast, below the cloud point, the polymer chains are hydrated and stretched. Thus, ionic permeability was markedly increased, perhaps five- to tenfold.

In these two examples, permeation of materials in shrunken and stretched polymers was reversed. This is probably related to differences in surface graft density as well as grafting inside pores.

Accordingly, it is possible to control interaction with materials by grafting a polymer onto the substrate surface and utilizing the change in surface properties, which respond to external stimuli. In fact, separation applications as described in the following section are now being re-evaluated.

3.3.3.6 Development of new thermoresponsive chromatography

Gewehr *et al.* [128] developed a system to separate polymers with size-exclusion chromatography (SEC) by employing porous glass beads on which PNIPAAm was affixed. As shown in Fig. 8, the elution time of dextrin ($C_6H_{10}O_5$) decreased as molecular weight decreased. Furthermore, elution time was time-dependent. Above the transition temperature of PNIPAAm, elution time slows discontinuously. This is because the pore size of the glass beads increases due to the coil-globular transition of PNIPAAm in response to temperature change, which causes changes in permeation of the solute.

Hosoya *et al.* [129, 130] developed an SEC substrate that could vary pore size with temperature variation by selectively treating the inner and

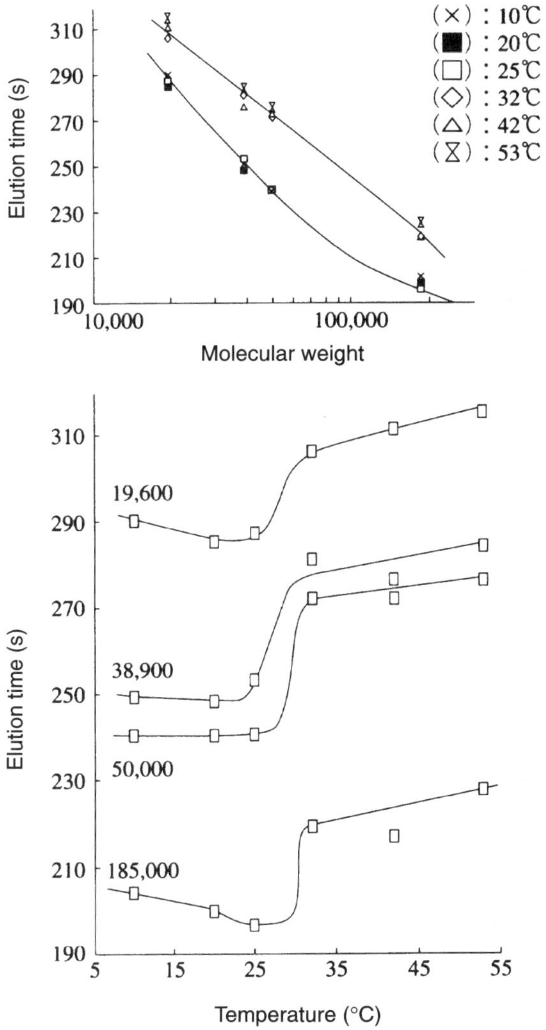

Fig. 8 Change in the elution time of dextrin with various molecular weights in response to temperature changes from the PNIPAAm-treated column (pore size 23.7 nm, flow rate 0.6 ml/min).

outer surfaces of polystyrene beads with PNIPAAm using different polymerization solvents. By changing the solvent, the inner and outer diameters could be changed. Using this substrate, the elution behavior of high-molecular weight polymers was shown to be controlled by changing the temperature. Furthermore, they also showed separation of the drug mixture above the transition temperature of PNIPAAm. This was likely due to the change in interaction between the amide group of PNIPAAm and the hydrophilic group of the drugs as a function of temperature.

Kanazawa *et al.* [131] prepared the HPLC substrate using amine-functional silica beads to which PNIPAAm was fixed at one end of the molecule. They made use of the solubility and nonsolubility change in PNIPAAm hydration and dehydration in response to temperature change of an aqueous solution. They attempted to control both hydrophilicity and hydrophobicity of a solid surface with this system. They also succeeded in achieving high-resolution separation of natural components using the change in hydrophobic interaction that accompanied surface changes. The materials studied included various steroids, peptide fragments, and proteins. If $\log P$ is used to evaluate the hydrophobicity where P is the distribution coefficient in an n-octanol/water system, which is used as the hydrophobicity parameter, the higher the hydrophobicity the slower the elution time as the temperature increases (Fig. 9).

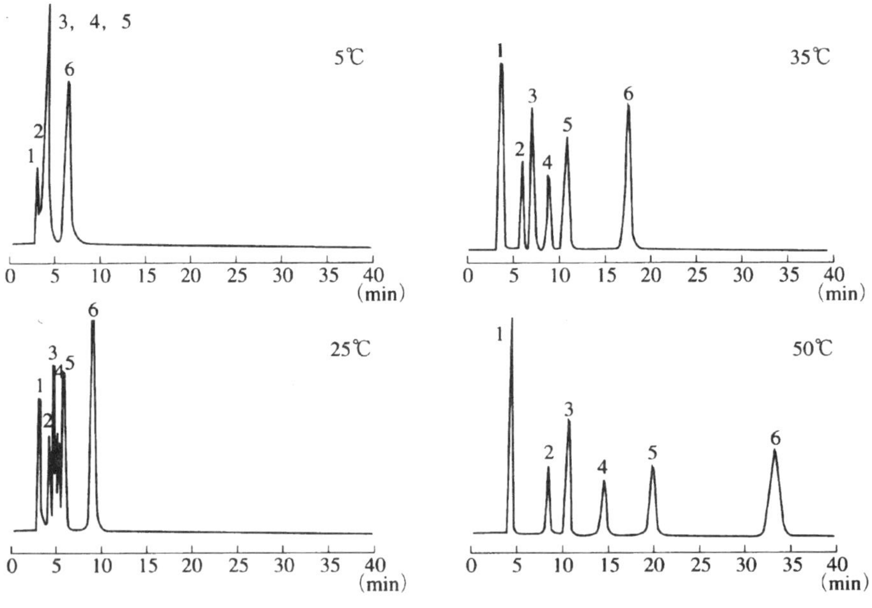

The carrier phase is water with peak 1, benzene (-); 2, hydrocortisone (1.61); 3, prednisolone (1.62); 4, dexametazone (1.83); 5, hydrocortisone acetate (2.30); and 6, testosterone (3.32). In the parentheses, hydrophobicity parameter (log *P*) is indicated.

Fig. 9 Elution behavior of steroids and benzene from the PNIPAAm treated column as a function of time [131].

It has been shown that increased hydrophobic interaction is the driving force in materials separation in aqueous systems as a result of increased hydrophobicity of surfaces following temperature increases. Furthermore, the authors applied the PNIPAAm-fixed surface to active field flow separation-cell adsorption chromatography using thermoresponsive hydrophilicity and hydrophobicity changes. They succeeded in changing the interaction of the surface with lymphocyte [132].

This chromatographic system has the advantages of: (1) maintaining the activity of physiologically active materials; (2) maintaining the survival rate and function of cells; and (3) avoiding environmental pollution as a result of replacing traditional systems, which use organic solvents (see Table 1).

3.3.3.7 Composite of enzyme and switch molecules

A composite of an enzyme or a physiologically active peptide that is treated with water soluble poly(ethylene glycol) shows a markedly increased stability in comparison with untreated materials. In the body, half life increased and there was a decreased antigenicity [133–136]. For example, Fuka *et al.* [137] treated trypsin with a single-, double-, and triple-chain PEG and studied resistance to hydrolysis by pepsin. They showed that the higher the PEG chain density the more resistance there was to pepsin degradation. Monfardini *et al.* [138] also prepared an agent similar to the one used by Fuka *et al.* that employed two amine groups, lysine and treated proteins. They reported increased stability of the proteins without activity loss.

The characteristics shown in Table 2 are expected to be observed if a composite of thermoresponsive PNIPAAm and enzymes is prepared. In addition to the properties of a protein or enzyme treated with water-soluble PEG, it is also possible to control dissolution-precipitation, adhesion-desorption to organs and cells, and on-off physiological activity

Table 1 Command responsive chromatography: characteristics of membrane separation.

High-resolution separation and purification in aqueous systems
Structure and function maintenance of physiologically active materials
Structure and function maintenance of cell membrane surface
Function maintenance of cells
Avoidance of environmental pollution by replacing the traditional organic solvent systems

Table 2 Characteristics of stimuli-responsive bioconjugate.

Stability control (structure and function)
Solubility control (dissolution and insolubilization)
On-off control of adhesion and desorption
On-off control of physiological activity
Functions as a barrier (recognition of foreign materials, avoidance of antigen-based problems)
Control of targeting (possibility of drug targeting)

as controlled by temperature. In fact, there have been several studies reported [139–146].

Takei *et al.* [147] introduced seven PNIPAAm (molecular weight 5000) to fibrinogen (molecular weight-340,000). The turbidity measurement of the aqueous solution indicated complete dissolution and transparency at low temperature. On the other hand, at higher temperatures, there was observed a phase transition phenomenon similar to PNIPAAm within a narrow temperature range and the composite became insoluble. Similarly, the albumin synthesized with PNIPAAm and different LCST copolymers exhibited soluble–insoluble changes near the phase transition temperature. When the temperature of the mixture of the two composites with different transition temperatures was increased, the fibrinogen treated with PNIPAAm of low LCST precipitated first at 28.5°C, followed by the albumin-PNIPAAm composite at 35°C. The composites did not interact with each other and were separable at their individual LCST temperatures [147]. If immunoglobulin (IgG) is modified by PNIPAAm, it is possible to control antigen-antibody interactions using temperature variation. In fact, Takei *et al.* [148] reported producing IgG-modified human albumin with PNIPAAm. As the amount of PNIPAAm increases, protein A, which recognizes the Fc part of IgG, reduces, while the bonding of the composite to human albumin is not sacrificed. The bonding of the composite to albumin disappeared above the phase transition temperature, indicating that antigen-antibody interaction can be controlled [148].

Matsuoka, Okano and others modified lipase with PNIPAAm and studied changes in fat digestion as a function of temperature [139]. The PNIPAAm-modified lipase showed changes in solubility that exhibited the same solubility characteristics of the PINPAAm chain in water as a function of temperature. They also showed that, when the modified enzyme became insoluble in water, enzyme reaction ceased. In addition, the precipitated modified enzyme can be recovered and used again (see

Fig. 10). Traditionally, solid enzymes in which the enzyme is fixed onto a solid substrate have been used in enzyme reactions. However, due to fixation of multiple locations, enzyme activity can be sacrificed. In fact, if the polymer was bonded to an enzyme at multiple locations, enzyme activity was reduced after repeated use [140]. On the other hand, if the polymer was fixed to the enzyme at only one terminal, the denaturation of the enzyme is almost eliminated. A new bioreactor can be made by performing enzyme reactions at a dissolution temperature of PNIPAAm, thereby increasing the temperature after completion of the reaction, recovering the enzyme, and separating it from the product [139, 140, 143–145]. The advantage of modifying the peptide at one end of the polymer is to bring the modification site nearer the active site of the peptide. Hoffman *et al.* demonstrated that it is possible to control the bonding of the substrate to the active site by changes of soluble/insoluble behavior in response to temperature changes [142, 144].

Such a concept can be applied to double-targeting drug delivery, which accumulates drugs only at the target. This can be achieved by

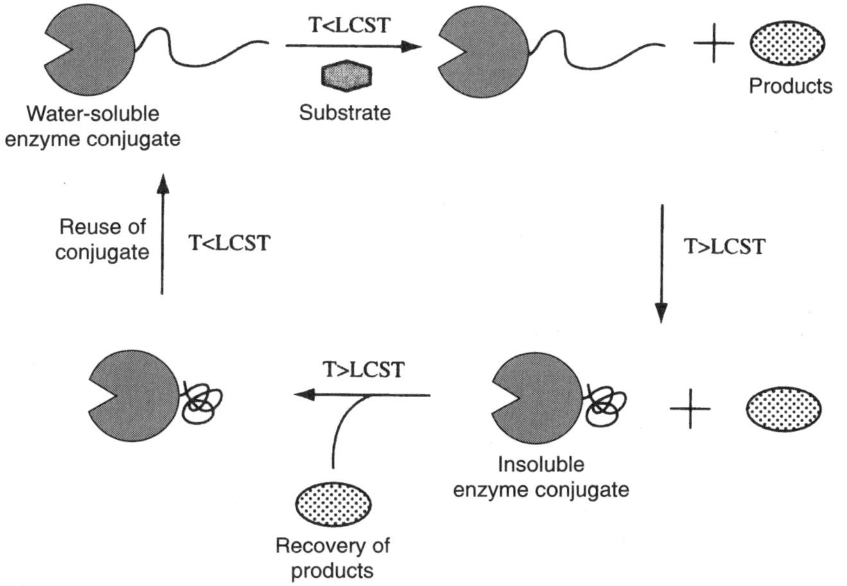

Fig. 10 Bioreactor system using an enzyme conjugator that exhibits solubility-insolubility in response to temperature changes.

modifying peptide-type drugs with PNIPAAm. Drug delivery is done by heating the target hypothermically.

Maeda and Umeno and Umeno *et al.* prepared a PNIPAAm oligomer with a light-bonded DNA intercalater, solaren, on one end [149, 150]. This solaren-PNIPAAm oligomer intercalates to the double helix if it is mixed with DNA. Afterwards, a DNA composite can be synthesized by UV irradiation and bonded to DNA. This DNA composite avoids being cut by the restriction enzyme as the degree of modification increases. Furthermore, this DNA composite maintains thermal sensitivity. Therefore, if it is desirable to separate the DNA component, it can be separated and recovered by mixing a solaren-PNIPAAm oligomer into the solution, irradiating it with UV, and heating.

3.3.4 Conclusions

Interactions between a solid surface that is covered by a gel layer and natural components has been reviewed. There have been many studies reported on attempts to inhibit interaction with natural components using neutral water soluble polymers. However, attempts have been made recently to control interaction actively by using stimuli-responsive polymers. Furthermore, a concept to use modified natural components has also been proposed.

REFERENCES

1 Cussler, E.L., Stokar, M.R., and Verberg, J.E. (1984). *AIChE J*, **30**: 578.
2 Freitas, R.F.S., and Cussler, E.L. (1987). *Chem. Eng. Sci.* **42**: 97.
3 Huang, X., Unno, H., Akehata, T., and Hirasa, O. (1987). *J. Chem. Eng., Jpn.* **20**: 123.
4 Huang, X., Unno, H., Akehata, T., and Hirasa, O. (1989). *Biotechnol. Bioeng.* **34**: 102.
5 Ito, A. (1986). *Sekiyu Gakkai-shi* **29**: 43.
6 Mishima, K., Ishida, O., Watanabe, T., Iwai, Y., and Arai, Y. (1989). *Proc. Soc. Chem. Eng., Jpn.*, F302.
7 Sakohara, S., Sakai, S., Maekawa, Y., and Asaeda, S. (1994). *Kobunshi Ronbunshu* **51**: 540.
8 Sakohara, S., Koshi, T., and Asaeda, S. (1995). *Kobunshi Ronbunshu* **52**: 155.
9 Sakohara, S., and Koshi, T. (1997). *Kobunshi Ronbunshu* **54**: 115.
10 Sakohara, S., Muramoto, F., Sakai, S., Yoshida, M., and Asaeda, M. (1991). in *Polymer Gels*, New York: Plenum Press, pp. 161–171.
11 Sakohara, S., Maekawa, Y., Tateishi, Y., and Asaeda, M. (1992). *J. Chem. Eng., Jpn.* **25**: 598.
12 Tanaka, T. (1981). *Scientific American* **244**: 110.

13 Sakohara, S., Muramoto, F., Sakata, T., and Asaeda, M. (1990). *J. Chem. Eng., Jpn.* **23**: 40.
14 Sakohara, S., Tateishi, Y., and Asaeda, S. (1995). *Kagaku Kogaku Ronbunshu* **21**: 547.
15 Sakohara, S., Tateishi, Y., and Asaeda, S. (1993). *Kagaku Kogaku Ronbunshu* **19**: 1192.
16 Kamitono, H., and Ogata, N. (1992). *Maku* **17**: 238.
17 Tanaka, T. (1978). *Phys. Rev. Lett.* **40**: 820.
18 Osada, Y., and Hasebe, M. (1985). *Chem. Lett.* **1285**.
19 Quinn, F.X., Kampff, E., Smyth, G., and McBrierty, V.J. (1988). *Macromolecules* **21**: 3191.
20 Gong, J.P., and Osada, Y. (1995). *Chem. Lett.* **6**: 449.
21 Manning, G.S. (1975). *J. Phys. Chem.* **79**: 262.
22 Honda, Y. (1954). *Ion Exchange*, Konan-do.
23 Kakibana, H., and Narita, K. (eds) (1960). *Advances in Ion Exchange*, Hirokawa Shoten.
24 Samuelson, O. (1960). *Ion Exchangers in Analytical Chemistry*,
25 Shimura, T., Ejiri, S., Yoshida, M., and Ishihara, H. (1948). *Experimental Methods in Biochemistry* 11: *Filteration by Gels*, Gakkai Publ. Center.
26 Schwarz, G. (1968). *Biopolymers* **6**: 873.
27 Schwuger, M.J. (1973). *J. Colloid Interface Sci.* **43**: 491.
28 Magny, B., Iliopoulos, I., Zana, R., and Audebert, R. (1994). *Langmuir* **10**: 3180.
29 Antonietti, M., Forster, S., Zisenis, M., and Conrad, J. (1995). *Macromolecules* **28**: 2270.
30 Starodubtzev, S.G. (1990). *Vysokomol. Soedin.* **31B**: 925.
31 Starodubtzev, S.G., Ryabina, V.R., and Khokhlov, A.R. (1990). *Vysokomol. Soedin.* **32A**: 969.
32 Khokholov, A.R., Kramarenko, E.Yu., Makhaeva, E.E., and Starodubtzev, S.G. (1992). *Macromolecules* **25**: 4779.
33 Osada, Y., Okuzaki, H., and Hori, H. (1992). *Nature* **355**: 242.
34 Osada, Y., and Ross-Murphy, S.B. (1993). *Sci. Am.* **268**: 82.
35 Okuzaki, H., and Osada, Y. (1994). *Macromolecules* **27**: 502.
36 Kokufuta, E., Zhang, Y.Q., Tanaka, T., and Mamada, A. (1993). *Macromolecules* **26**: 1053.
37 Kokufuta, E., Nakaizumi, S., Ito, S., and Tanaka, T. (1995). *Macromolecules* **28**: 1704.
38 Okuzaki, H., and Osada, Y. (1995). *Macromolecules* **28**: 4554.
39 Okuzaki, H., and Osada, Y. (1994). *Macromolecules* **27**: 502.
40 Okuzaki, H., and Osada, Y. (1994). *Chem. Mater.* **6**: 1651.
41 Gong, J.P., and Osada, Y. (1995). *J. Phys. Chem.* **99**: 10971.
42 Flory, P.J. (1953). *Principles of Polymer Chemistry*, Ithaca, New York: Cornell University Press.
43 Ising, E. (1925). *Pysik* **31**: 253.
44 Marcus, R.A. (1954). *J. Phys. Chem.* **58**: 621.
45 Khokhlov, A.R., Kramarenko, E.Y., Makhaeva, E.E., and Starodubtzev, S.G. (1992). *Makromol. Chem., Theory Simul.* **1**: 105.
46 Isogai, N., Gong, J.P., and Osada, Y. (1996). *Macromolecules* **29**: 6803.
47 Society of Polymer Science, Japan (ed.) (1990). Medical functional materials, vol. 9, Chapter 2, in *Polymeric Functional Materials Series*, Tokyo: Kyoritsu Publ.
48 Folkman, J., and Moscona, A. (1978). *Nature* **273**: 345.

49 Andrade, J.D., Lee, H.B., John, M.S., Kim, S.W., and Hibbs, J.B. Jr. (1973). *Trans. Am. Soc. Artif. Int. Organs* **19**: 1.
50 Lydon, M.J., Minett, T.W., and Tighe, B.J. (1985). *Biomaterials* **6**: 396.
51 Horbett, T.A., Schway, M.B., and Ratner, B.D. (1985). *J. Colloid Interface Sci* **104**: 28.
52 Ueda-Yuasa, T., and Matsuda, T. (1995). *Langmuir* **11**: 4135.
53 Nagaoka, S., Takiuchi, H., Yokota, K., Mori, Y., Tanzawa, H., and Kikuchi, T. (1982). *Kobunshi Ronbunshu* **39**: 165.
54 Nagaoka, S., Takiuchi, H., Yokota, K., Mori, Y., Tanzawa, H., and Kikuchi, T. (1982). *Kobunshi Ronbunshu* **39**: 173.
55 Mori, Y., Nagaoka, S., Takiuchi, H., Tanzawa, H., and Seikai, S. (1982). *Jinko Zoki* **11**: 971.
56 Tanzawa, H. (1986). *Jinko Zoki* **15**: 16.
57 Grainger, D.W., Nijiri, C., Okano, T., and Kim, S.W. (1989). *J. Biomed. Mater. Res.* **23**: 979.
58 Nijiri, C., Okano, T., Jocobs, H.A., Park, K.D., Mahommad, S.F., Olsen, D.B., and Kim, S.W. (1990). *J. Biomed. Mater. Res.* **24**: 1151.
59 Kishida, A., Mishima, K., Corretge, E., Konishi, H., and Ikada, Y. (1992). *Biomaterials* **13**: 113.
60 Desai, N.P., and Hubbell, J.A. (1992). *Macromolecules* **25**: 226.
61 Tseng, Y.C., and Park, K. (1992). *J. Biomed. Mater. Res.* **26**: 373.
62 Gombotz, W.R., Guanghui, W., Harbett, T.A., and Hoffman, A.S. (1991). *J. Biomed. Mater. Res.* **25**: 1547.
63 Amiji, M., and Park, K. (1992). *Biomaterials* **13**: 682.
64 Lee, J.H., Kopecek, J., and Andrade, J.D. (1989). *J. Biomed. Mater. Res.* **23**: 351.
65 Desai, N.P., and Hubbell, J.A. (1992). *Biomaterials* **13**: 505.
66 Desai, N.P., Hossainy, S.F.A., and Hubbell, J.A. (1992). *Biomaterials* **13**: 417.
67 Grainger, D.W., Knutson, K., Kim, S.W., and Feijen, J. (1990). *J. Biomed. Mater. Res.* **24**: 403.
68 Grainger, D.W., Okano, T., Kim, S.W., Castner, D.G., Ratner, B.D., Briggs, D., and Sung, Y.K. (1990). *J. Biomed. Mater. Res.* **24**: 547.
69 Han, D.K., Park, K.D., Ahn, K.D., Jeong, S.Y., and Kim, Y.H. (1989). *J. Biomed. Mater. Res., Appl. Biomater.* **23**: 87.
70 Park, K.D., Piao, A.Z., Jacobs, H., Okano, T., and Kim, S.W. (1991). *J. Polym. Sci., Part A Polym. Chem.* **29**: 1725.
71 Park, K.D., Okano, T., Nojiri, C., and Kim, S.W. (1988). *J. Biomed. Mater. Res.* **22**: 977.
72 Nojiri, C., Park, K.D., Grainger, D.W., Jacobs, H.A., Okano, R., Hoyanagi, H., and Kim, S.W. (1990). *ASAIO Trans.* **36**: M168.
73 Nagaoka, S., Kurumatani, H., Mori, Y., and Tanzawa, H. (1989). *J. Bioact. Compatible Polym.* **4**: 323.
74 Park, K.D., Kim, W.G., Jacobs, H., Okano, T., and Kim, S.W. (1992). *J. Biomed. Mater. Res.* **26**: 739.
75 Okano, T., Grainger, D., Park, K.D., Nijiri, C., Feijen, J., and Kim, S.W. (1988). in *Artificial Heart*, T. Akutsu, ed., Tokyo: Springer-Verlag, pp. 45–53.
76 Han, D.K., Jeong, S.Y., Kim, Y.H., Min, B.G., and Cho, H.I. (1991). *J. Biomed. Mater. Res.* **25**: 561.
77 Andrade, J.D., Nagaoka, S., Cooper, S., Okano, T., and Kim, S.W. (1987). *ASAIO* **10**: 75.
78 Amiji, M., and Park, K. (1993). *J. Biomater. Sci. Polym. Ed.* **4**: 217.

79 Llanos, G.R., and Sefton, M.V. (1993). *J. Biomater. Sci. Polym. Ed.* **4**: 381.
80 Tsuruta, T. (1996). *Adv. Polym. Sci.* **126**: 1.
81 Holmberg, K., Bergstrom, K., and Stark, M.B. (1992). in *Poly(ethylene glycol) Chemistry—Biotechnical and Biomedical Applications*, Chapter 19, J.M. Harris, ed., New York: Plenum.
82 Itoh, T., and Matsuda, T. (1989). *Jinko Zoki* **18**: 132.
83 Sawhney, A.S., Pathak, C.P., and Hubbell, J.A. (1993). *Biomaterials* **14**: 1008.
84 Pathak, C.P., and Hubbell, J.A. (1992). *J. Am. Chem. Soc.* **114**: 8311.
85 Matsuda, T., Itoh, T., Yamaguchi, T., Iwata, H., Kayashi, K., Uemura, S., Ando, T., Adachi, S., and Nakajima, N. (1986). *Trans. ASAIO* **32**: 151.
86 Matsuda, T., Nakajima, N., and Itoh, T. (1989). *Jinko Zoki* **18**: 405.
87 Hubbell, J.A., West, J.L., and Chowdhury, S.M. (1996). in *Advanced Biomaterials in Biomedical Engineering and Drug Delivery Systems*, N. Ogata, S.W. Kim, J. Feijen and T. Okano, eds., Tokyo: Springer, pp. 179–182.
88 Nakayama, Y., and Matsuda, T. (1995). *ASAIO Journal* **41**: M374.
89 Matsuda, T., Moghaddam, M.J., Miwa, H., Sakurai, K., and Iida, F. (1992). *ASAIO J.* **38**: M154.
90 Matsuda, T., Miwa, H., Moghaddam, M.J., and Iida, F. (1993). *ASAIO J.* **39**: M327.
91 Ikada, Y. (1984). *Adv. Polym. Sci.* **57**: 103.
92 Ikada, Y., Iwata, H., Horii, F., Matsunaga, T., Taniguchi, M., Suzuki, M., Taki, W., Yamagata, S., Yonekawa, Y., and Handa, H. (1981). *J. Biomed. Mater. Res.* **15**: 697.
93 Tamada, Y., and Ikada, Y. (1994). *J. Biomed. Mater. Res.* **28**: 783.
94 Matsuda, T., and Sugawara, T. (1995). *J. Biomed. Mater. Res.* **29**: 749.
95 Nakayama, Y., and Matsuda, T. (1994). *Jinko Zoki* **23**: 717.
96 New Energy, Industrial Technology Development Plan (1993). *The Report on the Development of High Functional Surface and the Research Trend.*
97 New Energy, Industrial Technology Development Plan (1994). *The Report on the Development of High Functional Surface and the Research Trend* (II)—*Investigation of the New Separation and Purification System Development.*
98 Nagasaki, Y., Honzawa, E., Kato, M., Kataoka, K., and Tsuruta, T. (1994). *Macromolecules* **27**: 4848.
99 Nagasaki, Y., Kazama, K., Honzawa, E., Kato, M., Kataoka, K., and Tsuruta, T. (1995). *Macromolecules* **28**: 8870.
100 Nagasaki, Y., and Kataoka, K. (1996). *Trends in Polym. Sci.* **4**: 59.
101 Nagasaki, Y., Tsujimoto, H., Honzawa, E., Kato, M., Kataoka, K., and Tsuruta, T. (1996). *Polym. Preprint, ACS* **36**: 59.
102 Takei, Y.G., Aoki, T., Sanui, K., Ogata, N., Sakurai, Y., and Okano, T. (1994). *Macromolecules* **27**: 6163.
103 Kawaguchi, H., Fujimoto, K., and Mizuhara, Y. (1992). *Colloid Polym. Sci.* **270**: 53.
104 Fujimoto, K., Mizuhara, Y., Tamura, N., and Kawaguchi, H. (1993). *J. Intelligent Mater. Systems Structures* **4**: 184.
105 Kawaguchi, H. (1994). *Kobunshi Kako* **43**: 542.
106 Yoshioka, H., Mikami, M., Nakai, T., and Mori, Y. (1994). *Polym. Adv. Technol.* **6**: 418.
107 Wang, K.L., Burban, J.H., and Cussler, E.L. (1993). *Adv. Polym. Sci.* **110**: 67.
108 Takei, Y.G., Aoki, T., Sanui, K., Ogata, N., Sakurai, Y., and Okano, T. (1995). *Biomaterials* **16**: 667.
109 Abuzakouk, M., Feighery, C., and O'Farrelly, C. (1996). *J. Immunol. Meth.* **194**: 211.

110 Yamada, N., Okano, T., Sakai, H., Karikusa, F., Sawasaki, Y., and Sakurai, Y. (1990). *Makromol. Chem., Rapid Commun.* **11**: 571.

111 Sakai, H., Doi, Y., Okano, T., Yamada, N., and Sakurai, Y. (1996). in *Advanced Biomaterials in Biomedical Engineering and Drug Delivery Systems*, N. Ogata, S.W. Kim, J. Feijen and T. Okano, eds., Tokyo: Springer, pp. 229–230.

112 Okano, M., Yamada, N., and Sakurai, H. (1991). *Soshiki Baiyo* **17**: 349.

113 Yamada, N., Sakai, H., Okano, M., and Sakurai, H. (1992). *Jinko Zoki* **21**: 206.

114 Okano, T., Yamada, N., Sakai, H., and Sakurai, Y. (1993). *J. Biomed. Mater. Res.* **27**: 1243.

115 Okano, T., Yamada, N., Okuhara, M., Sakai, H., and Sakurai, Y. (1995). *Biomaterials* **16**: 297.

116 Okuhara, M., Karikusa, F., Sakai, H., Kikuchi, A., Sakurai, H., and Okano, M. (1995). *Proc. 44th Annual Meeting of the Society of Polymer Science* **44**: 2786.

117 Okano, T., Kikuchi, A., Sakai, H., Okuhara, M., Ogura, F., and Sakurai, Y. (1995). *Proc. 2nd International Conf. on Cellular Eng.*, San Diego, California, p. 75.

118 Kikuchi, A., Okuhara, M., Karikusa, F., Sakai, H., Sakurai, Y., and Okano, T. (1996). *Proc. 5th World Biomaterials Congress, Transactions*, Toronto, Canada, p. 907.

119 Fukushima, K., Negishi, N., Kobayashi, M., Okano, M., and Nozaki, M. (1994). *Nikkeikai-shi* **14**: 636.

120 Takezawa, T., Mori, Y., and Yoshizato, K. (1990). *Bio/Technology* **8**: 854.

121 Takezawa, T., Mori, Y., and Yoshizato, K. (1991). *Oyo Saibo Seibutsugaku Kenkyu* **9**: 1.

122 Takezawa, T., Yamazaki, M., Mori, Y., Yonaha, T., and Yoshizato, K. (1992). *J. Cell. Sci.* **101**: 495.

123 Morra, M., and Cassinelli, C. (1995). *Polym. Prepr., ACS* **36**: 55.

124 Rollason, G., Davies, J.E., and Sefton, M.V. (1993). *Biomaterials* **14**: 153.

125 Kubota, A. (1994). *Jinko Zoki* **23**: 679.

126 Itoh, Y., Casolaro, M., Zheng, D.H., and Imanishi, Y. (1988). *Drug Delivery System* **3**: 391.

127 Okahata, Y., Noguchi, H., and Seki, T. (1986). *Macromolecules* **19**: 493.

128 Gewehr, M., Nakamura, K., Ise, N., and Kitano, H. (1992). *Makromol. Chem.* **193**: 249.

129 Hosoya, K., Sawada, E., Kimata, K., Araki, T., Tanaka, N., and Frechet, J.M.J. (1994). *Macromolecules* **27**: 3973.

130 Hosoya, K., Kimata, K., Araki, T., Tanaka, N., and Frechet, J.M.J. (1995). *Anal. Chem.* **67**: 1907.

131 Hanazawa, H., Yamamoto, K., Matsushima, Y., Takai, N., Kikuchi, A., Sakurai, Y., and Okano, T. (1996). *Anal. Chem.* **68**: 100.

132 Kikuchi, A., Sakurai, Y., and Okano, T. (unpublished data).

133 Harris, J.M. (ed.) (1992). *Poly(ethylene glycol) Chemistry*, New York: Plenum Press.

134 Inada, Y. (ed.) (1987). *Protein Hybrid: Advancement in Chemical Modification*, Kyoritsu Publ.

135 Inada, Y., and Maeda, H. (eds.) (1988). *Protein Hybrid, Sequel: Future Directions in Chemical Modification*, Kyoritsu Publ.

136 Inada, Y., and Wada, H. (eds.) (1990). *Protein Hybrid, Volume III: The State-of-the-Art Chemical Modification*, Kyoritsu Publ.

137 Fuka, I., Hayashi, T., Tabata, Y., and Ikada, Y. (1994). *J. Contr. Rel.* **30**: 27.

138 Monfardini, C., Schiavon, O., Caliceti, P., Morpurgo, M., Harris, J.M., and Veronese, F.M. (1995). *Bioconjugate Chem.* **6**: 62.

139 Matsukata, M., Takei, Y., Aoki, T., Sanui, K., Ogata, N., Sakurai, Y., and Okano, T. (1994). *J. Biochem.* **116**: 682.

140 Matsukata, M., Aoki, T., Sanui, K., Ogata, N., Kikuchi, A., Sakurai, Y., and Okano, T. (1996). *Bioconjugate Chem.* **7**: 96.

141 Ding, Z., Chen, G., and Hoffman, A.S. (1996). *Bioconjugate Chem.* **7**: 121.

142 Stayton, P.S., Shimoboji, T., Long, C., Chilkoti, A., Chen, G., Harris, J.M., and Hoffman, A.S. (1995). *Nature* **378**: 472.

143 Chen, G., and Hoffman, A.S. (1994). *Bioconjugate Chem.* **4**: 509.

144 Chilkoti, A., Chen, G., Stayton, P.S., and Hoffman, A.S. (1994). *Bioconjugate Chem.* **5**: 504.

145 Chen, G., and Hoffman, A.S. (1994). *J. Biomater. Sci., Polym. Ed.* **5**: 371.

146 Shiroya, T., Yasui, M., Fujimoto, K., and Kawaguchi, H. (1995). *Biointerfaces* **4**: 275.

147 Takei, Y.G., Aoki, T., Sanui, K., Ogata, N., Okano, T., and Sakurai, Y. (1993). *Bioconjugate Chem.* **4**: 341.

148 Takei, Y.G., Matsukata, M., Aoki, T., Sanui, K., Ogata, N., Kikuchi, A., Sakurai, Y., and Okano, T. (1994). *Bioconjugate Chem.* **5**: 577.

149 Maeda, M., and Umeno, D. (1996). *Proc. of the Symposium of the Materials Research Society, Japan*, B6, p. 74.

150 Umeno, T., Ihara, T., and Maeda, M. (1996). *Proc. 6th Biopolymer Symposium*, pp. 9–10.

Section 4

Transport and Permeation (Diffusion of Materials)

SHINGO MATSUKAWA AND ITARU ANDO

4.1 INTRODUCTION

Experimental and theoretical basis studies have attempted to understand the diffusion of molecules (solvents, polymeric materials etc.) in gels. Advances in instrumentation for diffusion coefficient determination have contributed significantly to this field. These techniques include the dynamic light scattering technique, nuclear magnetic resonance spectroscopy (NMR), and the forced Raleigh scattering technique. Application studies have been using the results of these basic studies. In this section, examples of recent studies will be discussed after briefly describing the fundamental theories and measurement methods regarding gels diffusion. In particular, those examples in which the NMR technique has been used will be referred to in detail.

4.2 THEORY OF MATERIAL DIFFUSION WITHIN POLYMER GELS

The diffusion coefficient D_0 of a particle in a solvent is expressed as follows by Einstein,

$$D_0 = kT/\zeta \tag{1}$$

where k is the Boltzmann constant, T the absolute temperature, and ζ the friction coefficient of the particle during transport. In the case of a macroscopic rigid sphere, the following Stokes equation applies:

$$\zeta = 6\pi R\eta \tag{2}$$

where R is the radius of the rigid sphere, and η the viscosity. From Eqs. (1) and (2),

$$D_0 = kT/6\pi R\eta \tag{3}$$

is obtained. In the case of polymer diffusion in a random coil state, the solvent within the coil travels together with the coil. If the fluid dynamic behavior is assumed to be approximated to the behavior of a rigid sphere with radius R_h, Eq. (3) can be written as

$$D_0 = kT/6\pi R_h\eta \tag{4}$$

and the relationship between R_h and the molecular weight M is expressed by

$$R_h \propto M^{\nu} \tag{5}$$

As ν is 0.6 in a good solvent, Eq. (4) becomes

$$D_0 = kT/6\pi M^{0.6}\eta \tag{6}$$

In the diffusion behavior of materials within gels, interaction between the diffusant and solvent dominates when network chain concentration is low. This behavior is similar to the diffusion in solution. When the network chain concentration increases, the restriction of diffusion by the network chain polymers becomes important. When the particle size R is smaller than the network size, the diffusion coefficient D of the particle in the gel is expressed as

$$D = D_0 \exp(-\kappa R) \tag{7}$$

where κ^{-1} is the fluid dynamic network size and is called the dynamic screening distance. This equation indicates that the diffusion is controlled by the ratio of the particle and network size. For diffusion of the probe polymer in a random coil state in a gel, Eq. (5) applies in the gel even if R_h is smaller than κ^{-1}. The diffusion of particles in a heterogeneous gel also can be analyzed by replacing κ^{-1} with the size of the hole in Eq. (7).

When R_h is large, the transport in the transverse direction of the probe molecule chain is highly restricted by the network chains. Only the movement along the chain direction becomes feasible. This is the so-called reptation region. The diffusion in a molten polymer is expressed by

$$D = M_e M^{-2} \tag{8}$$

where M_e is the molecular weight between the entanglement of the network polymers [1]. The dependence of M_e on network chain concentration and the effect of the chemical crosslinks of probe polymers in solvent-containing gels are not yet completely understood. However, it is known to be reciprocally proportional to the second power of the molecular weight of the probe polymer [2, 3].

When the dominant interaction between the solvent and network chains is a steric restriction, the movement of the solvent in a homogeneous gel can be explained by the free volume model. By focusing on an isolated solvent molecule, its diffusion coefficient is expressed by Eq. (1). The friction coefficient ζ can be expressed by the change of the volume ratio f as,

$$\frac{\zeta}{\zeta_r} = \exp\left[B\left(\frac{1}{f} - \frac{1}{f_r}\right)\right] \tag{9}$$

where ζ_r and f_r are the ζ and f at the reference concentration of the solvent, respectively. Also, B is the minimum hole size, which is necessary for the transport of solvent and can be regarded as unity (or 1). If f is assumed to increase proportionally to the volume fraction v of the solvent,

$$f = f_r + \beta(v - v_r) \tag{10}$$

where β is a proportionality constant and v_r is the v at the reference concentration. Fujita and others derived the free volume model using Eq. (10) in a region of sufficiently low concentration of solvent [4–6]. It has been shown that the experimental results of the diffusion coefficient of solvents in polymer solutions and within gels can be explained in a wide temperature range [7–10]. From Eqs. (9) and (10), we obtain

$$\frac{\zeta}{\zeta_r} = \exp[-(v - v_r)/f_r(f_r/\beta + v - v_r)] \tag{11}$$

Here, if we choose the reference concentration at $v_r = 1$, the diffusion coefficient of the solvent in a gel is expressed as

$$D = D_0 \exp[c/|cf_0 - f_0^2/\beta|] \tag{12}$$

where D_0 and f_0 are the diffusion coefficient and free volume ratio of the solvent not including the solvent, respectively, and c is the volume concentration of the polymer.

Thus far, the discussion focused on cases in which interactions among the probe polymer, solvent, and network chains are controlled by steric restrictions. However, in real systems, hydrogen bonding, hydrophobic interaction, or coulombic forces may control diffusion behavior. These effects must also be considered.

4.3 THE DIFFUSION COEFFICIENT MEASUREMENT METHODS

The classic method for measuring the diffusion coefficient within a gel is to measure the amount of the exclusion of the solute or to measure the amount of permeation through the gel. For example, Haggerty *et al.* immersed polyacrylamide, which contains rhodamine pigment, into a solution with no pigment. They determined the diffusion coefficient by measuring the time-dependent concentration of the excluded pigment [11]. Then by plotting the amount of solute passing through a gel from a solution that contains the solute and then into pure solvent, a straight line is obtained at steady state. From the time lag, obtained by extrapolating from this straight line to zero concentration, the diffusion coefficient can be obtained [12]. There are also methods in which a rod-like gel is immersed and the process of a solute diffusing within a gel by micro FTIR [13] is observed, as well as a method that analyzes time-dependent changes with ESR imaging [14].

Measurement of diffusion coefficients by light scattering has been widely adopted due to advancements in recent laser light technology and computers. Using the dynamic light scattering technique, the translational diffusion coefficient for a polymer in a dilute solution can be obtained, whereas for a gel the cooperative diffusion coefficient can be obtained by measuring the time-dependent autocorrelation function of the scattered light. This cooperative diffusion coefficient expresses the propagation rate of the concentration fluctuation by the cooperative movement of the polymer chain as a whole. In order to determine the translational diffusion coefficient of a probe polymer in a gel, the measurement usually is made by adjusting the refractive indices of the solvent and network and eliminating the scattering from the network chains [15, 16] . Even when there is scattering from the network chains, it is possible to determine the diffusion coefficient of the probe polymer separately by determining the relaxation time distribution by Laplace transformation of the time correlation function [3].

These methods measure the mutual diffusion coefficient that originates from the concentration difference or concentration fluctuation as a driving force. The pulsed magnetic gradient spin echo (PGSE)-NMR technique or forced Raleigh scattering technique measure the time-dependent average distance of the molecule that is tagged in the nuclear magnetic or photochemical excited state. Therefore, the self-diffusion coefficient can be obtained by these techniques.

In the PGSE-NMR technique, the pulsed magnetic gradient is applied in the middle of the pulse sequence of the spin echo technique and then tagged by the phase lag of the magnetic vector according to the coordinate. After a certain time, the magnetic gradient opposite the first magnetic gradient is applied and phase lag recovers. In the echo signal, the phases of the magnetic vectors are aligned in all coordinates. If the nuclear coordinates change during the period of two magnetic gradients the phases do not align and the spin echo signal reduces.

The damping (A/A_0) of the signal intensity is expressed as follows,

$$\frac{A}{A_0} = \exp(-KD) \tag{13}$$

where K is the parameter which relates to the distance between the magnetic gradient pulses, duration and intensity, and D is the diffusion coefficient. The quantity D can be determined from the signal intensity changes, which correspond to the variation of K. However, it is important to recognize that the obtained information is the diffusion coefficient in the time scale of the magnetic gradient period. Using an ordinary spectrometer, a magnetic gradient period from several milliseconds to several seconds can be applied. However, the effect of intramolecular displacement will not appear in this time scale. Further, signal intensity decreases due to the diffusion by translational movement. When the diffusional movement is spatially restricted the D observed by the magnetic gradient period changes [17, 18]. There is no need to artificially tag the molecule in the PGSE-NMR technique or limit the refractive index of the solvent. Thus, it is possible to measure diffusion in a heterogeneous, opaque gel. In addition, different diffusion coefficients can be simultaneously determined by taking advantage of the difference in resonance frequencies in each component of the system [19, 20].

In the forced Rayleigh scattering technique, the diffraction lattice of the excited state is formed by the interference of the two laser fluxes. A probe laser then follows the disappearing process of this lattice [21]. The disappearance process of this lattice reflects the lifetime of the excited state τ and molecular diffusion D. A plane wave laser light with a wavelength λ is divided into two and they cross each other at an angle q to observe the interference pattern with a period Λ. In the constructive interference area, the probe will be excited and the striation of the excited state can be observed. Period Λ of the striation is expressed as

$$\Lambda = \frac{\lambda}{2\sin(\theta/2)} \tag{14}$$

The striation of this excited state acts as a diffraction lattice for the laser light that is not absorbed by the probe. Consequently, the contrast of the striation of the excited state weakens and the intensity of the diffracted light decreases. The time-dependent diffraction intensity $I(t)$ is given by

$$I(\mathrm{t}) \propto \exp\left[-2t\left(\frac{1}{\tau} + \frac{4\pi^2 D}{\Lambda^2}\right)\right] \tag{15}$$

Here, if τ is known in advance, D can be directly obtained from Eq. (15). If it is unknown, τ and D are determined from the measurements with varied Λ. As the striation period can be as narrow as 1 μm, even slight movement and slow diffusion can be measured. The measurement area can also be small. For measurement purposes, a molecule that can be excited by a laser light as a probe or a probe that is tagged by a chromophore must be used. Studies on polystyrene diffusion partially labeled by 4-bromomethyl-azobenzene in gelatin [22] and spiropyrane in polystyrene gel [23] have been reported.

4.4 EXAMPLES OF INVESTIGATION

4.4.1 Solvent Diffusion in Gels

Diffusion of a solvent in a gel will be the basis for understanding the diffusion of particles or polymers in the gel. By investigating diffusion of a solvent, knowledge on the interaction between network chains and the solvent can be obtained.

Several researchers, including Matsukawa, measured the diffusion coefficient D of HDO (included in a minute amount in heavy water) in a poly(N,N-dimethylacrylamide) gel. The gel was swollen to equilibrium by heavy water with polyethylene glycol as the probe molecule. They measured the D using the pulsed magnetic gradient spin echo (PGSE) ^{1}H NMR technique and obtained the relationship between the degree of swelling q and the diffusion coefficient D of HDO [19, 20]. The solvent diffusion in the gel suddenly reduces when the degree of swelling decreases. As well, it is independent of the molecular weight of the probe molecule. Here, the degree of swelling q is the weight of the swollen gel, not the weight of the polymer in the gel.

As the weight fraction of the polymer is approximately equal to the volume fraction, it has the following relationship with the volumetric concentration c:

$$q = c^{-1} \tag{16}$$

From Eqs. (12) and (16),

$$D = D_0 \exp[q^{-1}/|q^{-1}f_0 - f_0^2/\beta|] \tag{17}$$

where D_0 is the diffusion coefficient of heavy water without the polymer and is 2.2×10^{-5} cm^2 s^{-1}. From this value and Eq. (17), the theoretical value (solid line) as shown in Fig. 1 was obtained using the least square method. The theoretical curve is in good agreement with the experimental results. As the degree of swelling decreases, the free volume decreases and the diffusion of the heavy water molecules is restricted. Similar results have been reported by several researchers on the water in poly(N,N-isopropylacrylamide) gel (see Fig. 2) [20, 24]. Yasunaga and Ando swelled poly(methacrylic acid) gel with water and measured its diffusion constant D_{H_2O} by (PGSE) ^{1}H NMR techniques as shown in Fig. 3 [25]. The diffusion constant of the water is proportional to the $\frac{1}{3}$rd power of the degree of swelling. As the network size increased linearly, the mobility of the water increased accordingly. In polyelectrolyte gels, water hydrates the ionized networks and thus the diffusion behavior of the water is complex. In order to study the diffusion behavior of water in detail, they measured the spin-lattice relaxation time (T_1) and spin-spin relaxation time (T_2) of protons (see Fig. 4). As the degree of swelling decreases, the relaxation time shortens in proportion to T_1 and T_2 and the mobility of the water is restricted.

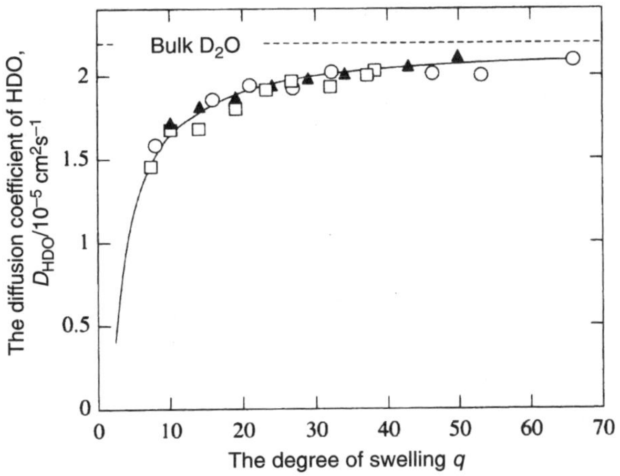

Poly(ethylene glycol) with molecular weight of 4250 (□), 10,890 (▲) and 20,000 (○) are included in the gel as probe molecules. Temperature 30°C.

Fig. 1 The relationship between the diffusion coefficient of HDO (included in a minute amount of heavy water) D_{HDO}, in a poly(N,N-dimethylacrylamide) gel which is swollen by heavy water and the degree of swelling q [19].

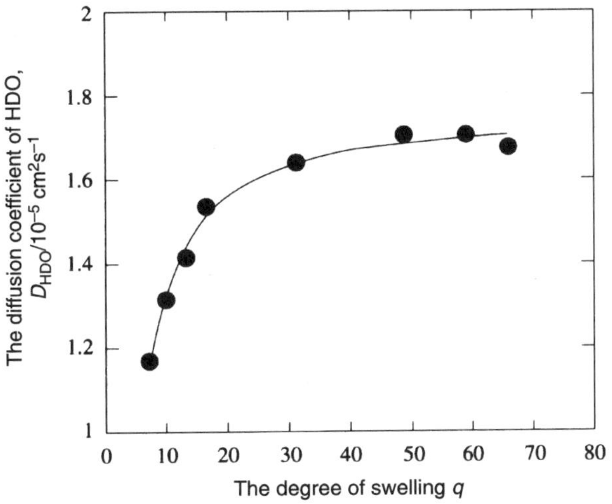

Temperature: 23°C

Fig. 2 The relationship between the diffusion coefficient, D_{HDO}, and the degree of swelling, q, of HDO (included in a minute amount of heavy water) in a poly(N,N-isopropylacrylamide) gel that is swollen by heavy water [24].

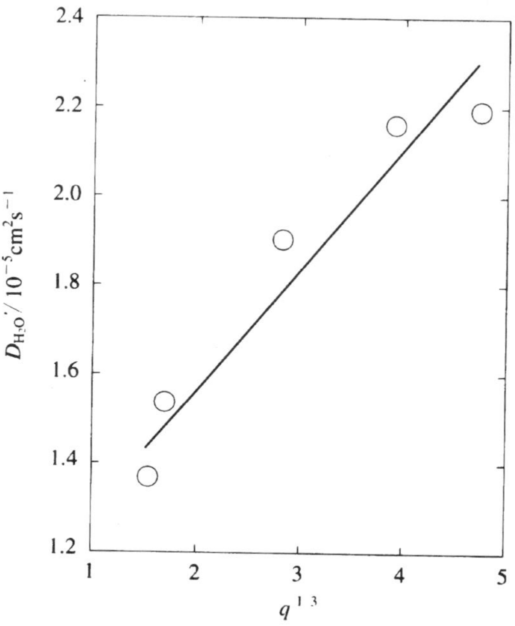

Temperature: 27°C; q is the degree of swelling

Fig. 3 The relationship between the diffusion coefficient of the water D_{H_2O}, and $q^{1/3}$ of the water in a poly(methacrylic acid) gel that is swollen by water [25].

The diffusion coefficient, which is determined by the (PGSE) ^{1}H NMR technique, is due to the translational motion of the center of gravity of molecules. On the other hand, information or water movement obtained by magnetic relaxation includes both rotational and translational movement. Therefore, they have different meanings. In the case of water, the spin-lattice relaxation is controlled by the dipole-dipole interaction.

T_1 has the following relationship with D_{H_2O} [26, 27]

$$\frac{1}{T_1} = \left(\frac{a}{b}\right)^2 \left(\frac{\gamma^4 h^2}{12\pi^2 b^4 D_{H_2O}}\right)\left(1 + \frac{3\pi N b^6}{5a^3}\right) \tag{18}$$

where a is the Stokes-Einstein radius of water cluster; b is the proton distance of water molecules; γ is the magnetogyric ratio; h is the Planck constant; and N is the number of protons of water in 1 cm^3. Using Eq. (18), a is calculated from Figs. 3 and 4 as shown in Fig. 5. It is found that

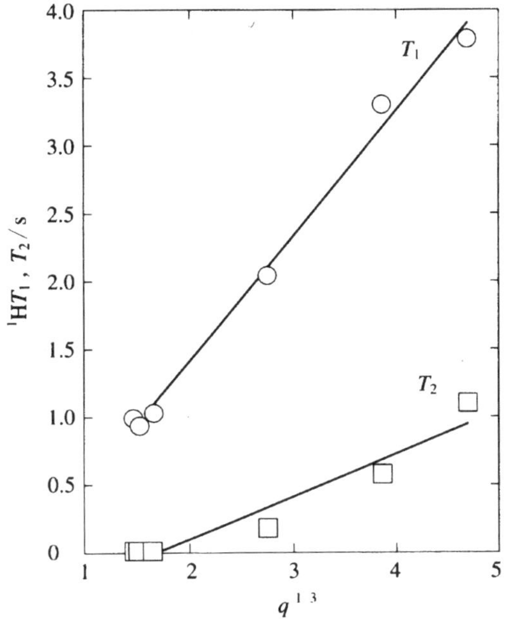

Temperature: 27°C; q is the degree of swelling

Fig. 4 The relationship between the spin-lattice relaxation time (T_1) or spin-spin relaxation time (T_2) and $q^{1/3}$ of the water in a poly(methacrylic acid) gel that is swollen by water [25].

molecular movement becomes easy and cluster size is reduced as network size increases.

The average square distance $\langle x^2 \rangle$ of the diffusant which diffuses in a homogeneous medium with the diffusion constant D is expressed as,

$$\langle x^2 \rangle = Dt \tag{19}$$

When the network structure of a gel is heterogeneous and the gel consists of small cells with microspace, the diffusion of the solvent is restricted spatially and will not follow Eq. (19). Thus, the observed diffusion coefficient reduces over time. Tanner [7] and Meerwall [18] studied such a limited diffusion using a model with parallel blocking walls with distance a and permeability p. They derived the relationship between the spin echo intensity $A(t)$ when the pulsed magnetic gradient with

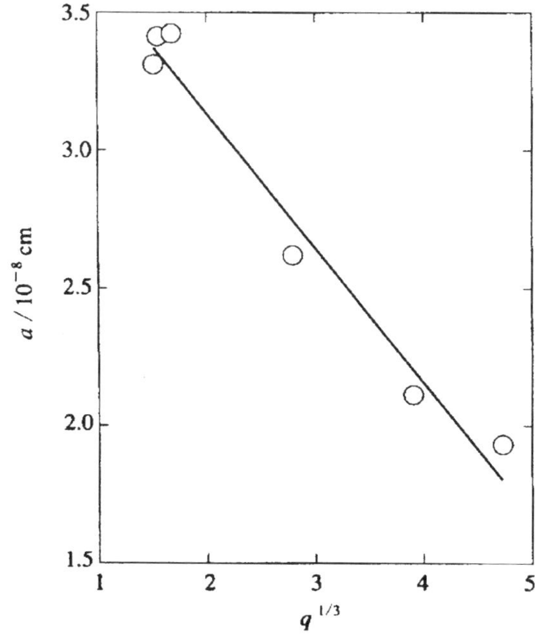

Fig. 5 The relationship between the Stokes–Einstein radius a and $q^{1/3}$ of the water cluster in a poly(methacrylic acid) that is swollen by water [25].

intensity is q and duration time δ and the structure of the blocking walls is considered.

$$\frac{A(t)}{A(0)} = \exp\left[-\frac{\theta^2 D_0 t}{a^2}(\sin^2\alpha + A)\right] \times \frac{2}{\pi^2 a^2}\left\{1 - \cos\pi d + 2\sum_{n=1}^{m}\left[\frac{1-(-1)^n\cos\pi d}{\left(1-\frac{n^2}{a^2}\right)^2}\exp\left(-\frac{n^2\pi^2 D_0 Bt}{a^2}\right)\right]\right\} \tag{20}$$

where $A(0)$ is the spin echo intensity without magnetic gradient, t the diffusion time, γ the magnetogyric ratio, D_0 the diffusion coefficient within the blocking wall, $\theta = \gamma g\delta a$, $d = (\theta/\pi)|\cos\alpha|$ and α the angle of the blocking wall with respect to the magnetic gradient.

Other researchers have measured the diffusion coefficient of water in a potato starch gel and reported the results as shown in Fig. 6 [28, 29]. The diffusion of internal water is spatially restricted due to the formation of compartments by the starch polymer and the apparent diffusion coefficient is consequently reduced. They estimated the diffusion coefficient within this compartment D_0, the size of the compartment a, and the blocking efficiency of the wall p, using Eq. (20) to obtain the results for the gel with various starch concentrations and storage times. The quantity D_0 was found to be smaller than pure water and it increased with increasing storage time. In addition to the polymer that forms the gel wall, the starch molecule that is dissolved in the compartment reduces D_0. In addition, as the starch gel ages, the starch molecules aggregate at the wall.

4.4.2 Diffusion of Probe Molecules in Gels

The homogeneous materials in a gel are considered to be dissolved in the solvent in the gel. The solvent and solute interact with the polymer networks in a complex manner. Tokita *et al.* measured the diffusion coefficient D of the water used as a solvent and other probe molecules like ethyl alcohol, glycerin, and poly(ethylene glycol) (molecular weight 200) using the (PGSE)-^{1}H NMR technique. To represent the size of the gel network, correlation length ξ, which is proportional to the $\frac{3}{4}$th power of the polymer concentration, will be used. Figure 7 is obtained by plotting the ratio of this correlation length with the size R of the probe molecule on the

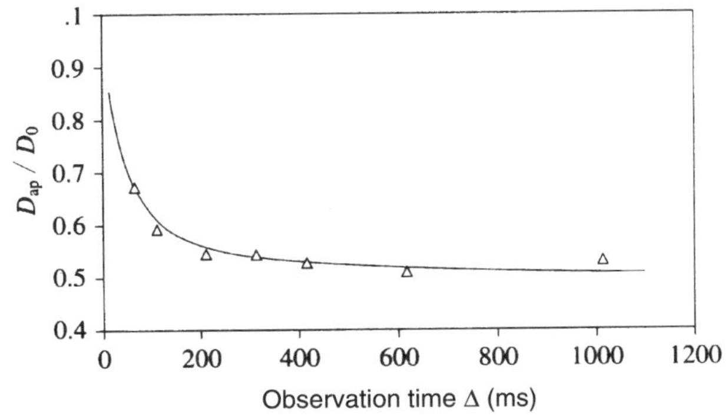

D_{ap} is the apparent diffusion coefficient and D_0 the diffusion coefficient of the water in the compartment

Fig. 6 The relationship between D_{ap}/D_0 and elapsed time [28].

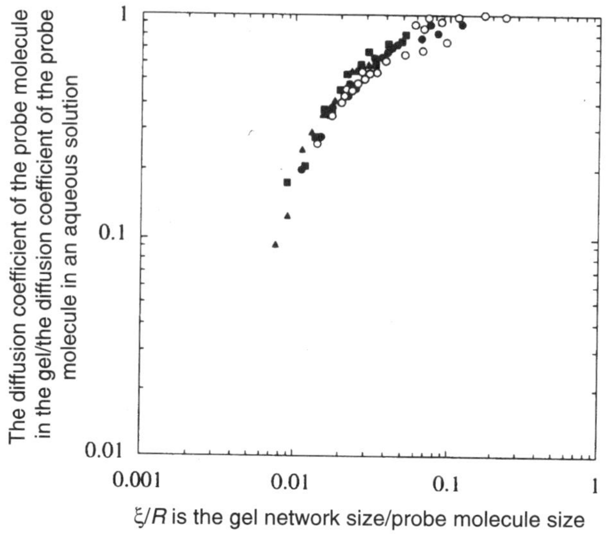

Fig. 7 The scaling plot of the diffusion coefficient D of the probe molecule in a poly(acrylamide) gel against $X = \xi/R$ [30].

abscissa with the diffusion coefficient D of each probe molecule in the gel as well as the diffusion coefficient in a solution D_0. The size R of the probe molecule is said to be proportional to the $\frac{1}{3}$rd power of the molecular weight M. Judging from the fact that the results of each probe molecule were on the same curve, gel diffusion is suggested to be independent of the type of probe molecules and may instead be determined by ξ/R. Further, an anomalous behavior near the phase transition temperature as shown in Fig. 8 is observed in the permeation study of a gel membrane made of poly(N,N-isopropylacrylamide) [31]. This anomalous behavior reflects the concentration fluctuation of networks or the change in interaction between the network and the probe molecule or water.

The diffusion behavior of a globular protein in a gel is important not only in applications involving the controlled release of a protein using a gel or for fixing an enzyme, but also for understanding the relationship between the diffusion behavior of materials of various shapes vis-à-vis the different structures of gels. Cameron *et al.* prepared an amylopectin gel in a capillary tube, and immersed one end into an albumin solution (see Fig. 9). The distribution of the albumin in the gel after a certain time is measured using the FTIR technique [13].

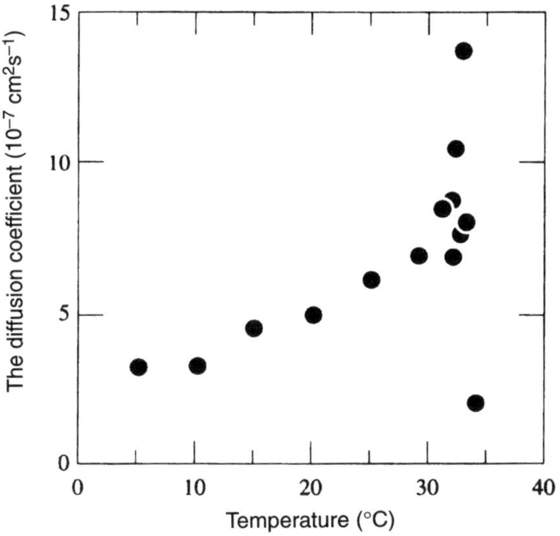

Fig. 8 The temperature dependence of the diffusion coefficient of Ponsor 3R in a poly(N,N-isopropylacrylamide) gel [31].

The diffusion coefficient increased as the concentration of the albumin increased when the albumin concentration was within 1% in the 4.5% amylopectin gel. When the albumin concentration was 2.5%, the diffusion coefficient increased at an increased salt concentration. It is

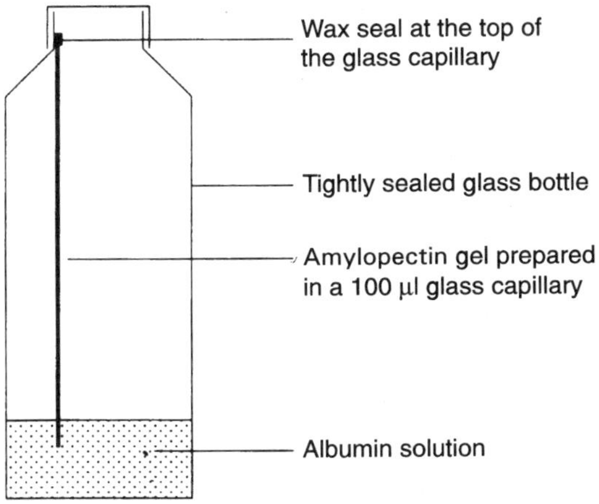

Fig. 9 The diagram for the diffusion coefficient measurement device for the albumin in amylopectin gel.

thought that the diffusion was accelerated by the interaction among albumin molecules caused by electric charges. Using an albumin solution of salt concentration 0.1 M NaCl, where interaction due to the electric charge decreases, the diffusion coefficient of albumin is measured as shown in Fig. 10 at various amylopectin concentrations.

This is the relationship between the dynamic shielding distance κ^{-1} of the gel network and the network concentration c:

$$\kappa^{-1} = c^{u} \tag{21}$$

The constant u expresses the network concentration dependence of the dynamic shielding distance and takes a value of 0.5–1.0 depending on the property of the network polymer. From Eqs (21) and (7), we obtain

$$D = D_0 \exp(-Rc^{-u}) \tag{22}$$

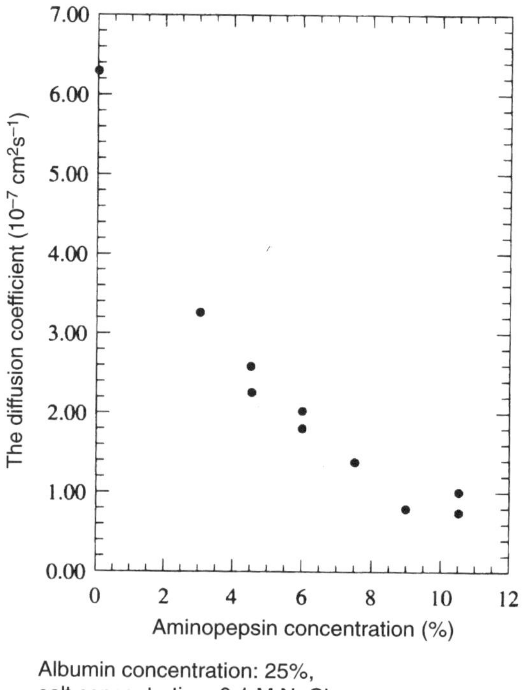

Fig. 10 The dependence of the diffusion coefficient of albumin on the amylopectin concentration [13].

The result of Fig. 10 is replotted in Fig. 11 in order to analyze the relationship between the amylopectin concentration and the diffusion concentration of albumin using Eq. (22). From the slope of this graph, $u = -0.93$ is obtained. However, the degree of aging and subsequently the degree of heterogeneity changes depending on the preparation method of amylopectin and the value u changes.

When the probe polymer is smaller than the network size, the diffusion behavior of the probe molecule can be expressed by Eq. (22) in a similar manner as for a spherical particle. In this case, the fluid dynamic radius R_h is used in place of the size of the particle R. Several researchers measured the diffusion coefficient of poly(ethylene glycol) (PEG) in a poly(N,N-dimethyl-acylamide) gel swollen to the equilibrium state by heavy water using the (PGSE)-^{1}H NMR technique. They obtained the relationship between the degree of swelling q and the PEG diffusion coefficient D_{PEG} (see Fig. 12) [19, 20].

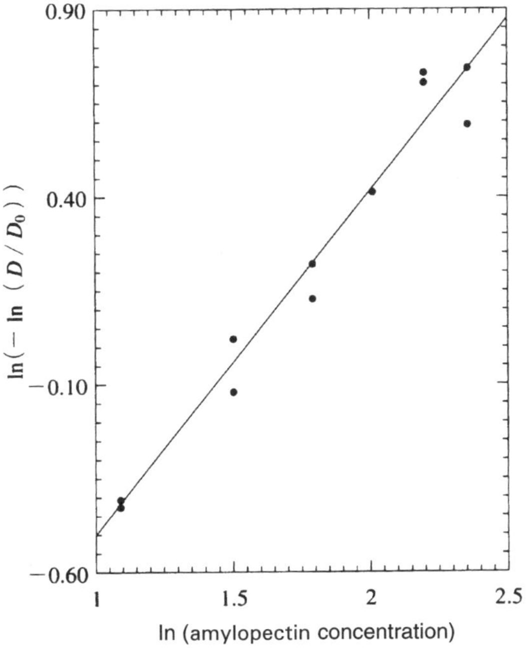

Albumin concentration: 2.5%,
salt concentration: 0.1 M NaCl

Fig. 11 A log–log plot of the diffusion coefficient of albumin and amylopectin concentration.

The quantity D_{PEG} is restricted in the gel for any molecular weight. We will attempt analysis of these results using Eq. (22). Here D_0 is the diffusion coefficient of PEG where there is no restriction by the network chain. To obtain this value from the diffusion coefficient in a PEG solution, the changes of the friction coefficient of the probe molecule in the solution and gel ($\zeta_{gel}^{polym}/\zeta_{soln}^{polym}$) need to be considered.

This can be estimated from the change in the frictional coefficient ($\zeta_{gel}^{solv}/\zeta_{soln}^{solv}$) or from the change in the diffusion coefficient of the solvent D_{HDO}^{neat}/D_{HDO}). Thus, the following equation applies:

$$\frac{\zeta_{gel}^{polym}}{\zeta_{soln}^{polym}} = \frac{\zeta_{gel}^{solv}}{\zeta_{soln}^{solv}} = \frac{D_{HDO}^{neat}}{D_{HDO}} \tag{23}$$

Using this relationship, D_0 can be expressed by using the D_{PEG}^{soln} of 1% PEG solution as

$$D_0 = \frac{D_{PEG}^{soln} D_{HDO}}{D_{HDO}^{neat}} \tag{24}$$

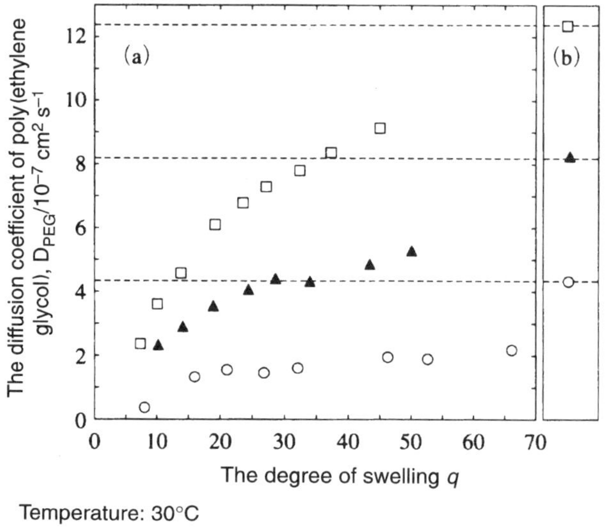

Fig. 12 (a) The relationship between the diffusion coefficient of poly(ethylene glycol) D_{PEG} with molecular weight of 4250 (□), 10,890 (▲), and 20,000 (○) and the degree of swelling in a poly(N,N-dimethylacrylamide) gel, and (b) D_{PEG} in 1% solution [19].

From Eqs. (22) and (24), we further obtain,

$$\ln\left(-\ln\frac{D_{\mathrm{PEG}}D_{\mathrm{HDO}}^{\mathrm{neat}}}{D_{\mathrm{PEG}}^{\mathrm{soln}}D_{\mathrm{HDO}}}\right) = u\ln q + \ln R_{\mathrm{h}} \tag{25}$$

R_{h} has the following relationship with the diffusion coefficient in a solution, $D_{\mathrm{PEG}}^{\mathrm{soln}}$,

$$R_{\mathrm{h}} = \frac{kT}{6\pi\eta_{\mathrm{s}}D_{\mathrm{PEG}}^{\mathrm{soln}}} \tag{26}$$

By inserting this equation into Eq. (25), we obtain

$$\ln\left[\left(-\ln\frac{D_{\mathrm{PEG}}D_{\mathrm{HDO}}^{\mathrm{neat}}}{D_{\mathrm{PEG}}^{\mathrm{soln}}D_{\mathrm{HDO}}}\right)D_{\mathrm{PEG}}^{\mathrm{soln}}\right] = u\ln q + \ln\left(\frac{kT}{6\pi\eta_{\mathrm{s}}}\right) \tag{27}$$

A result is obtained by using the measured diffusion coefficient, calculating the left-hand side term of Eq. (27) and plotting it against ln q. (Also see Fig. 13.) The slope of the straight line is $-\frac{3}{4}$ and $u = -\frac{3}{4}$. The network concentration dependence of the dynamic shielding distance changes by the properties of networks of a gel and solvent. For the flexible network chains in a good solvent, $u = -\frac{3}{4}$ and this is the same as the static correlation length ζ [32] and the experimental results are in good agreement.

When the probe polymer is larger than the network size, the Stokes-Einstein (S-B) type diffusion as shown in Eq. (7) becomes difficult and the polymer diffuses by reptation. In this case, the diffusion coefficient can be expressed by Eq. (8). Pajevic *et al.* [3] measured the diffusion coefficient D_{t} of polystyrene (PS), which is included in the gel as a probe molecule, by using dynamic light scattering. The gel used was poly(methyl methacrylate) gel, which is composed of 12.5% (v/v) polymer and is swollen by toluene. The results are shown in Fig. 14 [3] where D_0 is the diffusion coefficient of PS at each molecular weight (M_{p}) in toluene and $D_0 \approx M_{\mathrm{p}}{}^{-0.6}$.

The PS with molecular weight < 80,000 shows the same molecular weight dependency as that of toluene. It is thus thought that PS diffuses in the S-E type diffusion with the same fluid mechanics radius R_{h} as in solution. In the case of PS with molecular weight 100,000, the diffusion is more restricted as molecular weight increases. The slope of the straight line is -1.2 and $D_{\mathrm{t}}/D_0 \approx M_{\mathrm{p}}^{-1.2}$. Thus, the molecular weight dependence of D_{t} is $D_{\mathrm{t}} \approx M_{\mathrm{p}}^{-1.8}$. This is close to the molecular weight dependence in

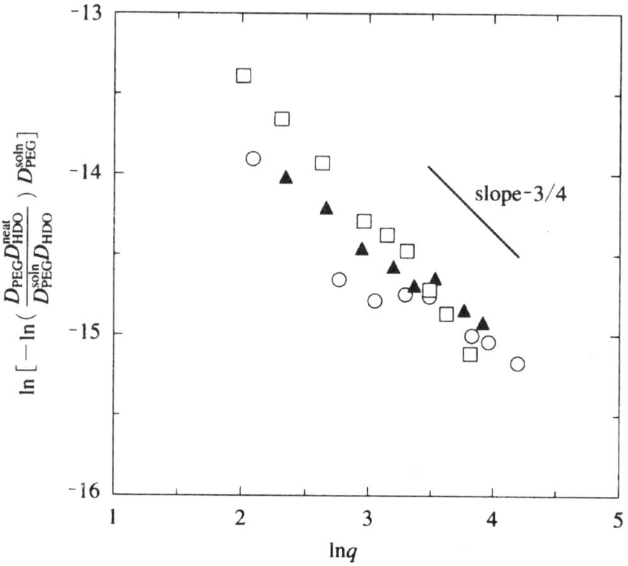

q is the degree of swelling, temperature: 30°C

Fig. 13 The retardation of the diffusion coefficient of poly(ethylene glycol) with molecular weights of 4250 (□), 10,890 (▲), and 20,000 (○) in a poly(N,N-dimethylacrylamide) gel.

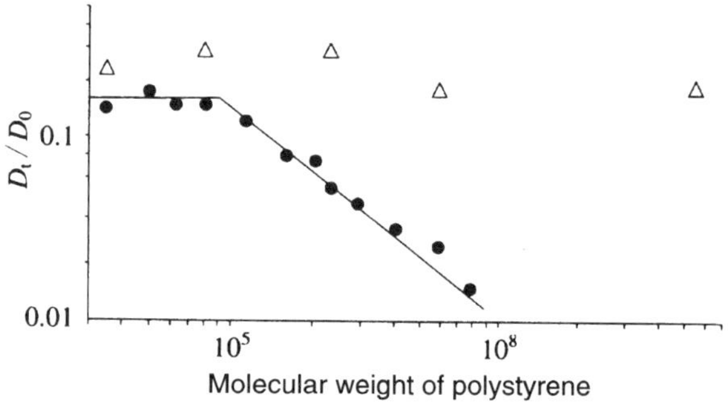

Network chain concentration: 12.5%, temperature: 4.6°C

D_0 is the diffusion coefficient of polystyrene in a toluene solution and Δ are the results in the poly(methyl methacrylate) solution at the same concentration.

Fig. 14 The molecular weight dependence of the diffusion coefficient D_t of polystyrene in the poly(methyl methacrylate) gel that is swollen by toluene.

the reptation region. It is thought that the probe polymer chains diffuse due to movement along the chain direction. The diffusion coefficient of PS in the PMMA solution at the same concentration is shown in Fig. 14 using the symbol Δ. In this case, the molecular weight dependence as seen in the MMA gel cannot be observed.

Matrix polymer entanglement is readily disentangled in a PMMA solution. In contrast, the chemical crosslinks stabilize the entanglement and S-E type diffusion is suppressed. As a result, diffusion by reptation occurs.

Rotstein *et al.* included polystyrene (PS) as a probe polymer in poly(vinyl methyl ether) that is with a network chain concentration of 0.235 g/ml and is swollen by toluene. They measured the diffusion coefficient of PS, D_{PS}, using the dynamic light scattering technique and obtained the results as shown in Fig. 15 [15]. The slope of this plot is −2.8. Similarly, the slope for the gel with network chain concentration of 0.200 g/ml was −2.7. These values for molecular weight dependence are larger than expected in a reptation region of −2. This was explained as due to the distribution of network size and the fact that the probe polymer localized near the entropically stable large networks. When the ratio r of

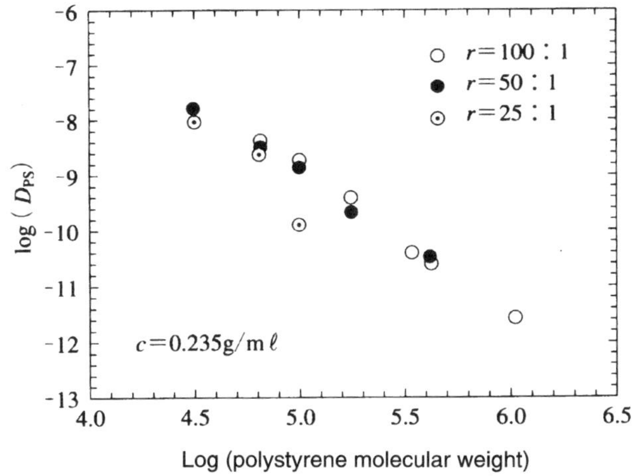

Network chain concentration: 0.235 g/mol; temperature: 30°C;
r : the ratio between vinyl methyl ether and the crosslinking agent.

Fig. 15 The molecular weight dependence of the diffusion coefficient of polystyrene D_{PS} in a poly(vinyl methyl ether) that is swollen by toluene [15].

VME and the crosslinker divinyloxybutane becomes small (i.e., the fraction of the crosslinker is high), it tends to reduce D_{PS} slightly. To clarify this effect, the diffusion coefficients of PS in the PVME solution were compared with those of the same concentration as the gel network concentration (see Fig. 16).

The DPS of large molecular weight PS in a PVME solution with small molecular weight is larger than the DPS in the gel. However, when the molecular weight of PVME becomes large, there is little difference in these systems. When the relaxation time of PVME is shorter than that of PS, PVME behaves like a viscous solution towards PS. In contrast, when the relaxation time of PVME is longer than that of PS, PVME behaves like the gel networks.

The difference in diffusion behavior of probe polymers has been investigated where the network formation of the host polymer is due to physical entanglement such as in a semidilute solution or covalent cross-linking as in a gel. Aven and Cohen [16] used polystyrene (PS) as a probe polymer. They studied the diffusion coefficient of PS ($M_w = 4140$, 7620, and 14,100) in various concentrations Φ swollen by tetrahydrofuran (THF) and PDMS ($M_w = 26{,}500$) solution using a dynamic light scattering technique.

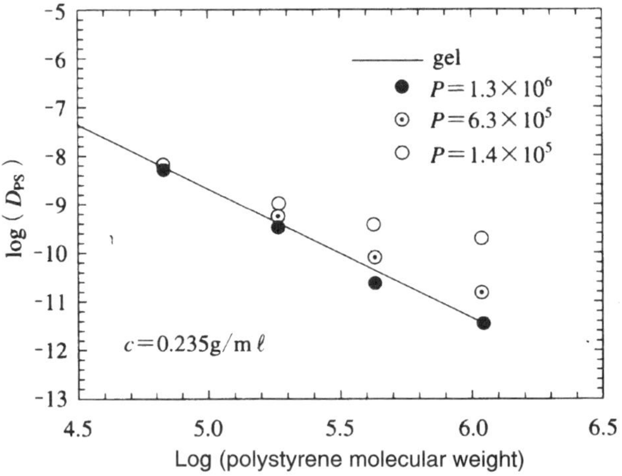

Network chain concentration: 0.235 g/ml; temperature: 30°C;
P: the molecular weight of poly (vinyl methyl ether).

Fig. 16 The molecular weight dependence of the diffusion coefficient of PS, DPS, in poly(vinyl methyl ether) toluene solution [15].

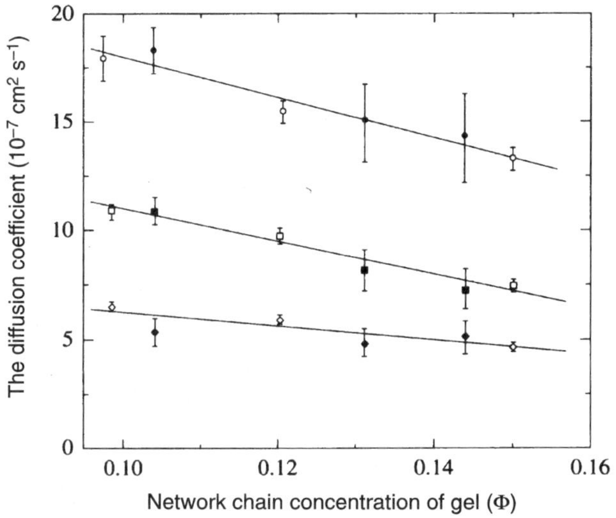

Temperature: 30 °C; molecular weight of polystyrene: 4140 (○,●), 7620 (□,■) and 14,100 (◇,◆).

Fig. 17 The network chain concentration (Φ) dependence of the diffusion coefficient of polystyrene in polydimethylsiloxane solutions (○, □, ◇) and gels (●, ■, ◆) using tetrahydrofuran as solvent [16].

The extrapolated value $D(0)$ to the PS concentration of zero is plotted against Φ (see Fig. 17) [16]. For PS with any molecular weight, there are no visible differences between the PS in the PDMS gel and the solution. Thus, the chemical crosslinks have almost no influence on the diffusion of the probe polymer. The fluid dynamic radius obtained from the diffusion coefficient of PS in the THF dilute solution was either smaller or the same as the dynamic correlation length ξ calculated from the concentration of PDMS. In the region where R_h/ξ is greater, a difference between the gel and solution is expected to be observed. Unfortunately, preparation of such samples is difficult and no experimental results have yet been reported. The extrapolated value of PS of $M_w = 4140$ and 7620 at $\Phi = 0$ equals approximately that of the diffusion coefficient of PS in THF. However, for PS with $M_w = 14{,}100$, the extrapolated value to $\Phi = 0$ is half that in THF. For the PS with $M_w = 14{,}100$, $R_h \geq \xi$ at $\Phi > 1$, it is inappropriate to extrapolate the value of $R_h \ll \xi$ when approaching $\Phi = 0$.

As material diffusion is influenced significantly by gel structure, knowledge about the structure can be obtained by observing the diffusion behavior of the probe material. Yoon *et al.* used poly(2-vinyl pyridine) (P2VP) as a probe polymer for a solution of gelatin randomly labeled by 4-methylbromoazobenzene. They measured the time-dependent changes of the diffusion coefficient D by forced Rayleigh scattering when quenched from 65°C to 5.7°C (see Fig. 18) [22]. During the first 10 h of quenching, D quickly reduced and then approached a constant value.

The restriction of the diffusion of P2VP is regarded to be due to network formation and reduction of the channel size between crystallites. By analyzing the results in Fig. 18, the gelation process of gelatin was considered. Also, in the temperature range of 0–50°C, the diffusion coefficient D of P2VP was measured after keeping a constant temperature for 24 h (see Fig. 19). At a gelation temperature of >33°C, D gradually decreased as the temperature decreased due to the reduction in viscosity of the gelatin solution. Below 33°C, the reduction in viscosity is sudden. Therefore, it is thought that the network was formed at 33°C and the amorphous portion decreased as the temperature decreased, resulting in

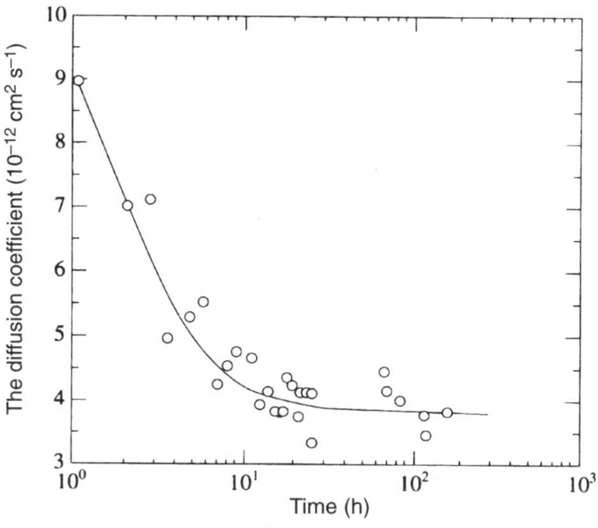

Fig. 18 Time-dependent change of the diffusion coefficient of poly(2-vinyl pyridine) with the molecular weight 96,000 in gelatin gel when quenched from 65 to 5.7°C [22].

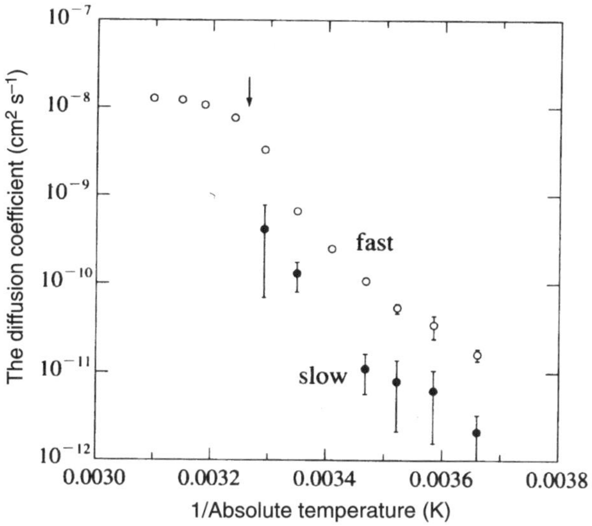

Below the gelation temperature of gelatin (indicated by an arrow), both fast and slow diffusion was observed. Network chain concentration: 10wt%.

Fig. 19 Temperature dependence of the diffusion coefficient of poly(vinyl pyridine) with molecular weight of 33,000 in a gelatin gel.

reduction in channel size. Below 33°C, a slower diffusion component was also observed and its fraction increased with the reduction of temperature. This slow component was thought to be due to the P2VP, which was interacting with the crystalline portion of the gelatin. The results of the measurement of D after maintaining a constant sample temperature for 24 h were the same regardless of increases or decreases in the measurement temperature. Thus, the structure formation of gels does not exhibit hysteresis and is thermally reversible.

REFERENCES

1 de Gennes, P.G., (1986). *Macromolecules* **19**: 1245.
2 Numasawa, N., Kuwamoto, K., and Nose, T. (1986). *Macromolecules* **19**: 2593.
3 Pajevic, S., Bansil, R., and Konak, C. (1993). *Macromolecules* **26**: 305.
4 Fujita, H. (1961). *Adv. Polym. Sci.* **3**: 1.
5 Fujita, H., and Kishimoto, A. (1961). *J. Chem. Phys.* **34**: 393.
6 Fujita, H., Kishimoto, A., and Matsumoto, K. (1960). *Trans. Faraday Soc.* **56**: 424.
7 Boss, B.D., Stejskal, E.O., and Ferry, J.D. (1967). *J. Phys. Chem.* **71**: 1501.
8 von Meerwall, E.D., Amis, E.J., and Ferry, J.D. (1985). *Macromolecules* **18**: 260.

9 Muhr, A.H., and Blanshard, M.V. (1982). *Polymer* **23**: 1012.
10 Yasuda, H., Lamaze, C.E., and Peterlin, A.J. (1971). *J. Polym. Sci.* A2 **9**: 1117.
11 Haggerty, L., Sugarman, J.H., and Prudhomme, R.K. (1988). *Polymer* **29**: 1058.
12 Tokita, M. (1995). *Jpn. J. Appl. Phys.* **34**: 2418.
13 Cameron, R.E., Jalil, M.A., and Donald, A.M. (1994). *Macromolecules* **27**: 2708.
14 Schlick, S., Pilaar, J., Kweon, S.C., Vacik, J., Gao, Z., and Labsky, J. (1995). *Macromolecules* **28**: 5780.
15 Rotstein, N.A., and Lodge, T.P. (1992). *Macromolecules* **25**: 1316.
16 Aven, M.R., and Cohen, C. (1990). *Polymer* **31**: 778.
17 Tanner, J.E. (1978), *J. Chem. Phys.* **69**: 1748.
18 von Meerwall, E., and Ferguson, R.D. (1981). *J. Chem. Phys.* **74**: 6956.
19 Matsukawa, S., and Ando, I. (1996), *Macromolecules* **29**: 7136.
20 Yasunaga, H., Kobayashi, M., Matsukawa, S., Kurosu, H., and Ando, I. (1997). in *Annual Reports on NMR Spectroscopy*, vol. 34, G.A. Webb and I. Ando, eds., London: Academic Press, p. 39.
21 Aoyagi, K., and Segawa, Y. (1983). *Kotai Butsuri* **18**: 221.
22 Yoon, H., Kim, H., and Yu, H. (1989). *Macromolecules* **22**: 848.
23 Widmaier, J.M., Ouriaghli, T.E., Leger, L., and Marmonier, M.F. (1989). *Polymer* **30**: 549.
24 Tanaka, N., Matsukawa, S., Kurosu, H.J., and Ando, I. (1995). *Polym. Preprints, Jpn.* **44**: 1612.
25 Yasunaga, H., and Ando, I. (1993). *Polym. Gels and Networks* **1**: 83.
26 Abragam, A. (1961). *The Principles of Nuclear Magnetism*, Oxford: Clarendon Press, pp. 323–327.
27 Hubbard, P.S. (1963), *Phys. Rev.* **131**: 275.
28 Ohtsuka, A., Watanabe, T., and Suzuki, T. (1993). *Polym. Preprints, Jpn.* **42**: 2997.
29 Ohtsuka, A., Watanabe, T., and Suzuki, T. (1994). *Carbohydrate Polymers*, **24**: 95.
30 Tokita, M., Miyoshi, T., Takegoshi, K., and Hikichi, K. (1996), *Phys. Rev.* **E53**: 1823.
31 Matsunaga, T., Miyamoto, K., Nakamura, T., Tokita, M., and Komai, T. (1994). *Polymer Preprints, Japan* **43**(10): 3650–3651.
32 de Gennes, P.G. (1976). *Macromolecules* **9**: 594–598.

Section 5

Insolubility and Supportability (including Absorption of Oil)

HISAO ICHIJO

5.1 FIXATION (MICROBES, ENZYMES AND CATALYSTS INCLUDED)

5.1.1 Introduction

It is said that the first attempt at enzyme fixation to make better utilization was done by Grubhofer and Schleith when they covalently bonded carboxypeptidase and diazotase to diazopolyaminostyrene [1].

Later, many new substrates and a bonding method have been developed. In the late 1960s, the number of articles published on enzyme fixation was ≈20. That number had increased to over 500 by 1975 [2]. Chibata *et al.* used aminoacylase ionically bonded to DEAE-Sephadex to split dextro-levorotary (D-L) amino acids. In 1969, they pioneered the industrial use of fixed enzymes. This then encouraged the industrial production of isomerized sugar from glucose isomerase, 6-aminopenicilin by penicilinacylase, low sugar lactose by lactase. It also broadened the research on fixed enzymes. The fact that these examples of industrialized production took place and there was a sudden increase in the published research in the 1970s points to the development of fixed enzymes with practical applications.

The natural polymers that have been used include polysaccharides like cellulose, dextrin, starch, agar, carrageenan, arginic acid, chitin and their derivatives including proteins like gelatin, albumin and collagen, and tannic acid. Natural polymers are superior to synthetic materials in terms of cost and safety. Carrageenan and arginic acid have often been used as substrates for foods and medical drugs. Although the degree of cross-linking is low and enzymes tend to leak, they have been said to be excellent materials with respect to safety and the dispersibility of the matrix when fixing microbes.

Synthetic polymers can vary their structure or achieve chemical modification relatively easily. Thus, attempts have been made to determine optimum composition by changing the composition of grafts and experimental conditions. A typical polymer is a polyacrylamide that is used for inclusion. Despite problems such as increased temperature from the heat of polymerization and the toxicity of the monomer, it has long been used as a substrate because the polymerization process has been understood and handling is easy.

Fixation of biocatalysts using natural and synthetic polymer gels has been extensively studied, to the point that developing focus is necessary here. In this section, unique gel substrates such superfine poly(vinyl alcohol) (PVA) gels and stimuli responsive gels will be described with respect to fixation while we will also take advantage of the substrates' characteristics.

5.1.2 Superfine Poly(vinyl alcohol) Fibers

One of the most important considerations in choosing a fixation substrate is the ease of making appropriate shapes that are suitable for individual applications. In traditional reactors such as column and tank, the majority of currently used fixed biocatalysts are pellet-shaped. Fibrous substrates have large surface areas and are easily changed into various shapes. They are porous, have unusual cross-sectional shapes, and are hollow at the microscopic level. Strings, cloth, filter papers, nonwoven, etc. are all used [3].

5.1.2.1 Preparation and characterization of superfine fibers

Completely saponified PVA is uniformly dissolved under pressure. Then, a component that causes microphase separation in the formed fiber, though it mixes stably with the completely saponified PVA, partially saponified PVA or poly(ethylene oxide), is mixed.

Upon dry spinning this solution in heated air, a fiber with a sea island structure is formed. After spinning, the fiber that has been stretched and heat treated is washed by water. Additives with poor molecular orientation are dissolved and removed, leaving a superfine fiber that consists of completely saponified PVA. When observed by an electron microscope, prior to microfibrilation the fiber has a diameter of approximately 20 μm, which then is reduced to submicrometers, thereby achieving sufficient microfibrilation [4].

Various functional groups can be readily introduced by acetal reaction. When acetal reaction is attempted, crosslinking treatment is also achieved to control the degree of swelling into 3 ~ 4.

Figure 1 compares the pore size distribution of micropores of various substrates using the methanol adsorption technique. The activated charcoal shows a large surface area at $>100\ m^2/g$. However, many of the pores consist of micropores with radii <1 nm. The ion-exchange resin (IER) similarly contains micropores. On the other hand, the microfibrilated fibers exhibit a surface area of $200\ m^2/g$, which is as large as 10–20% of

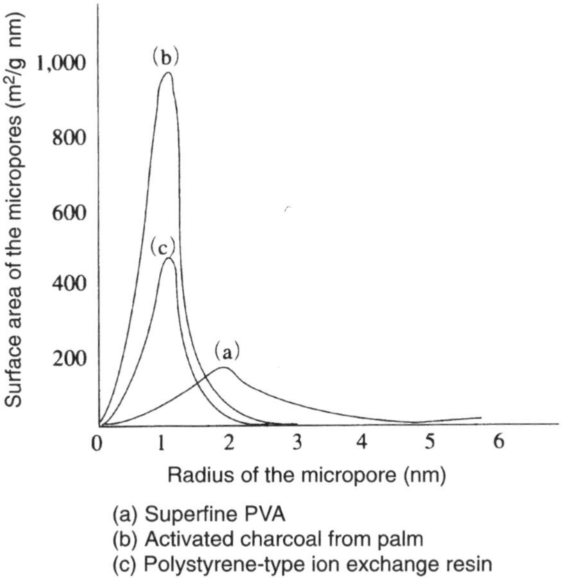

Fig. 1 Distribution curve of the micropore diameter of the fixation substrate.

activated charcoal and their surface consists of relatively large pores in comparison with the activated charcoal.

Many biocatalysts, including enzymes, are polymeric and thus diffusion and fixation are difficult in micropores. Thus, materials like superfine fibers, which possess relatively large micropores, are suitable for biocatalyst fixation substrates.

5.1.2.2 Mechanism of enzyme fixation by superfine fibers

Materials that exhibit an isoelectric point in the acid (e.g., glucosidase) can be fixed by aminated, superfine fibers. On the other hand, those with an alkaline isoelectric point can be fixed by sulfonated superfine fibers (see Table 1) [5]. When the nitrogen content in the fiber increased, the amount of enzyme fixation increased linearly proportionately and was almost saturated at 1 wt%.

Accordingly, fixation by this fiber is likely due to static electric attraction. However, there are enzymes that can be fixed more on dimethylaminated superfine fibers (SFF) than on trimethylaminated SFF. Thus, fixation using only static charge interaction is difficult to explain.

According to Giles *et al.* isothermal adsorption curves can be classified into four types depending on the initial shape of the adsorption isotherms [6]. The isothermal adsorption curve of invertase on aminated, superfine fibers follows that of Langmuir's. When the initial concentration is <100 mg/dl, the rate of fixation is easily explained. However, at a higher initial concentration, the difference between the

Table 1 Adsorptive fixation of various enzymes by three PVA superfine fibers.

Name of biocatalysts	Isoelectric point	The degree of adsorption		
		Dimethylaminated superfine fiber	Trimethylaminated superfine fiber	Sulfonated superfine fiber
Invertase	3.8	90	73	0
Glucosidase	4.3	98	98	27
Typhoid bacillus, β-galactose	4.6	92	92	15
β-Glycosidase	7.0	16	7	98
Lipase	7.3	65	2	100
β-Galactosidase from string beans	8.4	30	14	100

model and the experimental value widens. This is the case because as fixation proceeds the outer surface is covered by the enzyme and then the enzyme molecules must travel a longer distance through the entangled fibers to find available adsorption sites [7].

5.1.2.3 Fixation by superfine fibers and characteristics of fixed enzymes

The adsorptive fixation rate of enzymes by superfine fibers is much faster than ordinary PVA fibers. When the amount of added enzyme is 2000 U/g-fiber, 93% of the enzyme is adsorbed 5 min after the superfine fibers are added. By contrast, merely 9% is adsorbed during the same time with ordinary PVA fibers, and even after 4 h, only 26% is adsorbed. Accordingly, enzyme fixation ability improves drastically by increasing the surface area [4].

Even when compared with other ion-exchange materials, the enzyme (lipase) fixation ability of superfine fibers is extremely high (see Table 2). Of superfine fibers, a much higher fixation capability is possible with ion-exchange fibers (IEF) that have been dehydrated by heat, than with commercial polystyrene-type ion exchange resins. If the size of the lipase used is estimated based on the molecular weight [8], the radius is calculated to be 2 nm. Thus, a fiber surface having a pore radius that is larger than this value is useful. While the surface area of ion-exchange resin and fiber is extremely large, the effective surface area for adsorption with a superfine fiber that has a pore radius of 2 nm is larger. This might be why the fixation ability of superfine fibers is extremely high [9, 10]. However, considering that the effective surface area for adsorption is not necessarily proportional to the amount of enzyme adsorption, interaction between the substrate material and the enzyme might play a role, an

Table 2 Fixation of enzymes by various substrates.

Substrate	PVA superfine fiber	PVA type ion exchange fiber	Polystyrene type ion exchange resin
Ion exchange capacity (meq/g)	0.94	1.3	4.24
Total surface area (m^2/g)	217	515	410
Surface area of pores with radii more than 2 nm (m^2/g)	182	94	18
Amount of fixed enzyme (%)	66	5	5

interesting observation especially in relation to stability (to be described in a later section) [11].

In the case of invertase, the amount of fixation is linearly proportional to the initial concentration. At the initial concentration of 200 mg/dl, the adsorbed amount was as high as 800 mg/1 g of fiber. By adjusting salt concentration, adsorption that is higher than the equivalent weight of fiber is obtained.

In general, increased enzymatic activity shows that fixation has increased, approaching an asymptotic value. The superfine fiber-enzyme complex has not reached the asymptotic value even at a high amount of fixation at 800 mg/g-fiber. If it is applied to a system to which the substrate can be supplied quickly, further increased activity can be expected. From the amount of fixation of β-galactosidase (from string beans) by the three previously mentioned cationic exchange materials, residual activity in solution or UV absorption measurements demonstrate that the amount of fixation of the superfine fiber is much larger than was IEF or IER. This is also considered to be due to the surface characteristics of superfine fibers. Although the amount of fixed enzyme is 3–8 times that of another substrate, activity is much higher [12].

Change in final pH at fixation depends largely on the type of enzyme. The final pH of glucosidase shifted to an acidic profile at 0.7 units by fixation on dimethyl-aminated superfine fiber. This is probably due to static repulsion between the proton and the amine groups. However, in the case of invertase or β-galactosidase from the typhoid bacillus, there was no shift in the final pH.

Continuous hydrolysis of galactose is attempted by passing galactose solution through the column where aminated fiber, to which β-galactosidase from the typhoid bacillus is fixed, is packed. After 12 days, there were few problems other than the plugging of the column. During the entire period, the flow rate was nearly constant and there was no sudden decrease in activity. The half-life was approximately one month [13].

5.1.2.4 Thermal stability of the fixed enzyme on superfine fibers

Traditionally, the thermal stability of a fixed enzyme has been evaluated based on measurements of enzyme activity. In order to correlate the steric structure and changes in activity, there have been attempts made to compare enzyme activity and thermal stability by measuring the small energy that contributes to the maintenance of the structure by using differential thermal analysis (DSC).

Figure 2 shows the DSC thermograms of invertase that are free or fixed on fibers. The peak temperature of the thermogram corresponds to the thermal denaturation temperature (T_d). The T_d of the free and fixed enzymes are 72 and 77°C, respectively. It is possible that the thermal stability of invertase increased due to fixation on the fiber surface. The change of the higher-order structure started at 63°C for the free enzyme and 67°C for the fixed enzyme. Both processes completed at 76 and 83°C, respectively.

If the relative activity and denaturation enthalpy of the heat-treated (heated to the specified temperature of 1°C/min followed by quenching in ice water) free and fixed enzymes are plotted against the heat treatment temperature, there is good agreement between the activity measurement and the DSC one. Thus, it has been demonstrated that DSC can be used to evaluate the thermal stability of enzyme activity. If the thermal treatment temperature exceeds 60°C, the activity of the free enzyme is quickly lost. Although 30% activity is lost at 70°C, 94% of the original activity is lost at 72.6°C. In contrast, the activity of the fixed enzyme is high. Even when

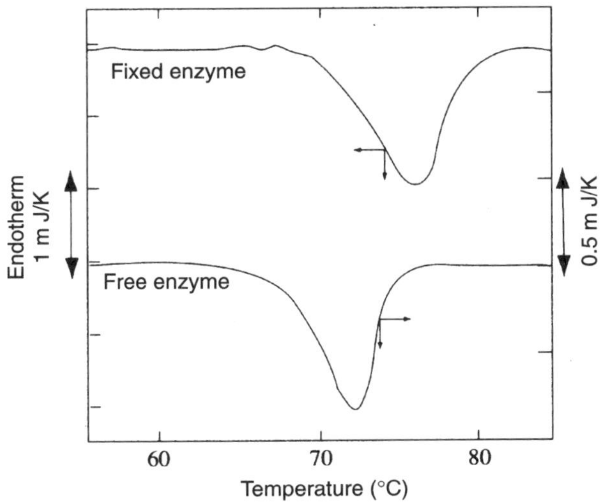

Fig. 2 The DSC thermograms of the free and the fixed enzyme on a PVA superfine fiber.

it was heated to 75 and 77°C, 50% and 25% activities were maintained, respectively. Upon heating to 70°C, 3.5% of the fixed enzyme shows a change in higher-order structure (see Fig. 2). However, it appears that this change is reversible and thus the fixed enzyme heated to 70°C did not lose any activity [14].

It has already been reported that the thermal stability of invertase improves upon enzyme concentration increase and inclusion fixation into PVA gel membrane [15]. Thermal stability improves upon fixation. Multi-functional alcohol also contributes to improvement in stability [11, 16]. The PVA is a kind of multifunctional alcohol and invertase is a typical polysaccharide having about a 50%-saccharide chain [17] . The thermal stability of the enzyme is probably improved by the stabilizing effect of the higher-order structure due to interaction between the hydroxyl groups and the water molecules included in the PVA and the polysaccharide chain.

In order to improve and enhance the function of a fixed biocatalyst, that is a substrate component, material design capable of controlling the microenvironment of the biocatalyst (interaction among the substrate, catalyst, and solvent) is required. If this is possible, it will be feasible to develop new biofunctions.

From DSC and activity measurements using glucosidase (GOD), approximately the same results have been obtained (see Fig. 3) [18]. Komori *et al.* measured the residual activity of the microencapsulated GOD from A. niger in polyurea by interfacial polymerization after keeping the sample at a specified temperature for 10 min. They reported that the thermal stability of GOD was improved by fixation in a capsule [19]. Despite the difference in the heat treatment method, the trend of the data is the same and the results on the free enzyme were nearly the same (see Fig. 3). Glucosidase from A. niger is a protein that contains about 16% saccharide chain [20]. Similar to invertase, it is possibly stabilized by interaction of the PVA-saccharide chain-water.

The thermal stability of the fixed enzyme increases as the pH approaches the isoelectric point or fixation density increases. However, the way it improves depends on the type of enzyme (see Fig. 4).

5.1.2.5 Fixation of microbes on superfine fibers

Using superfine fibers from different functional groups, the amount of fixed enzymes due to ionic bonding was compared. For an aminated fiber, a large amount of yeast is adsorbed and fixed quickly and the surface of

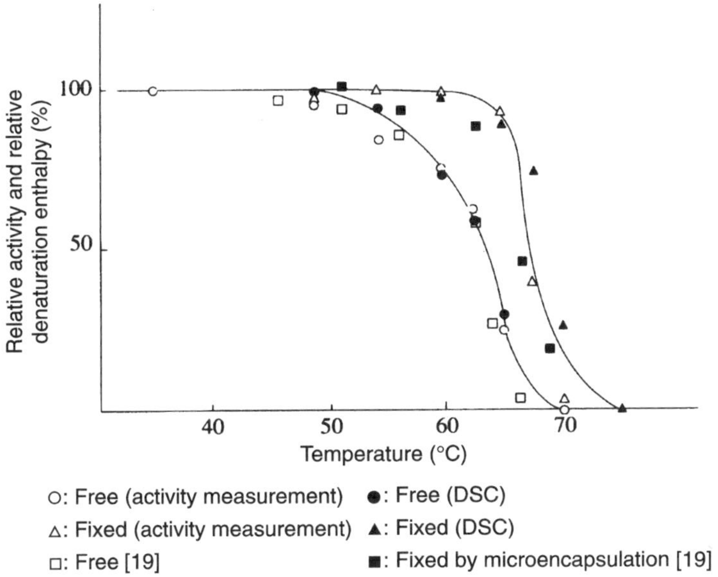

Fig. 3 Thermal stability of glucosidase that is free or fixed on a PVA superfine fiber.

the fiber is covered by a large number of yeast particles (see Fig. 5). In contrast, there was no fixation onto a sulfonated fiber. Both weight measurement and electron microscopy observation did not reveal adsorption of the microbe. Many microbes suspended in water are said to possess negative ions from the dissociated carboxylic acid or phosphoric acid [21]. It is also possible that fixation onto the superfine fibers is due to static interaction.

It has been reported that the amount of the fixed yeast using ion exchange resin is <130 mg per 1 g of resin [22]. Thus, it is clear that the amount of fixation onto the aminated surface is extremely high. When a small amount of yeast was fixed onto an aminated fiber and cultured in glucose solution, the fixed yeast that was found prior to culturing had grown substantially, completely covering the fiber surface [23]. This is a very interesting result because it indicates that mass production of cultured cells is possible.

If a 20 wt% glucose solution that contains yeast is passed through a column packed with dimethyl-aminated fiber onto which the yeast is fixed, alcohol was continuously produced.

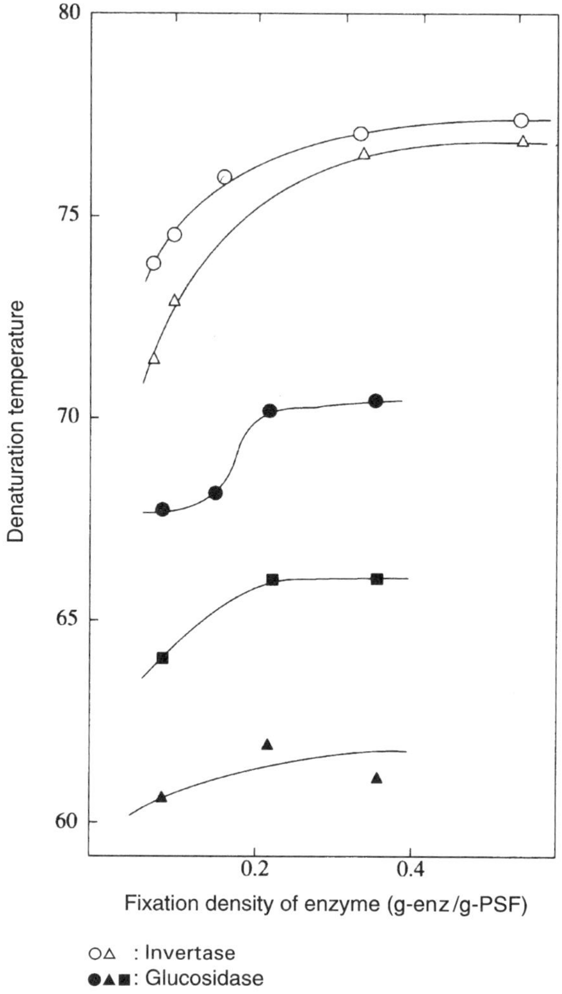

Fig. 4 Influence of the pH and fixation density of the fixed enzyme on thermal denaturation temperature.

Superfine fibers quickly fix a large quantity of microbes. A composite substrate that combines the oxygen enrichment function of silicone hollow fibers and the high fixation capability of superfine fibers is expected to be a material for future development. A silicone membrane with an oxygen enrichment effect and a superfine fiber that adsorbs a large amount of biocatalyst have also been developed.

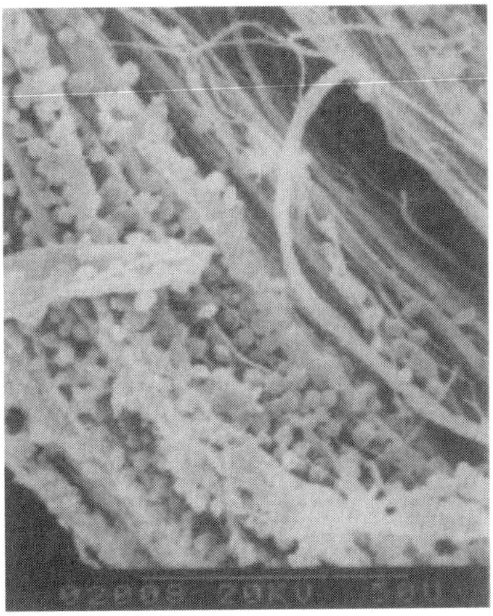

Fig. 5 An electron photomicrograph of yeast fixed onto an aminated PVA superfine fiber.

If activated sludge is fixed onto an oxygen enrichment module made of a silicone hollow tube covered with a superfine fiber (see Fig. 6) [24, 25], and air is passed through the hollow tube, air enriched with oxygen will reach the surface of the fiber. Hence, digestion by the fixed aerobic microbe accelerates. Comparing with the traditional activated sludge method, when the sludge charge is less than 1, the treatment capacity of the current module is 2–3 times higher and the replacement of the used sludge can be kept at 25%. Furthermore, the separation of sludge is unnecessary. Thus, it has been demonstrated that this module can be used as a high efficiency device for waste water treatment.

5.1.3 Fixation Using Stimuli-Responsive Polymers

There have been attempts to control activities in response to environmental changes using property and structural changes of a biocatalyst that is fixed onto an environment (stimuli)-responsive polymer substrate. For example, if a maleic acid/styrene copolymer is adsorbed onto a polystyrene microcapsule in which an enzyme is fixed, the polymer chains spread at $pH > 5$ due to the dissociation of carboxylic groups. As a result, the permeability of the microcapsule increases and the apparent activity of the

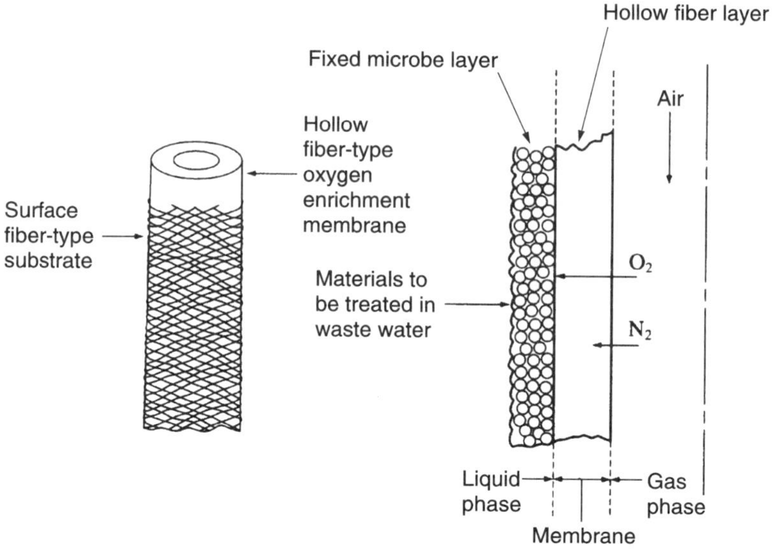

Fig. 6 Oxygen enrichment type fixation composite material.

enzyme also increases. However, at pH < 5, the activity is barely observed due to suppression of dissociation [26]. Another example demonstrates improving productivity by utilizing thermoresponsive swelling and shrinking (pumping effect) [27, 28]. This is done by fixing an enzyme to a thermoresponsive polymer, poly(isopropylacrylamide) (PNIPAAm) gel. Another method has also been studied to separate and recover the enzyme by aggregating the thermoresponsive polymer using increasing temperatures after the enzyme reaction without loss in activity; this is done by bonding PNIPAAm with the enzyme [29]. There have been many studies of PNIPAAm. If a PNIPAAm-grafted substrate is immersed into a suspension of cells and heated above the transition temperature, the cells adhere via hydrophobic interaction caused by polymer dehydration. After the cells are cultured, the grafted chains are hydrated by lowering the temperature and the cultured cells can be naturally desorbed. Thus, the recovery of cultured cells that were unharmed by a proteolytic digestive enzyme is possible [30]. A biochemomechanical system is proposed using fixed urease on the copolymer of NIPAAm and acrylic acid where the gel swells by dissociation of acrylic acid only when the matrix urea is provided [31]. Shown in the following is the result of enzyme activity changes that accompany temperature changes for the fixed enzyme on two kinds of thermoresponsive polymer gels [32, 33].

5.1.3.1 Fixation of gels by thermoresponsive polymers

An enzyme can be fixed by adding the NIPAAm monomer, a saccharide digestive enzyme, amyloglucosidase, and chemically fixing. Or it can also be achieved by irradiating with γ-radiation on a mixture of a thermoresponsive polymer, poly(vinyl methyl ether) (PVME), and an enzyme, and then fixing the enzyme during crosslinking.

Figure 7 shows the temperature dependence of the activity of a free enzyme and the enzyme when included as part of PNIPAAm and PVME gels. The activity of the free enzyme in the temperature range 20–50°C increases as temperature increases. In contrast, the enzyme included in the PNIPAAm gel drastically reduces its activity ≈30°C and there is almost no activity above the volumetric phase transition temperature. Park and Hoffman reported that there is little difference in the activity at a certain constant temperature above and below the transition when the gel is subjected to higher and lower temperatures, leading to an increase in apparent activity as a result of the acceleration of the supply of the matrix

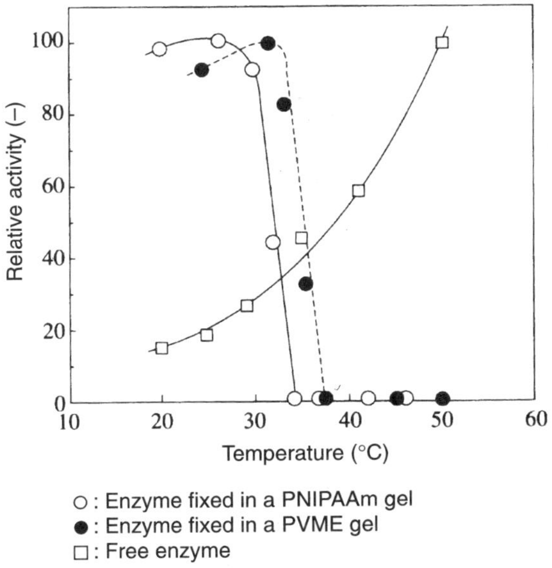

Fig. 7 Temperature dependence of the activity of free and fixed enzymes.

and removal of the products [27]. However, in this experiment, there is a clear difference in activity above and below the transition temperature.

The activity of the enzyme fixed in the PVME gel also depends on temperature in a similar manner as with the enzyme fixed in the PNIPAAm gel. Below the phase transition temperature of PVME, activity increases slightly upon increase in temperature. However, near the phase transition temperature, the activity suddenly decreases and there is almost no activity above the transition temperature.

Figure 8 shows the time variation of the glucose concentration produced during repeated temperature changes of above and below the volumetric phase transition temperature. In the case of PNIPAAm, there is no enzyme activity above the phase transition temperature at 37°C.

However, below this temperature—at 30°C—the amount of glucose produced quickly increased, confirming the progress of the enzyme reaction. If the enzyme reaction is done above and below the phase transition temperature, the activity of the fixed enzyme can be reversibly controlled. Accordingly, the loss of enzyme activity above the phase

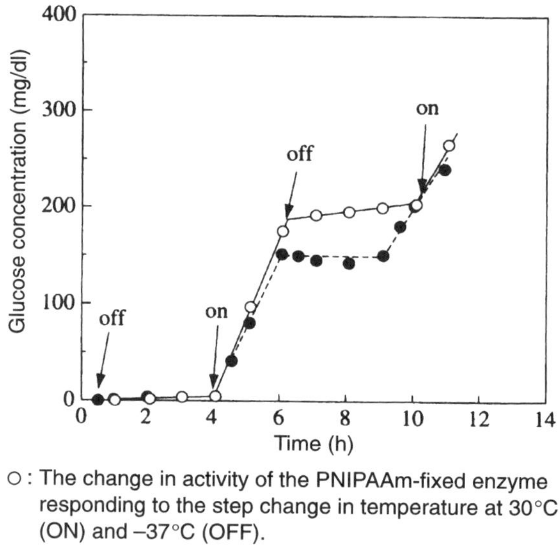

○ : The change in activity of the PNIPAAm-fixed enzyme responding to the step change in temperature at 30°C (ON) and –37°C (OFF).

● : The change in activity of the PVME-fixed enzyme responding to the step change in temperature at 32°C (ON) and –42°C (OFF).

Fig. 8 Activity changes of enzymes in response to reaction temperature.

transition temperature is probably not due to permanent loss of the activity but rather to the difference in the structure and property of the polymer gel at these temperatures.

5.1.3.2 Diffusion of reaction products through gel membranes

Figure 9 illustrates the temperature dependence of the diffusion of glucose through PNIPAAm and PVME gel membranes. Below the volumetric phase transition temperature, diffusion of glucose is rapid. The rate of diffusion slows near the transition temperature and eventually nearly halts above this temperature. Above the phase transition temperature, the PNIPAAm molecules shrink and the gel becomes more tightly structured. It also becomes hydrophobic and loses compatibility with glucose and maltose, leading to reduction in diffusion of both the raw material and the reaction product. This is probably why decreased enzyme activity occurs.

Upon calculation of the diffusion coefficient based on Fick's law of diffusion, it was 1.7×10^{-6} (cm^2/s) at 28.5°C. However, the coefficient decreased to 0.78×10^{-6} (cm^2/s) at 32.6°C and almost no diffusion was observed at 42°C. Reduction in apparent activity is related to the

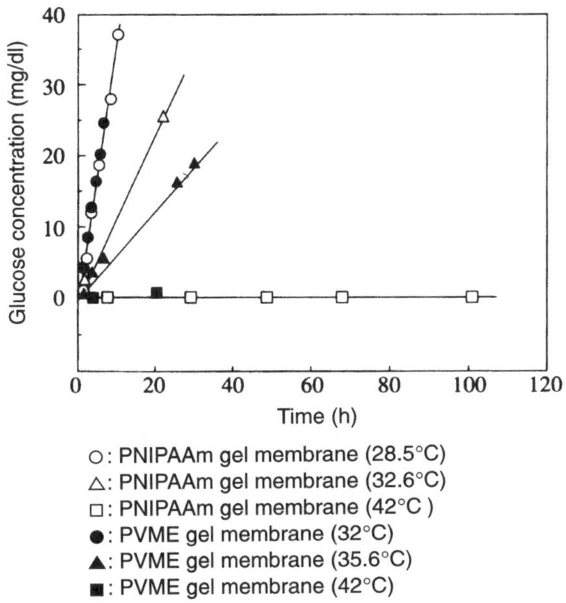

Fig. 9 Diffusion of glucose through thermoresponsive polymer gel membranes.

temperature dependence of the diffusion coefficient, with the final result a drastic reduction in raw material and reaction product above the phase transition temperature.

The amount of glucose passing through a PVME gel shows the same trend as for the PNIPAAm gel when temperature dependence is considered. Below the volumetric phase transition temperature of the PVME gel at 38°C, glucose diffusion is high. Around the transition temperature, it decreases and almost no diffusion takes place above the transition temperature.

5.1.4 Conclusions

To date, various substrate and fixation methods have been investigated. There will be further activity to control artificial biocatalytic reactions in the future. For this purpose, it is necessary to control interaction between the artificial material and the biocatalyst, that is, the enzyme's microenvironment and conformation both need to be controlled. Thus, molecular and material design will be essential in developing new substrates that possess tailored functions and properties.

Gels that contain large amounts of solvent are soft and therefore possess properties and functions that are similar to those of a naturally occurring organism, human or other.

Gels are attracting attention as high performance materials that have properties similar to those of natural bodies, and the ability to create environments similar to those in natural bodies, and to provide materials that model bodily functions. This will probably be the focus of fixation substrate research. However, those gels widely used today do not possess specific structures. If structural regularity is added to them, materials of higher performance can be expected. Superfine fiber will be useful to add structural regularity. In addition to high porosity and fibrillation of superfine fibers, it is possible to expect different reactivity or selectivity if a highly oriented polymer (such as liquid crystal) is used.

Fixation substrate functions depend largely on molecular level structure and properties such as chemical and stereospecific structure, the structure and property of supramolecular materials, and macroscopic properties like high porosity. In addition to the progress already made in both synthesis of functional polymers such as stimuli-responsive and molecular recognition types and development of composite manufacturing methods, if morphological characteristics can be utilized, enzyme reactions will be possible anytime they are needed.

5.2 GELATION AGENTS FOR OILS

KENJI EI AND OHHOH SHIRAI

5.2.1 Introduction

Treatment of oil spills, leaking oils, and domestic waste oils has become very important in recent years for many reasons, including concerns that treatment quickly address environmental pollution issues. If it were possible to add a small amount of oil gelation agent to solidify oil into a jelly, this would contribute significantly to saving the environment. In industrial wastes, there are liquid wastes (e.g., used organic solvents) that cannot be disposed of as they are. It is possible to use these wastes as fuels if the addition of a gelation agent quickly changes the used organic solvent into solid. It is also then easily disposable. Accordingly, the development of oil gelation agents is desirable for environmental protection.

The word "gel" has been used to date in an ambiguous manner. However, Almdal *et al.* recently proposed the following two conditions for the use of the word gel [34]. They are: (1) the material consists of more than two components and one of them is a liquid that exists in large quantity; and (2) the material has soft, solid, or solid-like shapes. Although these are qualitative descriptions, quantitative definitions can be given by using dynamic modulus and other properties. The oil gels described in this section are gels that conform to this definition.

Gels can be classified into thermoreversible gels, which become sols upon heating and return to gel when cooled, and thermoirreversible gels, which will not become sols once they become gels. For example, agar from seaweed forms jelly, which is a thermoreversible gel. On the other hand, a polyacrylamide made from acrylamide with a crosslinker, methylenebisacrylamide, forms a thermoirreversible gel. Thermoreversible gels form a gel structure by weak secondary bonding such as hydrogen bonding, hydrophobic bonding, or coordination bonding. Thus, those secondary bonds break upon heating and the gels return to flowing sols. The other thermoirreversible gels form strong network structures via covalent bonding. This covalent bonding will be broken upon heating and the formed gels will not return to sols. For this reason, a thermoreversible gel is called a physical gel and a thermoirreversible gel is a chemical gel. In this section, gelation agents for oils will be described. The authors understand the "oil gelation agents" to be compounds, which can gel liquids other than water with simple operations such as heating or cooling. The word "agent" means a compound that can gel a liquid by adding $<10\%$ by weight of the agent.

Gels can be formed not only by polymers but also by small molecular weight compounds, albeit there are a small number of examples. For small molecular weight gelation agents a relatively small amount is needed to form a gel, they dissolve quickly upon heating and gel readily upon cooling, and the formed gels are thermoreversible and thus by repeating heating and cooling, they change into solutions and gels. It is these small molecular weight gelation agents that are appropriate for use as the aforementioned "gelation agents for oils." In this section, low molecular weight gelation agents for oils will be introduced and driving force and gelation mechanisms will be described.

5.2.2 Amino Acid-Type Oil Gelation Agents

There are about ten agents used to gel organic compounds that have been reported by researchers other than the authors. They are, as shown in chemical formula **1–10**, 1,2,3,4-dibenzylidene-D-sorbitol (**1**) [35], 12-hydroxystearic acid (**2**) [36], N-lauroyl-L-glutamic acid-α, γ-bis-n-butylamide (**3**) [37], spin-labeled steroid (**4**) [38], cholesterol derivatives (**5**, **6**) [39–41], dialkylphosphoric acid aluminum (**7**) [42], phenolic-type cyclic oligomer (**8**) [43], 2,3-bis-n-hexadecyloxyanthracene (**9**) [44], and cyclic depsipeptide (**10**) [45]. The gelation phenomena of these compounds were discovered mostly by accident. Compound **1** is reported to be the oldest gelation agent for oils and is an excellent agent for a wide variety of organic solvents. Although compound **1** is not used as the gelation agent, it is nonetheless used as an additive for polypropylene. Compound **2** is commercially available as the gelation agent for waste tempura oils.

The authors have also found that an L-alanine derivative **11** gels methanol or cyclohexane during synthesis of an amino acid derivative [46]. For example, 4 g of **11** and 1 liter of methanol are heated to dissolve compound **11**. When the homogeneous mixture was at 20°C it gelled. If cyclohexane is used, it takes only 2 g of compound **11** to gel the same amount. Accordingly, compound **11** is an excellent gelation agent for methanol and cyclohexane. Unfortunately, only these two solvents can be gelled.

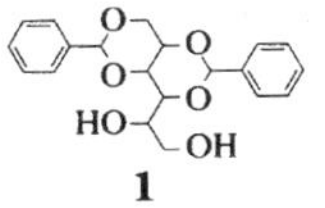

1

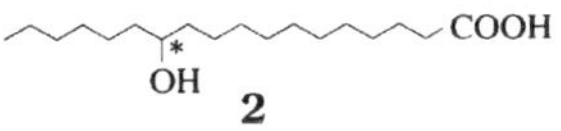

2

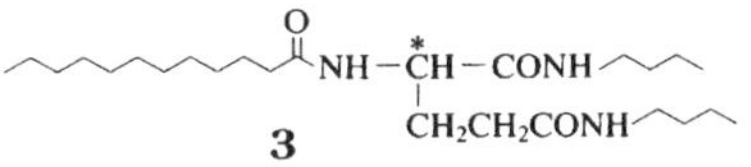

3

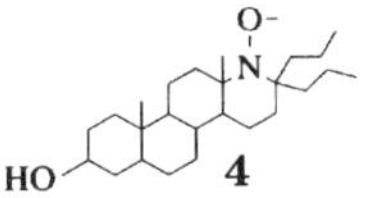

4

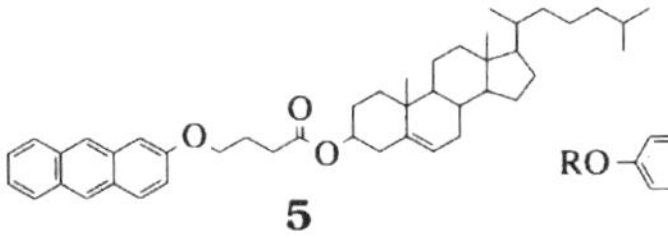

5

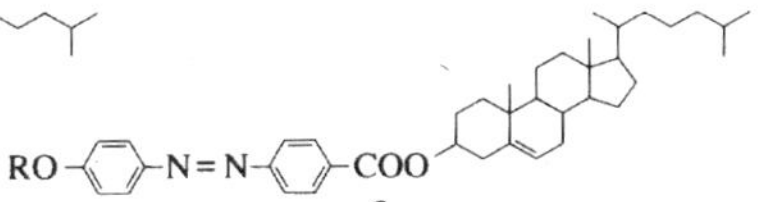

6

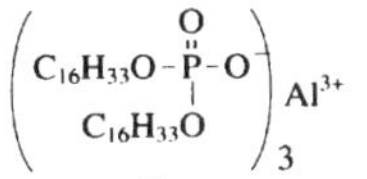

7

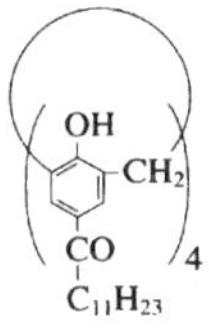

8

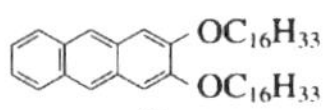

9

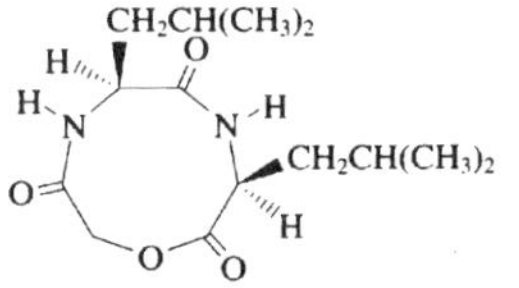

10

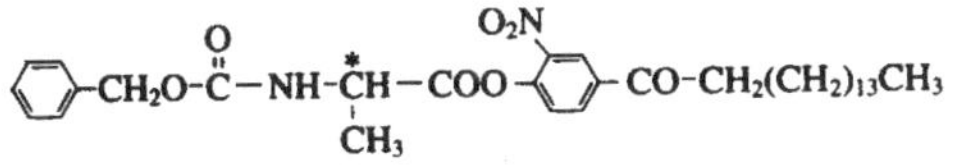

11

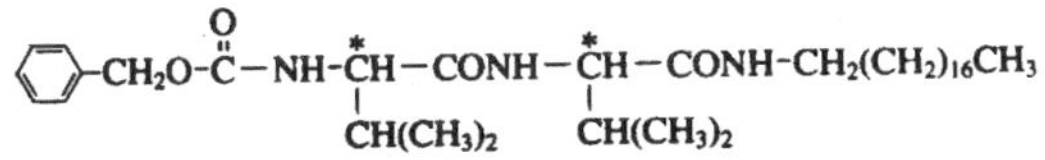

12

Structurally, nitro, carbonyl or long alkyl chains are essential and for amino acids this is limited to L-alanine or D-alanine. Hence the molecular structure required for gelation was simplified and a universal molecular structure was sought. As a result, a series of amino acid compounds, including Z-L-Val-L-Val-$NHC_{18}H_{37}$ (**12**) have been found to be excellent gelation agents for oils [47]. As shown in Table 1, compound **12** is an excellent gelation agent that can solidify hydrocarbons, various alcohols, ketones, esters, N,N-dimethylformamide, dimethylsulfoxide (DMSO), dioxane, carbon disulfide, aromatic compounds, mineral oil, vegetable oil, silicon oil, and a wide variety of solvents. For example, if 5 g of compound **12** is added to 1 liter of cyclohexane or acetone, a gel will be formed. Furthermore, the addition of 3 g to dimethylsulfoxide, kerosene, and salad oil is sufficient to cause gelling. However, amino acid derivatives whose structures are similar to compound **12** do not necessarily exhibit gelation ability. If the amino acid component is glycine, L-alanine, racemic D,L-valine, L-lysine, L-tert-lysine, L-phenylalanine, L-proline, or L-alanyl-L-alanine, these only crystallize and do not form gels.

Table 1 The minimum concentration of the compound **12** that is necessary to solidify various solvents at 20°C (g dm^{-3}) (gelation agent/solvent).

Solvent	Minimum gelation concentration	Solvent	Minimum gelation concentration
Cyclohexane	5	Chlorobenzene	4
Methanol	9	Nitrobenzene	4
Ethanol	15	Toluene	14
1-Propanol	15	DMF	18
2-Propanol	19	DMSO	3
1′-Butanol	15	Carbon tetrachloride	18
Ethylacetate	8	Turpentine	4
Acetone	5	Kerosene	3
2-Butanone	8	Heavy oils	8
Cyclohexanone	12	Silicone oil	4
1,4-Dioxane	18	Salad oil	2
Benzene	4	Soy oil	2

DMF; N,N-dimethylformamide
DMSO; dimethysulfoxide

In addition to compound **12**, Z-L-Val-$NHC_{18}H_{37}$, Z-D-Val-$NHC_{18}H_{37}$, Z-L-iso-Leu-$NHC_{18}H_{37}$, Z-D-Val-L-Val-$NHC_{18}H_{37}$, Z-L-Val-L-Leu-$NHC_{18}H_{37}$, Z-L-Val-β-Ala-$NHC_{18}H_{37}$, Z-L-Leu-β-Ala-$NHC_{18}H_{37}$, Z-β-Ala-L-Glu-$(NHC_{18}H_{37})_2$, were found to be gelation agents.

Compound **12** in cyclohexane gives rise to FTIR bands of NH stretching mode at 3290 cm^{-1}, C=O stretching mode of urethane at 1690 cm^{-1}, and amide stretching mode at 1640 cm^{-1}. The NH and C=O groups of the urethane and amide bonds, respectively, are bonded through hydrogen bonding.

Non-hydrogen-bonded NH or C=O hardly existed. Figure 1 is the sol-gel phase diagram prepared by plotting the minimum gelation concentration at each temperature. The upper portion of each curve is the gel phase and the lower portion is the sol phase. In general gelation by low molecular weight gelation agents shows the temperature dependency as shown in Fig. 1. It requires more gelation agent to gel a liquid at high temperature. From this sol-gel phase diagram, it was found that the change of enthalpy during the sol to gel transition of cyclohexane by compound **12** is $\Delta H = -8.73\,kcal\,mol^{-1}$ (20°C), indicating the formation of approximately 2.5 hydrogen bonds. In the case of compound **12**, there is one urethane bond and 2 amide bonds, allowing 3-intermolecular hydrogen bond formation. This consideration agrees with the results by

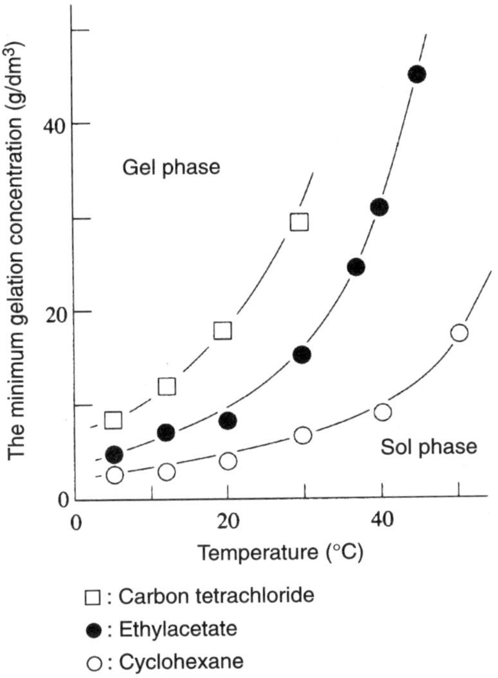

Fig. 1 The sol–gel phase diagram of compound **12**; relationship between the minimum gelation concentration and temperature.

FT-IR. The entropy change from sol to gel is $\Delta S = -15.9\,\text{kcal}\,\text{K}^{-1}\,\text{mol}^{-1}$ (20°C). Although gelation is disadvantageous from the entropic point of view, it seems enthalpy supplements this disadvantageous situation. As a whole, it is controlled by enthalpic phenomenon. Figure 2 indicates intermolecular hydrogen bonding for a dipeptide-type gelation agent like compound **12**. The structure of this intermolecular hydrogen bonding is the same as the antiparallel β sheet of proteins.

Figure 3 shows a TEM photograph of a dilute cyclohexane gel formed by Z-D-Val-$NHC_{18}H_{37}$. A fibrous structure with minimum diameter of 10–30 nm gathers to form a bundle of macroscopic association. This macroscopic association connects occasionally. It was found that the octadecyl group at the end of compound **12** is also important. If the length of the alkyl chain shortens, gelation ability is decreased. The van der Waals forces among the long alkyl chains are probably necessary to gather the string-like structures formed by intermolecular hydrogen

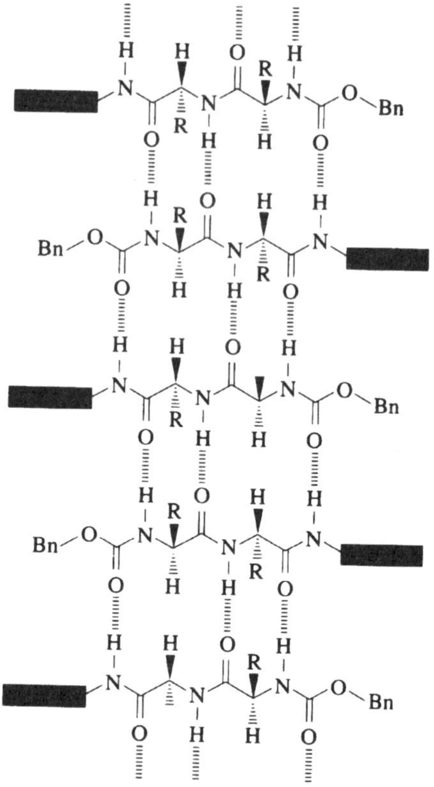

Fig. 2 Antiparallel β-type intermolecular hydrogen bonding formed by a dipeptide-type gelation agent for oils.

bonding and further formation of connected structure as shown in Fig. 3. The gelation mechanism with an amino acid compound is considered to be as follows. First, an associated structure is formed via intermolecular hydrogen bonding. This associated structure will then grow to a macroscopic association. Interactions such as van der Waals's forces form bundles of macroscopic structures and further result in a network structure, thereby restricting mobility and fluidity. The solvent included in this structure eventually causes gelation to occur.

The authors concluded from their studies on amino acid derivatives that it is necessary to find compounds to satisfy the following three conditions: (1) formation of macroscopic fibrous association by intermolecular interaction such as hydrogen bonding; (2) bonding and three

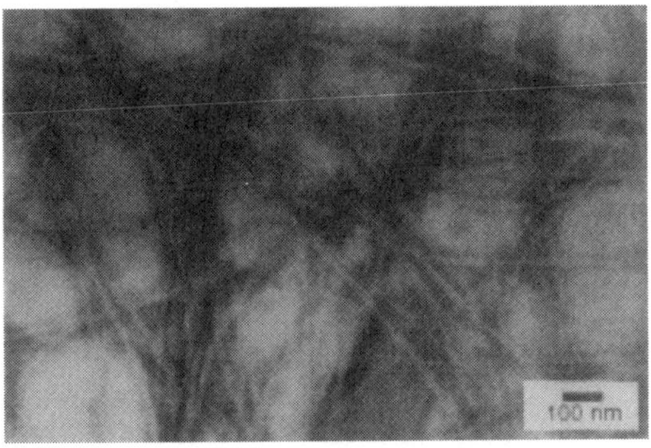

Fig. 3 Transmission electron photomicrograph (TEM) of a dilute cyclohexane gel by Z-D-Val-$NHC_{18}H_{37}$.

dimensionalization of fibrous association by forces such as van der Waals's forces; and (3) stabilization of the gel (which is quasistable) and inhibition of crystallization.

As a compound that satisfies conditions (1)–(3), the authors next identified a cyclic dipeptide derivative as a candidate. Cyclic dipeptide (2,5-diketopiperazine derivative) is a six-membered ring and possesses two amide bonds. Thus, through the hydrogen bonds shown in Fig. 4, it forms molecular assembly by condition (1). If the hydrogen bonding has

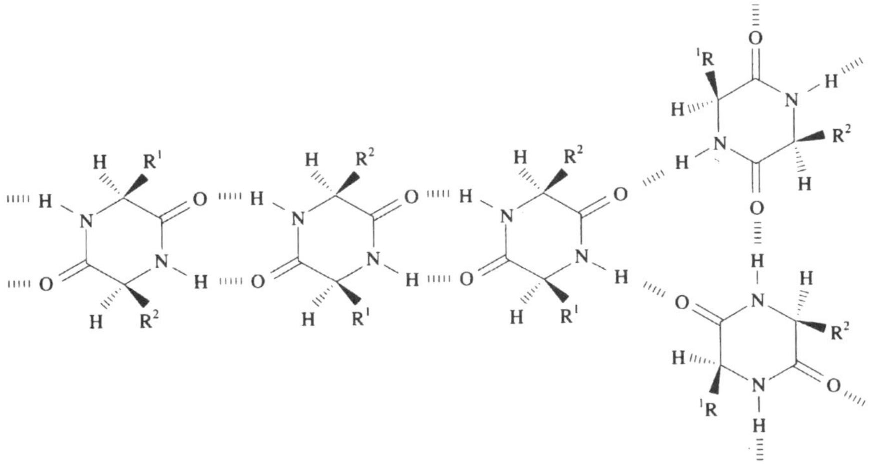

Fig. 4 Intermolecular hydrogen bonding formed by a cyclic dipeptide.

defects, fibrous association in a 3D manner will be formed as in condition (2). Finally, the random arrangement of R1 and R2 groups prevents crystals from forming and the gel structure will be stabilized, which satisfies condition (3). The authors synthesized various dipeptide derivatives and evaluated them for whether or not they act as gelation agents [48]. As a result, the cyclic dipeptides, **13–15**, whose chemical structures are shown here, were found to possess gelation characteristics. In general, the cyclic peptides with neutral amino acids such as L-valine, L-lysine, and L-phenylalanine and acidic amino acids such as L-glutamic acid-γ-ester, L-aspartic acid-β-ester are excellent gelation agents. Of those cyclic peptide-type gelation agents, cyclo(L-Asp(OR)-L-Phe), **15**, is an excellent gelation agent for oils. It can gel alcohols, esters, ketones, aromatic compounds, soy oil, and triolein.

Of the three conditions mentioned here, how to incorporate the third condition into the molecular design so as to prevent crystallization is most difficult.

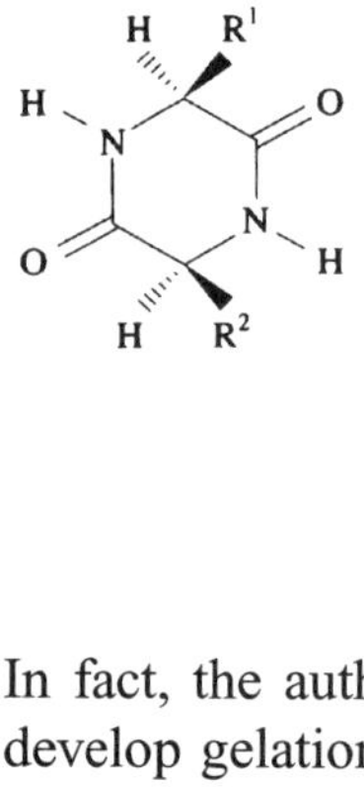

Cyclo(L-Val-L-Glu)
OR
13

$R^1 = CH(CH_3)_2$
$R^2 = CH_2CH_2COOC_2H_5$

Cyclo(L-Leu-L-Glu)
OR
14

$R^1 = CH_2CH(CH_3)_2$
$R^2 = CH_2CH_2COOCH_2CH_3$

Cyclo(L-Asp-L-Phe)
OR
15

$R^1 = CH_2COO$ 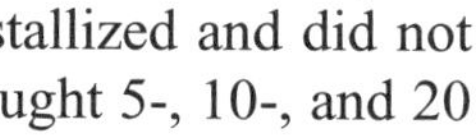
$R^2 = CH_2Ph$

In fact, the authors have synthesized many compounds in an attempt to develop gelation agents but the majority of them crystallized and did not form gels. To overcome this difficulty, the authors thought 5-, 10-, and 20 mers of amino acid oligomers to be potentially useful compounds. If the NCA monomer, an amino acid, is polymerized by elimination of carbon dioxide using an appropriate primary amine, an α-amino acid oligomer that possesses the initiator primary amine at the C-terminal of the molecule can be synthesized. This oligomer has rather large molecular weight of 1000 with broad molecular weight distribution, thus preventing crystallization. In fact, the chemical structures of α-amino acid oligomers shown here, **16–19**, have been found to be gelation agents for oils [49].

Table 2 lists the results of the gelation test for compounds **16–19** as gelation agents.

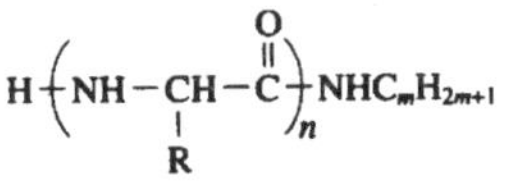

16 H-(L-Val)$_5$-NHC$_{18}$H$_{37}$ R=CHMe$_2$, n=5.0, m=18

17 H-(L-Ileu)$_5$-NHC$_{18}$H$_{37}$ R=CHMeEt, n=5.0, m=18

18 H-(L-Phe)$_5$-NHC$_{12}$H$_{25}$ R=CH$_2$Ph, n=5.1, m=12

19 H-[L-Glu(OMe)]$_4$-NHC$_{12}$H$_{25}$ R=CH$_2$CH$_2$CO$_2$Me, n=4.2, m=12

From the FT-IR and CD spectra, gels are formed when a β-sheet is constructed by intermolecular hydrogen bonding and gels are not formed when an α-helix structure is constructed. Random structure also does not lead to gelation. The construction of a β-sheet is essential for condition (1) to form macroscopic fibrous association via intermolecular hydrogen bonding. It is also thought to be related to condition (2), which calls for the formation of bonding among fibrous associations and 3D structures.

Table 2 Gelation test and minimum gelation concentration (g dm^{-3}) (gelation agent/solvent) for compounds **16–19** against various solvents at 25°C.

Solvent	16	17	18	19
Methanol	30	30	Insoluble	19
Ethanol	Insoluble	28	24	10
1-Propanol	29	17	26	13
Oleyl alcohol	Insoluble	High viscosity	High viscosity	9
Ethylacetate	High viscosity	15	10	26
Acetone	High viscosity	14	Insoluble	Solution
Chloroform	High viscosity	30	29	Solution
Carbon disulfide	High viscosity	High viscosity	30	High viscosity
DMF	9	11	Solution	Solution
DMSO	17	12	Solution	Solution
Benzene	Insoluble	Solution	18	29
Toluene	Insoluble	High viscosity	8	14
Chlorobenzene	25	15	9	30
Nitrobenzene	10	9	10	Solution
Soy oil	High viscosity	High viscosity	21	15
Renolic oil	High viscosity	Insoluble	Insoluble	High viscosity
Tricaprylin	18	27	30	7
Triolein	Insoluble	Insoluble	30	14

DMF; N,N-dimethylformamide
DMSO; dimethylsulfoxide

5.2.3 Two-component Gelation Agents for Oils

In the following, development of a two-component type gelation agent for oils using the concept of molecular lago is based on the aforementioned hypotheses (1)–(3).

Lehn *et al.* reported that the mixture of barbituric acid derivative and triaminopyridine derivative forms intermolecular hydrogen bonding, resulting in alloy-type co-crystals [50]. This system satisfies hypothesis (1). Careful molecular design is all that is needed to satisfy hypotheses (2) and (3) to form a gelation agent. The authors found that by synthesizing triaminopyrimidine derivative (**20**), which contains a 3,7-dimethyloctyl group and barbituric acid derivative (**21**) as shown in chemical formulae **20** and **21**, and combining them a two-component type gelation agent will result [51]. The combination of **20** and **21** could gel one of the most difficult solvents, chloroform, in addition to 1-propanol, benzene, ethylacetate, and acetone. In particular, both **20** and **21** dissolve very well. Simply by adding them to chloroform at room temperature, these compounds will naturally dissolve and cause gelation. In general, low molecular weight type gelation agents require dissolution in solvents at elevated temperature and cooling to room temperature to form gels. In contrast, the two-component type gelation agent can cause gelation simply by adding two components that dissolve at room temperature.

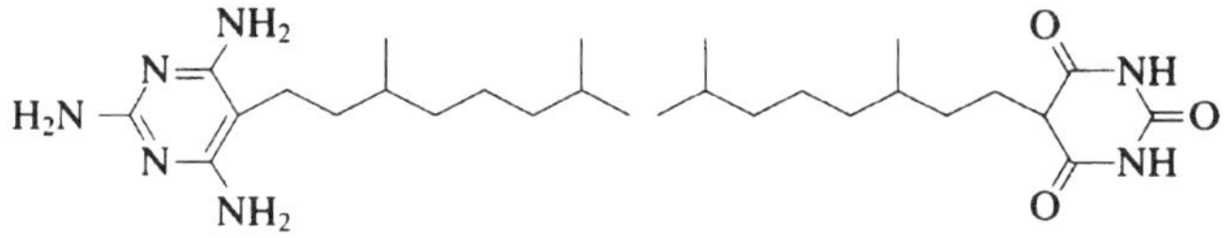

20 **21**

Figure 5 depicts the complementary hydrogen bonding structure of **20** and **21** that was predicted from the x-ray diffraction data. The TEM photograph of a dilute ethylacetate gel (stained by osmium tetraoxide) that used the same mixture is shown in Fig. 6. It shows occasional necking-like patterns. This necking is probably caused by the twisting of the tape-like assembly formed by supplemental hydrogen bonding.

5.2.4 Gelation Agent for Oils from Cyclohexanediamine Derivatives

Diamide **22** and diurea derivative **23** synthesized from trans-1,2-cyclohexanediamine exhibit excellent gelation ability, comparable to that of the

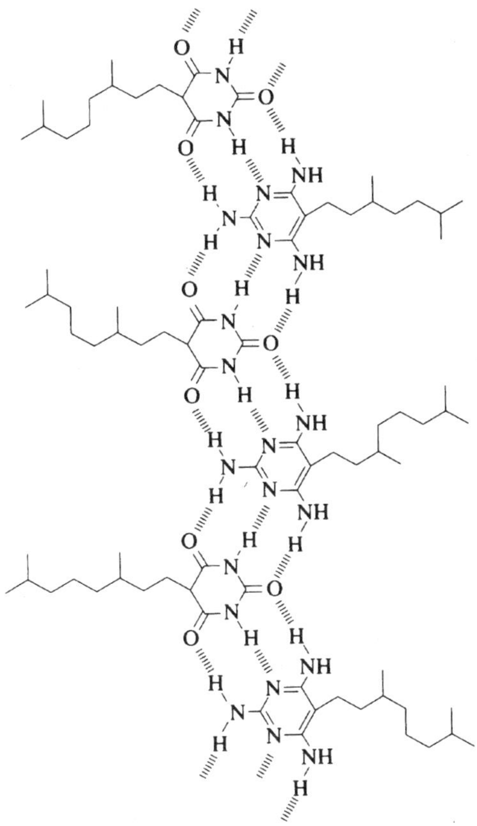

Fig. 5 Complementary intermolecular hydrogen bonding formed by compounds **20** and **21**.

dipeptide derivative **12**. These compounds, **22** and **23**, can gel a wide variety of organic solvents with just a small amount of compound [52, 53]. Table 3 lists the results of a gelation test for 27 kinds of solvents. The 1,2-cyclohexanediamine has one cis and two trans, (IR, 2R) and (1S, 2S), structures, totaling three isomers. The diamide synthesized from the cis-1,2-cyclohexanediamine and diurea did not show gelation ability. From molecular modeling, it can be seen that the two derivatives of the trans-form, **22** and **23**, are on the equatorial position. As shown in Fig. 7 (a), it is possible to form molecular assembly via hydrogen bonding. However, in the case of the cis-form, the derivatives exist at the axial and equatorial positions and thus cannot form intramolecular hydrogen bonding. This is

Table 3 Gelation test for and minimum gelation concentration (gdm^{-3}) (gelation agent/solvent) compounds **22** and **23** to various solvents at 25°C.

Solvent	22	23	Solvent	22	23
Hexane	6	Insoluble	Benzene	20	9
Cyclohexane	11	2	Toluene	12	7
Methanol	20	3	Chlorobenzene	22	13
Ethanol	33	4	Nitrobenzene	12	Solution
2-Propanol	40	5	Pyridine	25	2
Acetonitrile	5	12	DMS	11	2
Chloroform	Solution	Solution	DMF	10	2
Carbon tetrachloride	23	15	DMSO	12	5
Ethylacetate	8	Insoluble	Kerosene	7	11
Acetone	10	2	Gasoline	8	15
2-Butanone	15	2	Salad oil	6	30
Cyclohexanone	11	2	Soy oil	7	Solution
THF	Solution	3	Silicon oil	2	Solution
1,4-dioxane	12				

THE; tetrahydrofurane, DMA; N,N-dimethylacetoamide, DMF; N,N-dimethylformamide, and DMSO; dimethylsulfoxide

probably the reason why the cis-form does not exhibit gelation. Of the minimum gelation concentration amounts, it is seen that the urea derivative **23** gels with the least amount among those tested. This is because of the hydrogen bonding among the urea bonds as shown in Fig. 7 (b), indicating that the urea bond is also useful for molecular design of a gelation agent in a manner similar to that of the amide bond.

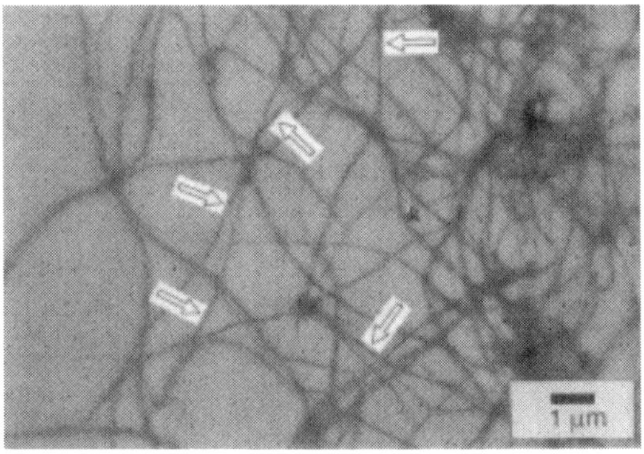

Fig. 6 Transmission electron photomicrograph (TEM) of a dilute gel of ethylacetate obtained by a mixture of compound **20** and 5-(2-ethylhexyl)barbituric acid.

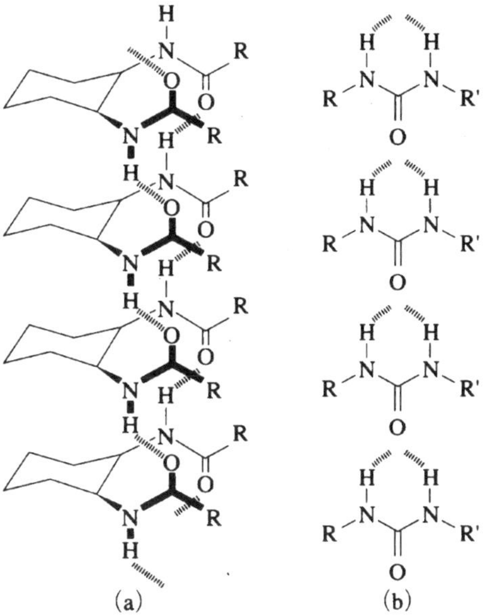

Fig. 7 Molecular assembly of diamide **22** (a) and intermolecular hydrogen bonds of urea bonds (b).

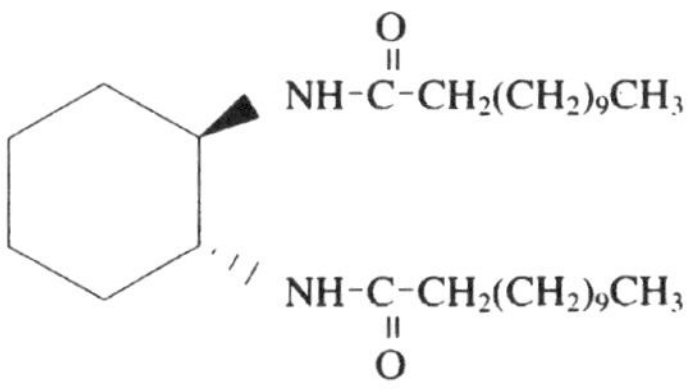

22

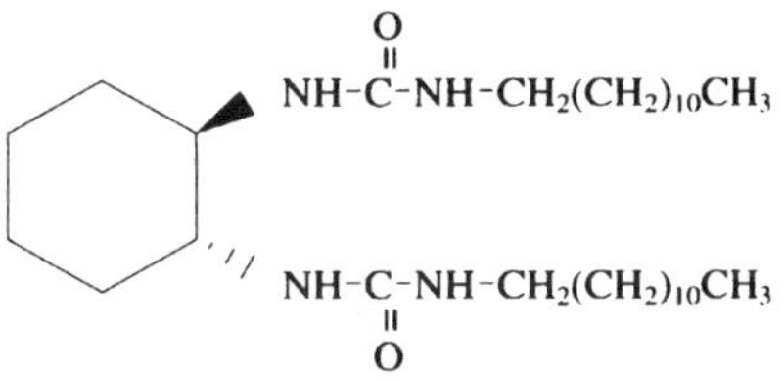

23

5.2.5 Conclusions

In this section, low molecular weight gelation agents related mostly to the authors' works have been described. There are only about 20 kinds of known low molecular weight gelation agents for oils. One reason for this is that gels formed by low molecular weight compounds are in metastable conditions and, thus, these compounds usually transfer in time to more stable crystals. The gelation mechanism of low molecular weight compounds involves first molecular assembly formation via intermolecular interaction, followed by growth into a macroscopic association. Then those associated structures further form bundles via interactions such as van der Waals's forces. Finally, these fibrous structures form networks that inhibit mobility, leading to loss of fluidity and resulting in a gel. Thus, it is possible to design a gelation agent or oils if a compound is molecularly designed so as to satisfy the following three conditions: (1) formation of macroscopic fibrous association by intermolecular interaction such as hydrogen bonding; (2) bonding and three-dimensionalization of fibrous association by forces such as van der Waals's forces; and (3) stabilization of a gel that is quasistable and inhibition of crystallization. The simpler the compound the more academically interesting and higher the potential for commercialization.

REFERENCES

1 Grubhofer, N. and Schleith, L. (1953). *Naturewissenshaften* **40**: 508.
2 Yamauchi, A. and Ichiju, H. (1986). *Energy Shigen* **7**: 141.
3 Ichijo, H. and Yamauchi, A. (1987). *Kobunshi Kako* **36**: 237.
4 Ichijo, H. *et al.* (1982). *J. Appl. Polym. Sci.* **27**: 1665.
5 Ichijo, H. (1992). *Hyomen* **30**: 684.
6 Giles, C.H. *et al.* (1960). *J. Chem. Soc.* 3973.
7 Ichijo, H. (1983). *J. Appl. Polym. Sci.* **28**: 1447.
8 Horiuchi, Y. and Imamura, S. (1977). *J. Biochem.* **81**: 1639.
9 Ichijo, H. *et al.* (1983). *Sen-i Gakkaishi* **39**: T-532.
10 Ichijo, H. *et al.* (1985). *Report on the Fiber and Polymeric Materials Institute* **147**: 31.
11 Kamidaira, H. (1981). *Sen-i Gakkai-shi* **37**: P-436.
12 Ichijo, H. *et al.* (1985). *Sen-i Gakkai-shi* **41**: T-303.
13 Ichijo, H. *et al.* (1986). *Sen-i Gakkai-shi* **42**: T-115.
14 Ichijo, H. *et al.* (1985). *Agric. Biol. Chem.* **49**: 3591.
15 Ichijo, H. *et al.* (1984). *Sen-i Gakkai-shi* **40**: T-317.
16 K. Gekko, (1982). *J. Biochem.* **91**: 1197.
17 Izuka, M. *et al.* (1978). *Agric. Biol. Chem.* **42**: 1207.
18 Ichijo, H. *et al.* (1989). *Agric. Biol. Chem.* **53**: 833.
19 Komori, T. *et al.* (1986). *J. Microencapsulation* **3**: 219.
20 Swoboda, B.E.P. and Massey, V. (1965). *J. Biol. Chem.* **249**: 2209.
21 Morizaki, H. (1986). *Interfaces and Microbes*, Tokyo: Gakkai Pibl. Center, pp. 13–14.
22 Krug, T.A. and Daugulis, A.J. (1983). *Biotechnol. Lett.* **5**: 159.
23 Ichijo, H. *et al.* (1987). *Sen-i Gakkai-shi* **43**: T-271.
24 Hirasa, O. *et al.* (1991). *J. Ferment. Bioeng.* **71**: 206.
25 Hirasa, O. *et al.* (1991). *J. Ferment Bioeng.* **71**: 376.
26 Kokufuta, E. *et al.* (1988). *Biotechnol. Bioeng.* **32**: 289.
27 Park, T.G. and Hoffman, A.S. (1988). *Appl. Biochem. Biotechnol.* **19**: 1.
28 Park, T.G. and Hoffman, A.S. (1989). *Biotechnol. Lett.* **11**: 17.

29 Matsukata, M. *et al.* (1994). *J. Biochem.* **116**: 111.
30 Yamada, N. *et al.* (1990). *Makromol. Chem. Rapid Commun.* **11**: 571.
31 Kokufuta, E. *et al.* (1994). *J. Biomater Sci.* **6**: 35.
32 Kokufuta, E. *et al.* (1992). *J. Chem. Soc., Chem. Commun.* **5**: 416.
33 Hirasa, K. (1994). *Report of the Materials Engineering Industrial Technology Institute* **2**: 483.
34 Almdal, K., Dyre, J., Hvidt, S. and Kramer, O. (1993). *Polym. Gels Networks* **1**: 5.
35 Yamamoto, S. (1943). *Kogyo Kagaku Zasshi* **46**: 779.
36 Tachibana, T., Mori, T. and Hori, K. (1980). *Bull. Chem. Soc. Jpn.* **53**: 1714.
37 Honma, M. (1987). *Gendai Kagaku* **54**.
38 Terech, P. and Wade, R.H. (1988). *J. Colloid Interface Sci.* **125**: 542.
39 Lin, Y., Kachar, B. and Weiss, R.G. (1989). *J. Am. Chem. Soc.* **111**: 5542.
40 Murata, K., Aoki, M., Nishi, T., Ikeda, A., and Shinkai, S. (1991). *J. Chem. Soc., Chem. Commun.* 1715.
41 Murata, K., Aoki, M., Suzuki, T., Harada, T., Kawabata, H., Komori, T., Ohseto, F., Ueda, K., and Shinkai, S. (1994). *J. Am. Chem. Soc.* **116**: 6664.
42 Fukasawa, J. and Tsutsumi, H. (1991). *J. Colloid Interface Sci.* **143**: 69.
43 Aoki, M., Nakashima, K., Kawabata, H., Tsutsui, S., and Shinkai, S. (1993). *J. Chem. Soc., Perkin Trans.* **2**: 347.
44 Brotin, T., Utermohlen, R., Fages, F., Bouas-Laurent, H. and Desvergne, J. (1991). *J. Chem. Soc., Chem. Commum.* 416.
45 Vries, E.J. and Kellogg, R.M. (1993). *J. Chem. Soc., Chem. Commun.* **238**.
46 Hanabusa, K., Okui, K., Karaki, K., Koyama, T., and Shirai, H. (1992). *J. Chem. Soc., Chem. Commun.* 1371.
47 Hanabusa, K., Tange, J., Taguchi, Y., Koyama, T., and Shirai, H. (1993). *J. Chem. Soc., Chem. Commun.* 390.
48 Hanabusa, K., Matsumoto, M., Miki, T., Koyama, T.O., and Shirai, H. (1994). *J. Chem. Soc., Chem. Commun.* 1401.
49 Hanabusa, K., Naka, Y., Koyama, T., and Shirai, H, (1994). *J. Chem. Soc., Chem. Commun.* 2683.
50 Lehn, J.-M., Mascal, M., DeCian, A. and Fisher, J. (1990). *J. Chem. Soc., Chem. Commun.* 479.
51 Hanabusa, K., Miki, T., Taguchi,Y., Koyama T., and Shirai, H. (1993). *J. Chem. Soc., Chem. Commun.* 1382.
52 Hanabusa, K., Yamada, M., Kimura, M., and Shirai, H. (1996). *Angew. Chem. Int. Ed. Engl.* **35**: 1949.
53 Hanabusa, K., Shimura, K., Hirose, K., Kimura, M., and Shirai, H. (1996). *Chem. Lett.* 885.

Section 6
Transparency (Optical Properties)

ATSUSHI SUZUKI AND TOSHIHARU TANAKA

6.1 TRANSMISSION OF LIGHT

6.1.1 Introduction

Gels are three-dimensional polymer networks that contain a solvent. The many examples we encounter every day include agar and jelly. Even our own eyes have a gel constituent. The many applications for synthetic polymer gels today have resulted in their use in many fields [1,2]. Superabsorbent materials, moisture and water retention materials, and sustained release drugs are among the applications for gels. These applications take advantage of the unique properties of gels, that is, their ability to hold water or solution while continuing to maintain shape. Most gels are used as superabsorbents in diapers and in moisture/water retention materials. The optical properties of gels are interesting and relate to the basic properties they possess in the raw state. Functionally, they also demonstrate volumetric changes that are triggered by light.

6.1.2 Polymer Gels and Transmission of Light

In general, transmissivity in both optical lenses and ordinary films is expressed by the intensity of the transmitting light/intensity of the incident light. The transmissivity T_r of light, passing through a film thickness d, with reflectivity R and extinction coefficient α, normal to the surface is given by Eq. (1) [3]:

$$T_r = \frac{(1 - R)^2 \exp(-\alpha d)}{1 - R^2 \exp(-2\alpha d)} \tag{1}$$

Obviously a polymer used for optical lenses has high transmissivity. Polymethyl methacrylate (PMMA) used for hard contact lenses has a transparency (transmissivity) of 92% [3,4]. Hydrogel lenses (poly(hydroxy-ethyl methacrylate) (PHRMA)) has over 98% and poly(vinyl pyrrolidone) (PVP) has over 97% transmissivity [4].

Loss of light originates either by absorption or scattering. Absorptions intrinsic to polymers include those due to electronic transition in the short wavelength region and atomic vibrations in the long wavelength region. The theory of light scattering can be classified into two theories, the electromagnetic theory and the fluctuation theory. The phenomenon represented by the electromagnetic theory is Rayleigh scattering, which explains the scattering phenomenon in air. On the other hand, the fluctuation theory explains light scattering in liquids and solids. Einstein's light scattering theory assumes that there is no fluctuation due to the heterogeneity of a polymer sequence. This fluctuation theory can be applied well to liquids like water. This may also explain the scattering phenomenon from the local and instantaneous heterogeneous structure due to the micro-Brownian motion of water molecules. In general, polymers that absorb in the visible region are colored by resonance and thus absorption of light. Absorption takes place at a particular wavelength. Therefore, if the reduction of light intensity is seen in a colorless, turbid polymer, it is due mostly to scattering rather than absorption. These theories have been proposed for polymers in general [5]. However, few experiments and theories specifically on the transparency of gels have been reported.

Light irradiated onto a gel is transmitted, absorbed, or scattered. An example of a material that uses the high transparency of a gel in its raw state is the contact lens. Contact lenses can be classified largely into hard and soft lenses. The majority of soft contact lenses are made of hydrogels. The material development of contact lenses began with the hard contact lens made of polymethyl methacrylate. Since then, wetting and gas permeability have been improved and new materials such as silicone-methacryl type and fluorinated polymers have actively been developed. Of the soft types, there are both water and nonwater types of lenses. The nonwater lenses are made of silicone rubbers or acrylic rubbers. Those containing water are made of hydrogels. Those with water content $< 40\%$ are low water content lenses and those $> 60\%$ are high water content lenses. In general, the term "soft lens" indicates these hydrogel lenses. Soft contact lenses are more comfortable in comparison with hard lenses. The materials used for hydrogel lenses are PHEMA and PVP [4]. The PHEMA type has hydroxyl groups and water content is $\approx 40\%$. It is used as the main component for low water content soft lenses. The PVP type has water content of 70% and is the main component of high water content soft lenses.

6.1.3 Application of Polymer Gels Utilizing Transparency Characteristics

An example in which gel transparency is useful involves an optical shutter that takes advantage of volumetric phase transition [6]. There are studies on light-induced volumetric phase transition of dyed thermoresponsive polymer gels [7–10]. Among the factors that cause volumetric changes in gels, light is useful for its speed, cleanliness, and simplicity. The gel is made of a thermoresponsive polymer and a photoresponsive dye. It changes volume discontinuously in water by irradiation of light. Molecular design of gels and transparency and kinetics studies have been made with applications to lens shutters, photosensors, and photoactuators in mind. The principle of volumetric phase transition by local temperature rise is close to being confirmed [7,8]. Using this principle, the intensity of the transmitting light can be discontinuously changed with a gel that consists of any polymer and dye and which shrinks thermally. In this section, we will describe phase transition, a function found from the study of gel characteristics, and its application to photoresponsive polymer gels using local heating.

It is possible to synthesize a copolymer gel using thermoresponsive N-isopropylacrylamide (NIPAAm) and a dye (a sodium salt of copper chlorophilin-3) and also to induce phase transition by visible light (see Fig. 1) [7]. A rod-shaped, submillimeter diameter gel was prepared using an argon gas laser at a wavelength of 488 nm as the light source. When the light was not irradiated, the gel exhibited linear volumetric changes in response to the temperature variation. However, if the temperature was changed as the light was irradiated, the swelling curve exhibited a sudden

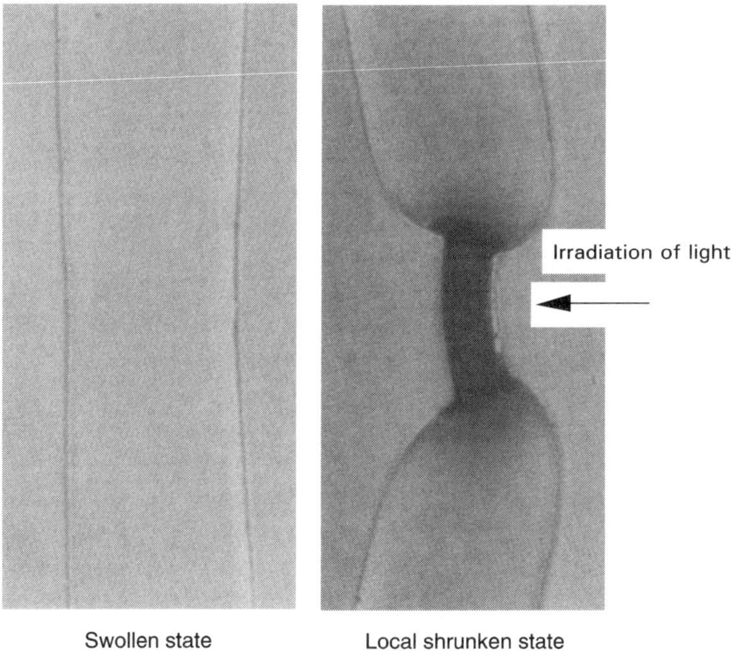

Fig. 1 Local volumetric phase transition of a dye-modified polymer gel by visible light irradiation.

change. If light intensity was even further increased, a discontinuous phase transition could be observed. Further, if the surrounding solvent's temperature was kept constant and the intensity of the irradiating light was increased slowly, the gel initially shrank gradually. However, it showed a discontinuous phase transition (shrinkage) at a certain light intensity (see Fig. 2(a)). The gel with higher dye concentration exhibited a smaller critical light intensity for phase transition at the same temperature to be induced. A similar experiment was performed with the copolymer between NIPAAm and porphyrin dye (sodium salt of protoporphyrin IX). This dye shows much higher specific molar absorptivity with the visible light at 488 nm than the chlorophyrin dye. Thus, phase transition is expected to occur at a lower light intensity. The relationship between light intensity and gel diameter at a constant temperature is shown in Fig. 2(b). Although the porphyrin dye was at 1/6 of the chlorophyrin concentration in the copolymer, the critical light intensity required to cause phase transition is nearly equal. Therefore, the minimum necessary

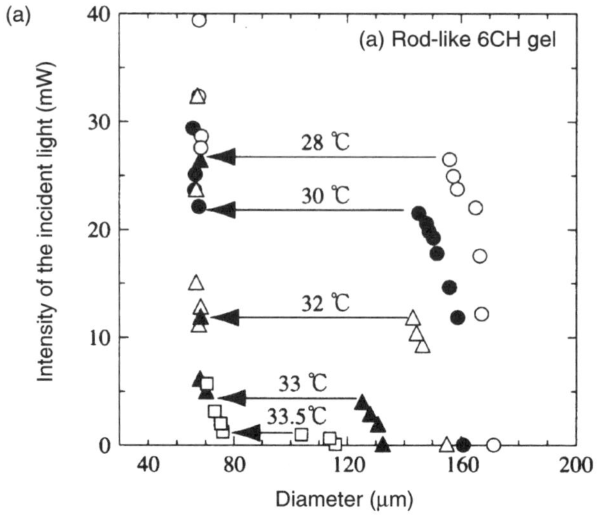

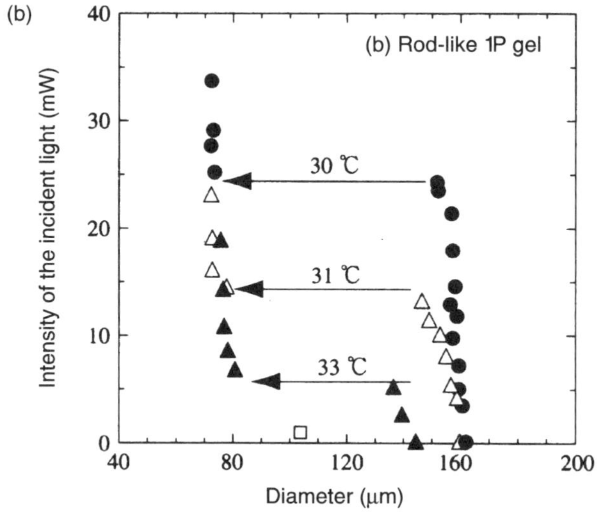

(a) 6CH (chlorophyrin) gel
(b) 1P (porphyrin) gel
6CH, 1P indicates the concentration of the dye, and the concentration of the dye in 6CH is six times that of the 1P.

Fig. 2 Relationship between light intensity and the diameter of the gel at a constant temperature.

light intensity needed to cause phase transition depends on the concentration and specific absorptivity of the dye. This phenomenon was caused by absorption of the irradiated light by the dye, resulting in localized heating, which then caused the gel to shrink. The effect of adding light to a gel can be explained theoretically with an equation. If the local temperature rise is proportional to the light intensity I_0 and also to the concentration of the dye, in other words to the volume fraction of the polymer networks ϕ, the temperature of the gel T will be higher than the surrounding temperature T_0. Thus, the gel temperature can be expressed as follows using a proportionality constant α:

$$T = T_0 + \alpha I_0 \phi \tag{2}$$

From the equation of the state of gel, gel temperature T can be expressed by the function of ϕ, $T_{gel}(\phi)$. Hence, the relationship between T_0 and ϕ is

$$T_0 = T_{gel}(\phi) - \alpha I_0(\phi) \tag{3}$$

The second term of Eq. (3) increases the unstable region in the temperature-volume diagram, induces the discontinuous phase transition, and at the same time, reduces the phase transition temperature. Figure 3 illustrates the theoretical curve as well as the effect of the second term. From this diagram, it can be seen that a discontinuous phase transition is induced by light irradiation. The phenomenon shown in Fig. 2 also can be qualitatively explained. The time-dependent measurement of the volume of this system is also measured.

Because thermal diffusion is sufficiently fast in comparison to the collective diffusion of the gel, the shrinking and swelling behaviors with on/off operation of light are governed only by network movement. Thus, by setting the starting time of the irradiation at zero time, the kinetics of shrinking or swelling can be observed. Unfortunately, the time variation of shrinking is rather complex and the mechanism has not yet been fully understood. The time necessary for a 100-μm diameter gel to shrink was 50 s.

Using the same principle, a submillimeter thick film-like gel was synthesized. The one side of the gel film was chemically affixed to a glass plate. Figure 4 shows the transmittance of light with varied intensity of incident light at a constant temperature for the two kinds of gel prepared with different dye concentrations. When the intensity of light was slowly increased, the transmittance decreased discontinuously at a certain intensity. Before and after this transition, transmittance was almost constant. As

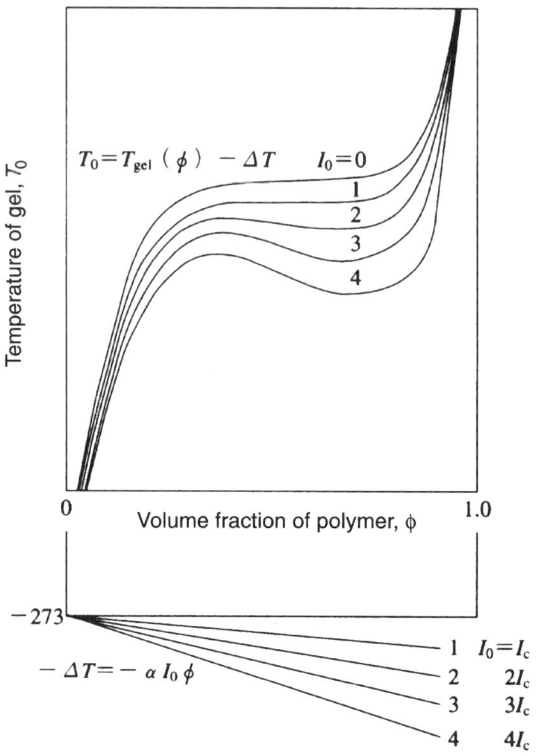

The intensity of light is from the top of the curves $I_0=0$, $I_0=I_c$ (arbitrary value), then increases twice, three times and four times.

The curve at $I_0=0$ is $T=T_{gel}$ (ϕ).

The straight lines express the influence of irradiation and the light intensity is proportional to the slope of the line.

Fig. 3 Conceptual diagram of theoretical swelling curves under light irradiation.

the temperature increased, the critical light intensity decreased. The 10CH gel with higher dye concentration shows lower transmittance in the swollen and shrunken phases. The intensity of light required for the phase transition is lower when the gels are compared at the same temperature. The dye concentration is proportional to the density ϕ of the polymer networks (α_0 is the proportionality constant). When the dye

(a)

(b)

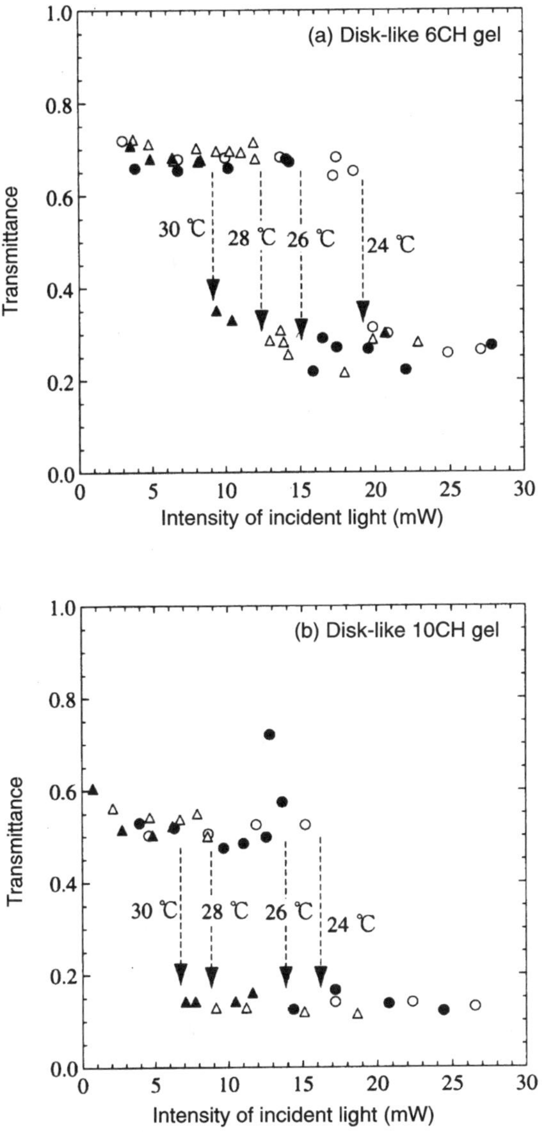

(a) 6CH gel (b) 10CH gel
Here, 6CH and 10CH express the concentration of the dye; 6CH gel indicates that the dye concentration is 0.6 times that of 10CH [7].

Fig. 4 Relationship between the light intensity and transmittance at a constant temperature.

concentration is sufficiently low and the film thickness is very thin, the intensity of transmitted light I is given by the Lambert-Beer equation:

$$\ln\left(\frac{I_0}{I}\right) = \alpha_0 \varepsilon \phi l \tag{4}$$

where ε is the specific molar absorptivity and l is the film thickness (this equation is equivalent to Eq. (1) when $R = 0$). Hence, the energy dissipated in the gel upon irradiation is approximated by the following equation:

$$I_0 - I = I_0\{1 - \exp(-\alpha_0 \varepsilon \phi l)\} \tag{5}$$

Assuming that the majority of this energy dissipated as heat, when $\alpha_0 \varepsilon \phi l \ll 1$, namely when the dye concentration is sufficiently low and the film is very thin, the right-hand side of Eq. (5) becomes $\alpha_0 \varepsilon \phi l \approx \alpha I_0 \phi l$. Therefore, if the heat capacity is assumed to be constant, the local temperature rise of the gel film is expressed by $\alpha I_0 \phi$ (α is a proportionality constant). This is nothing but the second term of Eq. (3). However, this discussion will not explain the results as already described. For example, since one side of the gel is fixed to the glass surface, l is inversely proportional to ϕ. Hence, according to Eq. (4), the transmittance (I/I_0) should be constant regardless of the intensity of the incident light. In order to explain the experimental results, it is important to consider the shrinking not only in terms of thickness but also in terms of width. Accordingly, study of gel thickness and restrictions become important.

In this section, the volumetric phase transition of the dye-modified, thermoresponsive polymer gel by irradiation of light has been reviewed. Using this principle, the amount of transmitted light can be reduced discontinuously by irradiating any gel made of any polymer that will shrink when heat is used. Furthermore, it is also possible to develop a gel that swells by irradiation of light and increases the intensity of the transmitted light discontinuously. In this case, it is necessary for the gel to swell discontinuously when light is not irradiated.

6.1.4 Transmission of Light and Structure of Gel Networks

Thus far, the properties of gels that are used in both raw and functional materials have been reviewed. Many of these phenomena depend on the properties of the gel structure itself. Gel structure is hierarchical from the macroscopic to the molecular level. Of the optical properties, transmission of light depends heavily on the submicrometer level structure, which is comparable to the wavelength of light. Gels exhibit geometrical patterns when volumetric changes, both swelling and shrinking, occur. These patterns are sensitive to the gel's shape, the condition of the gel before and after the phase transition, and mechanical restriction. As a result, many patterns appear [11,12]. Much of the pioneering work has been done in the size range of from millimeters to submillimeters.

Direct observation of polymer gel surfaces has been made recently underwater with atomic force microscopy (AMF) in order to elucidate surface morphology and stimuli responsiveness in the micrometer to several tens of nanometers range [13,14]. The gel used was disk-shaped and the bottom of the gel was chemically affixed to a glass plate. The surface morphology of the gel was determined by the properties of the networks (type and hardness), external environment, and restriction conditions [8].

Figure 5 illustrates the domain seen in transparent NIPAAm gel, which was synthesized at the temperature of ice water. The swollen gel surface did not divide into evenly sized domains, but rather elongated and shrunken ones, which seem to be more stable. The domain structure and its temperature variation are observed by synthesizing a gel at 40°C into different transparencies (heterogeneity). The swollen and shrunken phase domain size of the more heterogeneous NIPAAm gel synthesized at 40°C showed larger domain size than the highly homogeneous gel synthesized at very cold temperature. The higher temperature sample is also more sensitive to temperature changes and morphology changes as temperature increases. Although we must await a study regarding the relationship among these surface structures, bulk structures and various functions, it is expected that structural changes at the submicrometer level significantly influence gel functions. If the domain structure and properties of gels could be freely controlled, many applications would be able to be developed not only in the engineering field but also in medical and biological fields. The simplest example, use of the photo-responsive function of gels, is the control of light transmission via local temperature rise by a domain. By changing the domain size on the scale of the wavelength of light, it is possible to design a gel with intelligent light transmission.

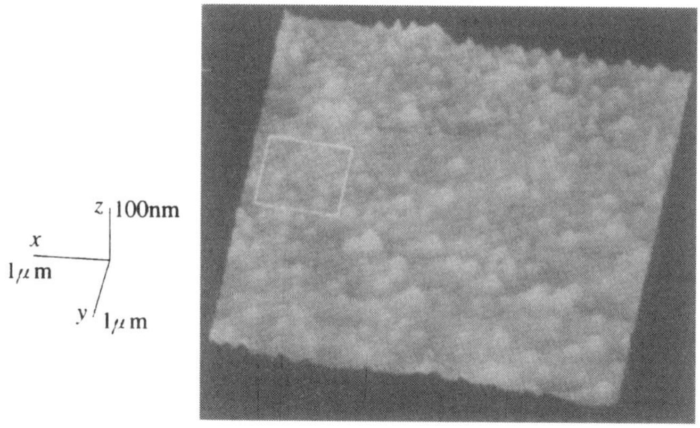

(a) 5×5 μm^2 sized image

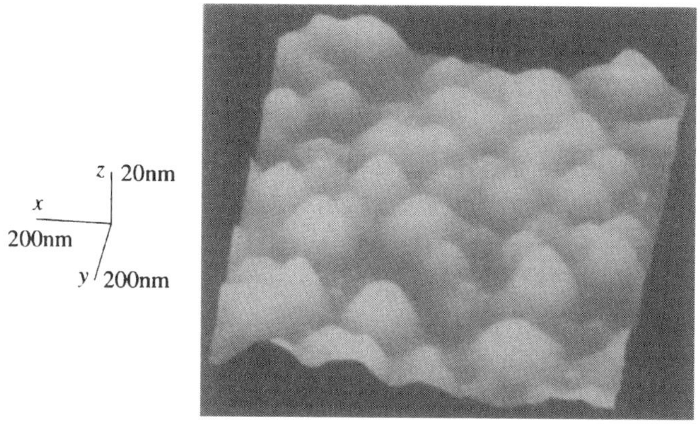

(b) Enlarged image of the white square of (a)

Fig. 5 AFM images recorded at 25°C from NIPAAm gel synthesized at ice temperature.

6.1.5 Conclusions

The use of gels in both raw materials and functional materials has accelerated in recent years. The molecular design and development of gels with stimuli-responsive functions have been established gradually. It is hoped that practical and useful approaches will be developed in the near future.

6.2 REPLACEMENT MATERIALS FOR THE VITREOUS OF HUMAN EYES

AIZO YAMAUCHI

6.2.1 Introduction

Many gels are found in nature—for example, the eggs of frogs. Others include those designed by nature for transparency, including the ocular media consisting of the cornea, lens, and vitreous materials of eyes. Figure 1 illustrates the transparent materials of the eye as indicated by the cross-sectional diagram of an eyeball and Table 1 lists structure and components [15,16].

It is perhaps not necessary to point out that of all the senses, vision demands both quality and quantity. If any disease process leads to loss of sight, life quality is seriously compromised. Thus, many attempts have been made to develop cures and methods to recover sight.

Loss of sight is caused by many things, with the following ones especially important:

(i) any opacity to any sight portions of the eye;
(ii) diseases that affect the ability of the retina and/or the optic nerve to recognize, transmit, or use visual stimuli; and
(iii) malfunctions in translating or relaying information to the brain.

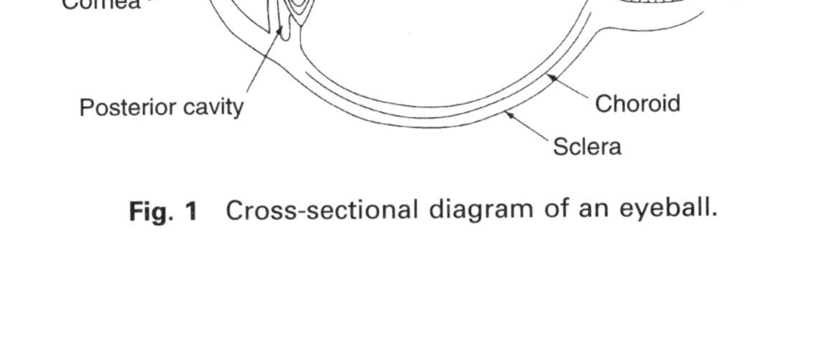

Fig. 1 Cross-sectional diagram of an eyeball.

Table 1 The structure and components of the eye.

The cornea	The lens	The vitreous
The outer layer is covered by conjunctiva. It consists of five layers of outer wall cells, Bowman's membrane, the parenchyma, Descemet's membrane, and inner wall cells. The majority is the real layer, which consists of collagen fiber and mucopolysaccharide. The fiber aligns parallel and the transparency is attained due to its size, which is smaller than the wavelength.	The lens is surrounded by a membrane that contains collagen. The structure of the cells in the forefront surface is fibrous and ordered. It takes α, β, and γ crystalline forms and the closer it approaches to the center of the lens the higher the density and refractive index. When the ordered structure becomes disordered, the system becomes cloudy, resulting in cataracts.	The vitreous is a composite gel structure, which consists of collagen fibers and hyaluronic acid. The water content is as high as around 99%. It maintains the shape of the eyeball, absorbs impact, maintains the balance of the aqueous humor in the eyeball, and regulates eye pressure. However, details are not known.

Of these, a relatively well-established cure exists only for opacity-based problems (i). Cataracts are easily cured by extracting the clouded lens and inserting a plastic intraocular lens. In any society with a large number of aged citizens, this technique is extraordinarily valuable.

Corneal clouding is handled by corneal transplantation from a donor, and as the cornea does not contain blood, compatibility problems can be minimized and thus the success rate is very high. Unfortunately, there are not enough donors. Although the study of hybrid materials (something between a biomaterial, a hyaluronic acid material, and synthetic polymer gels) that could serve as artificial corneas [17] is underway, this is not something available now to restore sight.

The vitreous situates behind the lens and occupies 80% of the volume of the eyeball. It is a gel that is 99% water, with collagen fiber, sodium hyaluronic acid, and a small amount of soluble protein and glycoproteins. The proposed structure is a composite of collagen fibers and coiled hyaluronic acid in a network. This structure operates as a stable hydrogel to maintain the mechanical and optical properties of the vitreous (humor) [18]. A structure in which hyaluronic acid is entangled with collagen fibers was proposed based on electron microscopy as shown in Fig. 2 [19]. This fiber is a one-dimensional (1D) rather than two-dimensional (2D) structure for the cornea and is three-dimensional (3D) for the lens.

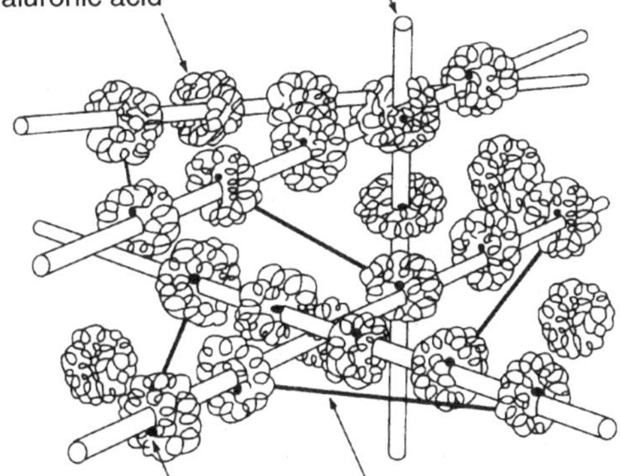

Fig. 2 Hypothetical diagram of the details of the vitreous [19].

In any case, this is a homogeneous system that handles light wavelengths and this is what maintains transparency. However, the composition, structure, and even the properties and functions of the vitreous are not yet well understood. Glaucoma, retinal detachment, opaque membranes, etc. can be treated medically but there are to date no absolute cures or ways to prevent these problems.

New instrumentation and methodologies for analysis of the eyes are being developed. Physical properties, for example, phase transition and thermal fluctuation measurements, have been explored by looking at biogels (cornea, lens, and vitreous) with dynamic laser scattering under a microscope [20–22]. It is therefore expected that our understanding of the mechanisms required to adjust eye functions and correct opacity of ocular media will be advanced (see this monograph, Volume 1, Chapter 3, Section 2-12, Optical Properties, page 390).

6.2.2 Development of Replacement Materials for the Vitreous

As shown in Table 2, attempts to replace vitreous materials have been active. Unfortunately, useful materials have not been found due to problems with biocompatibility and transparency. However, some materials (air, saline solutions or hyaluronic acid-aqueous solutions) are currently used for specific, appropriate illnesses. High molecular weight silicone oil also had been used until recently as a filling material for detached retinas. Due to the problem of silicone microparticles invading surrounding tissues and the possibility that silicone is carcinogenic, it is no longer used today.

The requirements for vitreous replacement materials include: (1) transparency; (2) similar refractive index and density to the vitreous; (3) no germs and nontoxic; (4) no antigen effect; (5) nonflammable; (6) no enhancement of cell growth; (7) stability in the body; (8) slow absorption; (9) no solubility in water; and (10) ability to inject. However, as the vitreous replacement material is just a filler for a short period of time during retinal or lens surgery, it only needs to satisfy conditions (1), (2), (3), (4), (5), and (10).

Table 2 History of replacement materials for the vitreous [23, 24].

Natural materials	1900	Saline solution	Andrews
	1906	The vitreous of domestic rabbit and calf	Deutschmann
	1911	Air	Elschnig
	1911	Humor under retina	Ohm
	1911	Marrow	Birch-Hirchfeld
	1928	The vitreous of a corpse	Hehmer
	1946	The freeze-dried human vitreous	Cutler
	1953	Hyaluronic acid	Paufique and Moreau
	1960		Widder
	1960	Collagen, hyaluronic acid (domestic rabbit)	Balazs
	1974	UV-treated collagen	Pruett
	1981	Purified hyaluronic acid	Schepens
Synthetic materials	1954	PVP[a] (animal)	Scuderi
	1958	Liquid silicone	Stone, Armaly
	1959	PVP[a]	Hayano *et al.*
	1966	Synthetic polypeptides	Oosterhuis
	1967	PAAm[a] gel	Muller-Jensen
	1968	PGMA[a] gel	Daniele
	1973	6-Fluorosulfur gas	Norton
	1973	Silicone rubber balloon	M.F. Refojo *et al.*
	1974	8-Fluorocyclobutane	Brubaker
	1977	PVA[a] gel (domestic rabbit)	Yamauchi *et al.*
	1979	Liquid silicone total replacement	Scott
	1984	PHEA[a] (domestic rabbit)	I.M. Chan, M.F. Refojo *et al.*
	1991	PMAGME[a] (domestic rabbit, animal)	T.V. Chirila *et al.*

[a]See the text for the abbreviation

Chirila *et al.* introduced synthetic polymers as replacement materials for the vitreous in their review article [24]. They also stated that, to date, an ideal material had not been found. However, synthetic hydrogels do not fragment during surgical manipulation and do not induce cytotoxic reactions and therefore they are potentially useful as a long-term substitute material, particularly because they show little degradation in and absorption by the body and they stay confined to a specific space without dissolving or dispersing.

- Poly(vinyl pyrrolidone) (PVP)
  Historically, PVP aqueous solution was clinically used as a blood substitute and it is therefore considered to be of low toxicity. It was used by Scuderi in 1954 for domestic rabbits and in 1959 Hayano and Yoshino used it clinically. However, there has been no clinical development since then.
- Poly(acrylamide) (PAAm)
  In 1968, Muller-Jensen *et al*'s group polymerized the monomer by injecting it into the eyes of domestic rabbits. Since it had a strong toxicity, they purified the monomer and repeated the same experiment in 1969. In 1973, Refojo and Zauberman used a crosslinked, fine PAAm hydrogel, and studied light scattering at the gel and body interface.
- Poly(glycerol methacrylate) (PGMA)
  Starting in 1965, Refojo *et al.* have reported on a series of *in vivo* and *in vitro* experiments using a hydrogel and dehydrated PGMA gels. They reported that the dehydrated gel causes wounds and swelling in the vitreous. In 1976, Hogen-Esch *et al.* used slightly crosslinked PGMA gel that contained 96% saline solution. They injected this into the vitreous of domestic rabbits and obtained good results. However, there have been no detailed reports since then.
- Silicone rubber
  Since 1973, Refojo *et al.* have been evaluating thin silicone balloons with a thin tube in place of silicone oil. They inserted this balloon into the vitreous of monkeys and domestic rabbits and injected saline solution into the balloon to inflate it. After six months, the balloon was transparent. However, temporary opacity of the lens and retinal hemorrhage were observed. This method might be useful only as a temporary measure during transplantation.

- Poly(2-hydroxyethyl acrylate) (PHEA)
  Chan, Refojo and others synthesized PHEA hydrogels, extracted the residual monomer, sterilized it, inserted it into a domestic rabbit's eye, and the extracted eyes were pathologically evaluated. They found that the PHEA gels are excellent for their transparency, high viscosity, non-absorbing properties, and ease of injection, and that they are compatible with the vitreous of animals. However, they pointed out that PHEA should not be used as a vitreous substitute material because it causes various complications.
- Poly(methyl-2-acrylamido-2-methoxyacetate) (PMAGME)
  The authors of this review, Chirila *et al.*, synthesized approximately 100 different polymers and copolymers of PMAGME with vitreous water contents and the appropriate gels were selected based on their properties. A gel treated with phosphate-buffered saline solution (PBS) that was first deaerated with perfluoropropane followed by adjustment by a phosphate salt in order to imitate the buffer function of vitreous humor materials was injected into the vitreous of a domestic rabbit. After the operation, inflammation was observed. After extraction of the eye, systematic analysis indicated retinal detachment and damage to the vitreous as well as shrinking of eye nerves. As a result, PMAGME hydrogels are known to be toxic to the nervous system. They are further evaluating interpenetrating networks (IPN) between less toxic polymers and hyaluronic acid or collagen.
- Poly(vinyl alcohol) (PVA)
  The PVA hydrogels developed by the author of this section are also introduced here and details of this work are given.

6.2.3 PVA Hydrogels as Replacement Materials for the Vitreous

As described before, many materials have been examined as substitutes for the vitreous. However, no ideal materials have been found. There is no material specifically designed as replacement material for the vitreous. Those used are selected either from materials already in use in other systems or from the literature.

As a result of our evaluation of vitreous replacement materials capable of satisfying the aforementioned conditions that need to be met, we chose the transparent, 98% PVA hydrogel [23,25–28], that had been obtained by irradiating a PVA aqueous solution with γ-rays. This hydrogel was selected for the following reasons.

(1) Because a stable polymer is used as the raw material, monomers with strong toxicity do not remain.
(2) The structure of the repeat unit is similar to ethyl alcohol and, thus, it is expected to be less toxic than other materials.
(3) The biocompatibility of PVA has been evaluated favorably in other medical applications.
(4) It is water soluble and stable in aqueous solutions.
(5) Crosslinking by γ-ray does not require an additive and thus the gel can be prepared using only PVA and water.
(6) Based on the mechanism of γ-ray crosslinking, the crosslink point is statistically distributed, leading to a transparent gel with other good properties.
(7) Because the gel contains more than 98% water (like the vitreous of domestic rabbits) it showed a similar refractive index.
(8) Autoclave sterilization can be used.
(9) A gel with 98% water content can be injected with a thin syringe needle and no fragmentation results.
(10) Finally, PVA is industrially readily available in large quantities and it is homogeneous in quality.

In order to determine the type of material, the properties of the desired gel, and the manufacturing conditions needed, hydrogel preparation conditions, such as the degree of saponification and polymerization, concentration of the aqueous solution, and irradiation doses of the γ-ray, were varied. Then the relationship between the structure and properties was evaluated by the swelling ratio, viscosity or elasticity, infrared spectroscopy, transmission electron microscopy, permeability of materials into the gel, and refractive index. A comparison of the properties of this created gel with those of a domestic rabbit's vitreous materials demonstrated that our synthesis was successful. The obtained gel was as transparent as albumen and fluidity permitted injection.

Specifically, a 7% aqueous solution of completely saponified PVA with molecular weight of 2000 was irradiated by γ-ray. After the unreacted PVA was eliminated by hot water, it was brought into equilibrium at 37°C. A 7% aqueous solution is used because 1–2% solutions will have large intermolecular distances, which become the cause of intramolecular crosslinking or localization of PVA molecules near the crosslink point. These then become microgels or heterogeneous gels. The equilibrated homogeneous gel, which is obtained by irradiating the 7% aqueous solution with a predetermined dose of γ-ray creates a material with water content similar to that of the vitreous and is injectable by needle.

By comparing the relationship between the swelling ratio and refractive index of this gel and the data on the vitreous of domestic rabbits, it was found that the appropriate dosage is approximately 0.6 Mrad and the water content is 80–85%. However, as further evaluation revealed that the refractive index increased due to permeation of proteins and polysaccharides, it was deemed better to use gels that have a swelling ratio of between 90 and 100.

The replacement PVA gel for the vitreous is sterilized and injected into the vitreous of a domestic rabbit from which about 0.5 ml of the vitreous material is extracted using an 18–24 gauge syringe needle. The uniqueness of this material is that it is adjusted so as to be injected by needle. This allows for easy replacement without resort to surgery on the eyeball. The domestic rabbit whose eye underwent replacement surgery was observed by ophthalmoscope, ocular tension, photograph of eyeground, and electroretinagram. Furthermore, it was confirmed that optically normal refraction was maintained and the information was relayed to the brain by examining the hemorrhage and opacity, photo-stimulus responsiveness, and clarity of the retinal blood veins. The ocular tension revealed that the artificial vitreous remained in the vitreous and shape was maintained.

Figure 3 shows the photograph taken when the artificial vitreous was injected and Fig. 4 illustrates the photograph six days after replacement. The retinal veins, which could not be seen upon injection, are clearly seen on the photograph taken six days later. It can be seen that the transparency and refractive index of the artificial vitreous and the domestic rabbit are the same, the gel shows excellent biocompatibility, and the gel is performing its optical function as an artificial vitreous.

Figure 5 shows the infrared spectra of the PVA gel before and after the injection along with the vitreous near the replacement gel. The spectrum of the gel six days after injection is similar to the spectrum of the vitreous. This implies that the vitreous humor penetrated into the gel. This gel might be optically the same as the vitreous and would

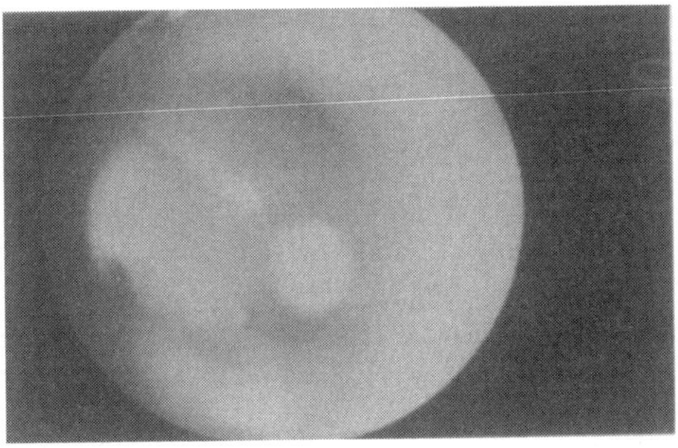

Fig. 3 A photograph taken during the injection of agel into the vitreous of a domestic rabbit.

therefore demonstrate excellent biocompatibility. When an *in vitro* experiment was performed with bovine serum using refractive index and UV absorption, it was found that the serum components penetrated the gel and the composition of the gel became the same as its surroundings in only 2–6 h.

From these results, the PVA hydrogel artificial vitreous implanted into a domestic rabbit is reconstructed like the one shown in Fig. 2 except that the collagen fiber is replaced with the PVA gel system. As a result, the requirement that an artificial vitreous possess the same transparency and refractive index, biocompatibility with the surroundings, and spatial filling

Fig. 4 An image of blood veins of a domestic rabbit six days after the replacement

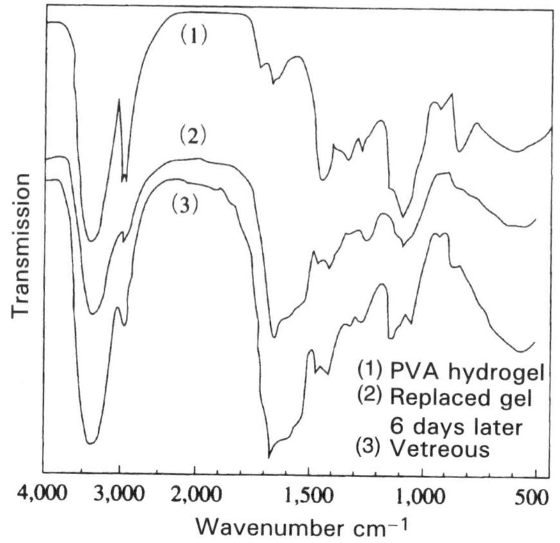

Fig. 5 The infrared spectra of the PVA used and the surrounding vitreous of a domestic rabbit.

in the vitreous as would be found in a naturally occurring vitreous has been satisfied.

6.2.4 Conclusions

The authors are continuing to improve PVA gels and to evaluate the retina and surrounding vitreous after injection in the hope that these gels will be used clinically.

6.3 COLORATION

EIJI NAKANISHI

6.3.1 Introduction

Colored light is visually detectable in wavelengths ranging from 380–780 nm. Any wavelength shorter than this is ultraviolet radiation and longer wavelengths are considered infrared radiation. Color recognition occurs when light not absorbed by a material or a substance is visually perceived either directly by transmission or indirectly by reflection. Therefore, if the color blue is absorbed, the material will look red and vice versa.

The history of dyes is old. There are today many synthetic dyes numbering perhaps as many as 7000 kinds. In general, these compounds possess a p-electron system that absorbs visible radiation. It is now possible to analyze the relationship between a dye color and its structure with the PPP-MO (molecular orbit) method [29]. It is possible today to synthesize molecularly designed dyes. Dyes possessing the ability to change in response to external stimuli are attracting attention. These are not simple dyes, but rather functional ones on which active research is being done. Readers are referred to the many monographs on functional dyes [30–36]. Research and development are ongoing in the areas of acid coloration of dyes, sublimation transfer, photoconductivity, sensitivity to semiconductor lasers, information storage dyes that use electrical charges, energy exchange dyes that use photoexchangeability or nonlinearity, and medical diagnosis dyes that are biosensitive. Of these functions, chromism will be introduced in detail in this section.

6.3.2 Chromism

Phenomenologically chromism is a color change in response to some external stimulus. The name will vary depending on the type of external stimulus. If color change results from polarity of a solvent, it is called solvatochromism. Photochromism is caused by light and electrochromism is caused by electric fields. Other stimuli such as heat, pressure, or magnetism provide the names thermochromism, piezochromism, and magnetochromism, respectively.

6.3.2.1 Solvatochromism dyes

The absorption spectrum of a dye is determined by the energy difference between the ground state and excited state, that is, the transition energy between the frontier orbital. If the polarity of the solvent is different, the corresponding dipolar moment of the orbital is varied. Thus, the absorption spectrum maximum shifts and the color changes. In general, nonionic dyes (which have polar structures in the excited state) shift the absorption maximum to a longer wavelength in order to stabilize the excited state in a polar solvent, which then causes a reduction in the energy difference. On the other hand, ionic dyes reduce ionic properties in polar solvents by hydrogen bonding, which then leads to a shorter wavelength shift. An example of this is pyridium phenolphthalein (**1**). In dye-solvent interactions, the charge transfer-based transition used will differ greatly depending on the type of solvent. There are even compounds that can evaluate the polarity of solvents.

Reversible color changes based on the acid-base equilibrium of a solution are called halochromism. In this category there are many compounds capable of changing the electronic structure of the conjugated system by proton dissociation, with the result a change in color. There are also compounds that show and eliminate color by the opening and closing of lacton ring such as in or phenolphthalein (**2**).

6.3.2.2 Photochromism dyes

Many compounds exhibit photochromism based on cis-trans isomerism or photoisomerism (such as tautomerism). When these are considered as dye materials, a few show complete reversibility and many accompany photodegradation. These are used for recording media and displays.

Azobenzene (**3**) has been known for a long time. This compound shows cis-trans isomerism upon irradiation of light. When UV light is irradiated, the planar, stable trans structure isomerizes into nonplanar, unstable cis structures. This reaction is reversible. If the visible light is irradiated, the cis structure returns to the trans structure. Another well-studied example is spiropyran (**4**), which exhibits ionic dissociation of a bond. When irradiated by UV light, ring opening occurs, leading to longer conjugation. This also accompanies an ionic structure, with coloration the result. When this compound was commercialized problems with aging and thermal stability were found. To help with these problems a fulgid derivative (**5**) was developed. This compound maintains thermal stability by the introduction of methyl groups. When the ring is open, the compound is colorless, whereas upon UV irradiation the ring structure opens and shows a red color.

Fulgid compounds have short conjugation, which is unfavorable for exhibiting deep colors. Therefore, in order to display various colors diarylethane (**6**) was developed.

6.3.2.3 Electrochromism Dyes

These dyes are used for electrochromic display. Many compounds show color changes by oxidation-reduction cycle. The biologen type (**7**) yields radical cations by reduction and displays brilliant color. In comparison to tungsten oxide, which shows a blue color by reduction, the response is faster. By changing the substituents, many colors can be obtained. Pyrazoline-type compounds (**8**), which possess a nitrogen heterocycle, can also display black.

1

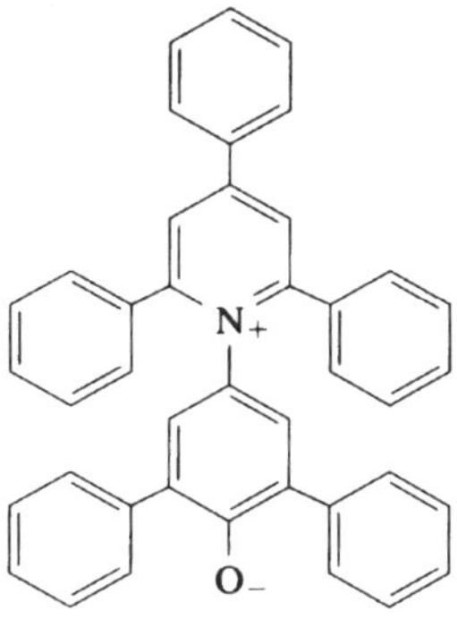

2

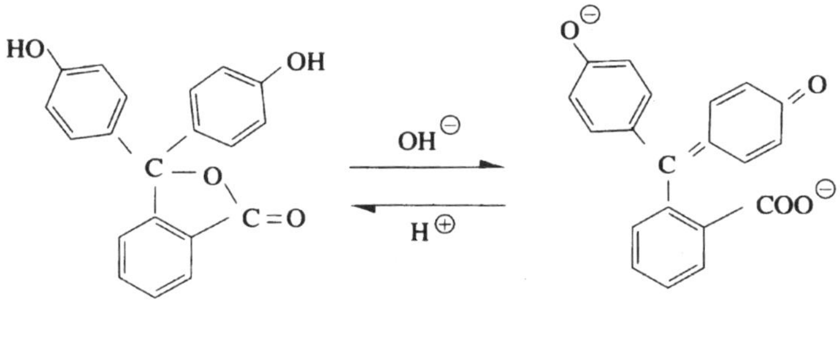

3

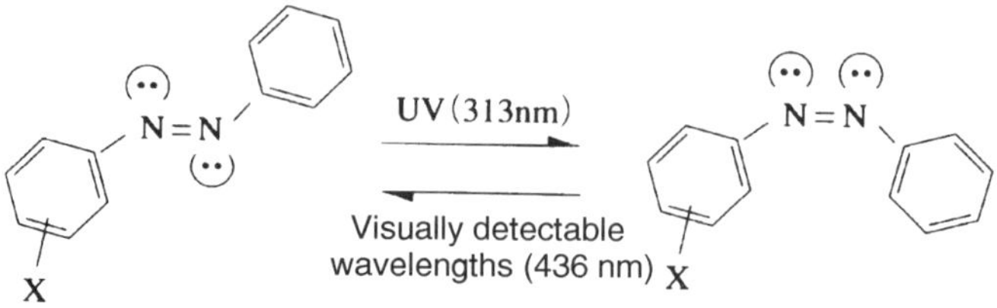

4

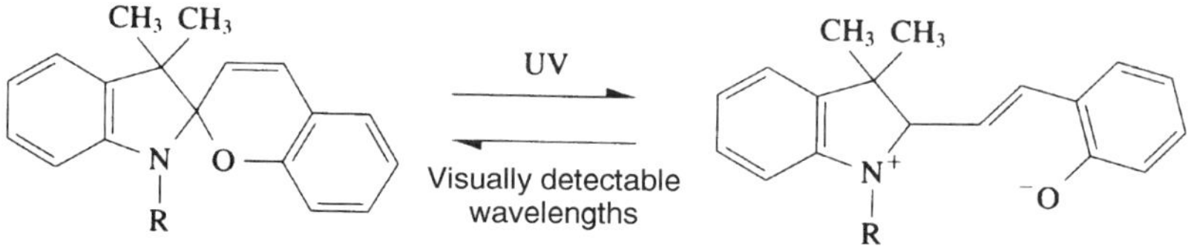

5

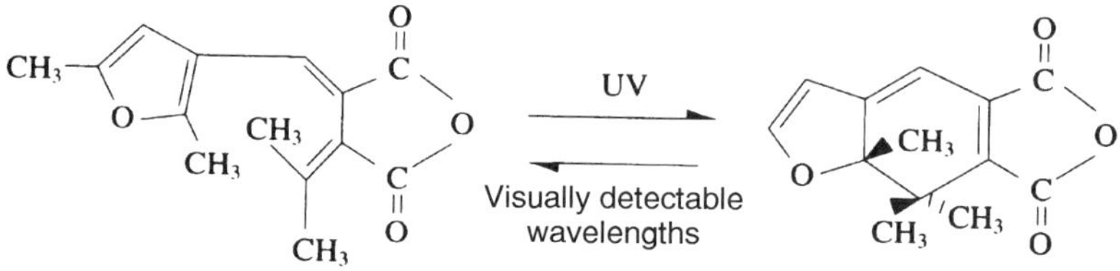

6

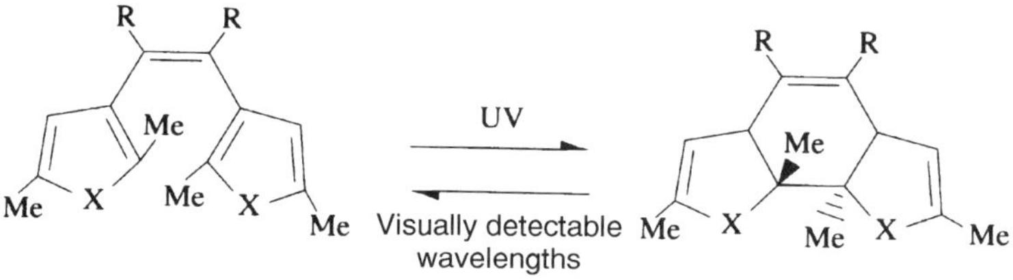

$X = O, S, Se, NMe$

$R = CN$, $R,R = (CO)_2O$

7

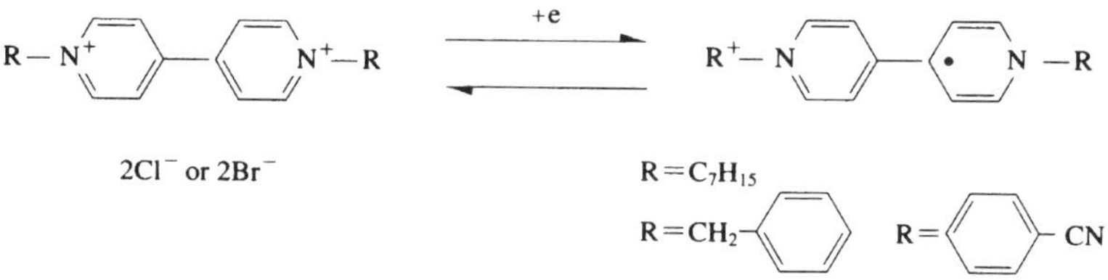

8

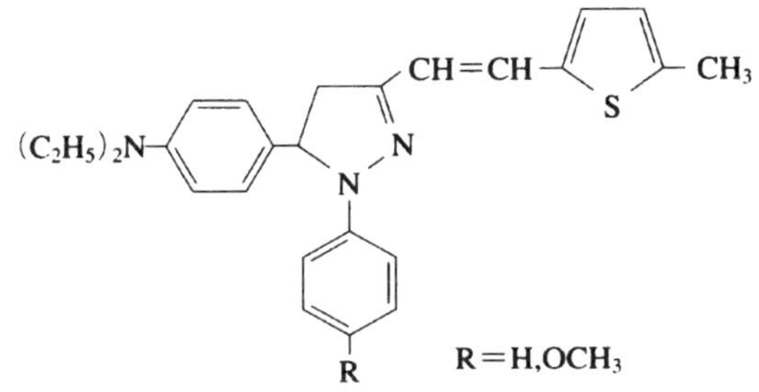
CH=CH
CH3
S
(C2H5)2N
N
N
R
R=H,OCH3

9

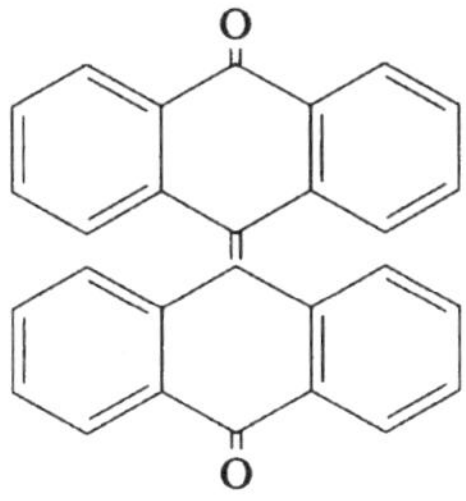
O
O

10

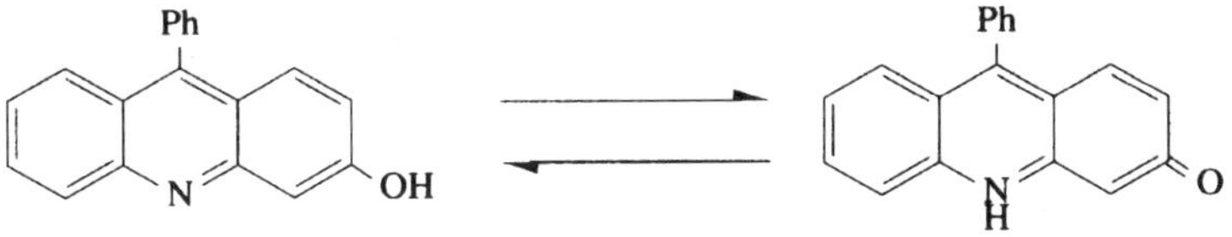
Ph
N
OH
Ph
N
H
O

11

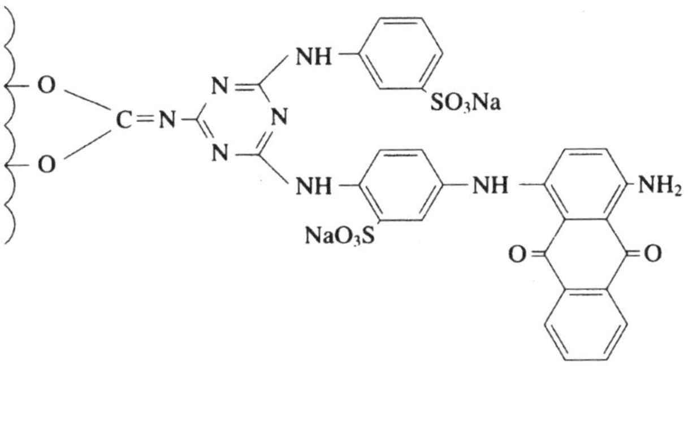
O
O
C=N
N
N
N
NH
SO3Na
NH
NaO3S
NH
NH2
O
O

12 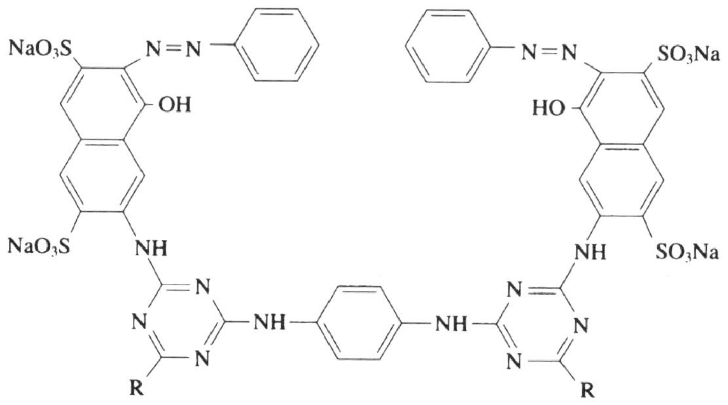

6.3.2.4 Other chromism dyes

Of the thermochromism dyes that indicate temperature, some of them (including biantron (**9**)) change color by twisting the double bonds and (**10**) which shows piezochromism by lactam-lactim tautomerism.

6.3.3 Application to Gels

Gels used as separation materials and polypeptides and which show halochromism will be described here.

6.3.3.1 Separation gels

Polysaccharides (such as agarose and dextrin) are used in chromatography. Introduction of dyes permits certain materials to be selectively separated.

If the Chibcron blue F3GA (**11**) [37] or Prussian red HE3B (**12**) are bonded to a gel for support, the proteins or enzymes that interact with the nicotinamide adenine nucleotide can be selectively adsorbed and separated because the structure of these dyes is similar to that of the nicotinamide adenine nucleotide [38]. There are also examples of selective separation using a dye that exhibits photochromism by polarity changes of the molecule.

If spiropyran is introduced into a gel, the polarity increases upon irradiation by UV and cytochrome C can be selectively adsorbed [39]. Also, by terminating the irradiation, the adsorbed molecules can be desorbed and this can be used to separate molecules. Since azobenzene increases its hydrophilicity upon UV irradiation, separation of hydrophobic protein has been attempted using this system [40].

6.3.3.2 Peptide gels

Because hydrogels are soft and do not harm the body, and they easily diffuse substances, many materials seen as biocompatible have been proposed. Polypeptides assume conformations based on types of residues and the environment in which they are used, which means their characteristics vary accordingly. The authors have been studying polypeptide gels made of poly(N-hydroxyalkyl-L-glutamine) with different lengths of the alkyl group [41–45]. Figure 1 shows the gel's preparation method. Peptide gels have been found to be superior to other gels in terms of biocompatibility, biodegradability, material diffusion, and mechanical properties. It also was found that the properties of the gel can be freely adjusted by controlling the conformation.

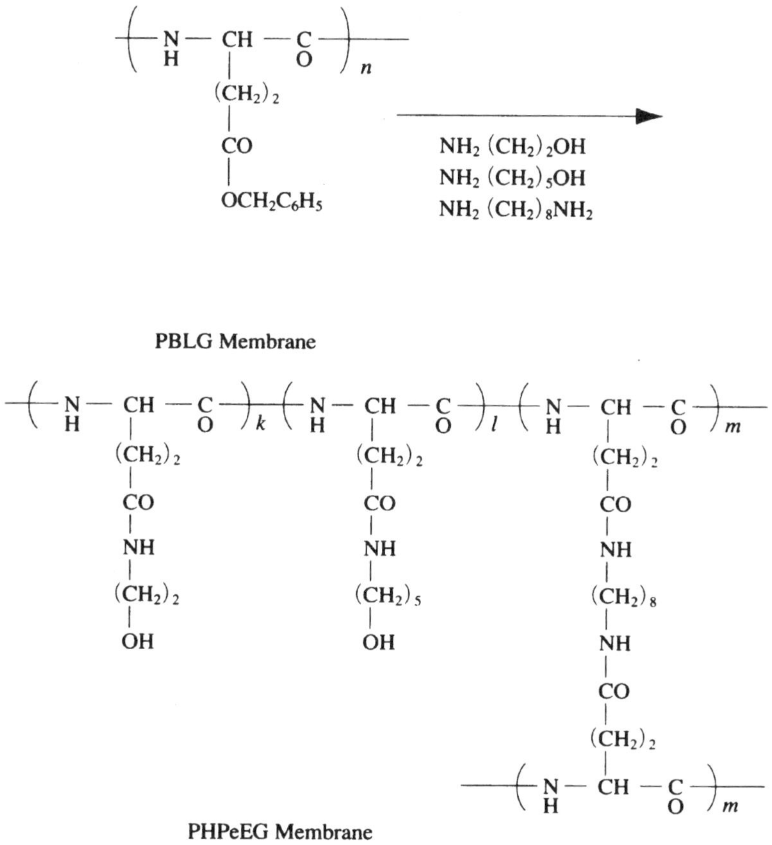

Fig. 1 The preparation method of PHPeEG gel.

Of all the natural amino acids, tryptophan (Trp) has the highest hydrophobicity and is the only compound that contains an indole ring. The indole ring of Trp has excess π-electrons and high electron density. Hence, it can be readily oxidized under acidic conditions to form various structures [46]. Further, Trp colors by reacting with aldehyde under acidic conditions. This is because the α-position of the indole ring undergoes various condensation reactions under acidic conditions, which often results in brilliant colors. Therefore, peptides that contain Trp are qualitatively and quantitatively determined using this coloration. Using this coloration of Trp, if a Trp derivative is dissolved in a strong acid like trifluoroacetic acid (TFA) or nitric acid and UV is irradiated onto the solution, the solution colors. We found that the solution shows reversible color change as a function of pH [47]. Additionally, the color depends on the type of acid and the pH region. A TFA-treated Trp derivative shows red $<$ pH 4.0 and is yellow $>$ pH 5.5. In contrast, the nitric acid-treated Trp is yellow $<$ 10.5 and red $>$ pH 12.5. Hence, Trp treated with strong acids was then purified and the coloration mechanism was studied spectroscopically [48,49]. As shown in Fig. 2, the Trp derivatives assume a three-ring structure upon treatment by strong acids. The Trp derivative shows the yellow color because the conjugation length

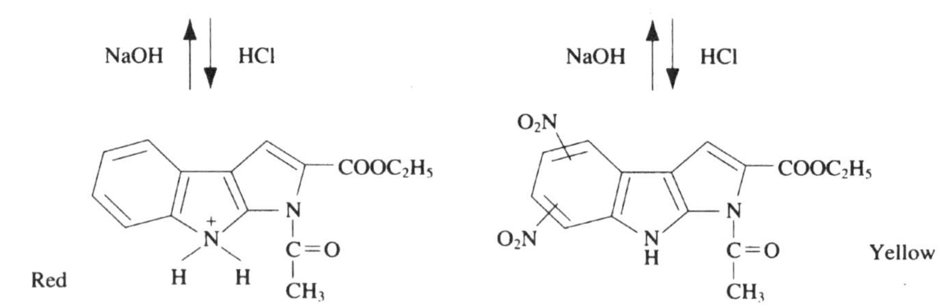

Fig. 2 Isomerism of tryptophan (Trp) derivatives by treatment with a strong acid.

increases by TFA treatment and the nitro groups are introduced into the indole ring. With TFA treatment, the compound possesses a positive charge at low pH and the electronic structure of the aromatic ring changes. With the nitric acid treatment, the conjugation length increases due to the negative charge at high pH. Color change is the result. The Trp treated with strong acids possesses charges and coloration takes place by dissociation.

Based on the knowledge obtained from these monomers, a polypeptide gel (EGT) is prepared from N-hydroxyethyl-L-glutamine (L-Trp) and pH responsiveness following strong acid treatment is evaluated. The TFA-treated hydrogel (EGT-T) showed reversible coloration from red and yellow in the pH range of 4.0–5.5 and the nitric acid-treated one showed reversible yellow and red colors in the pH range of 10.5–12.5. As shown in Fig. 3, the water content of the colored gels changed significantly in these pH ranges. Water content of the gels increased in the low pH range when treated by TFA and in the high pH when treated with nitric acid. Accordingly, we have succeeded in preparing materials that exhibit changes in both color and property simultaneously.

To make use of this function, microspheres were prepared. Certain peptides show specific interaction with certain molecules. Here, using a strong acid-treated EGT, a peptide microsphere with a particle diameter of from 0.1 to 150 μm is prepared. Interactions between EGT-T and warfarin which is an ionic drug, and between EGT-N and metallic ions have been studied.

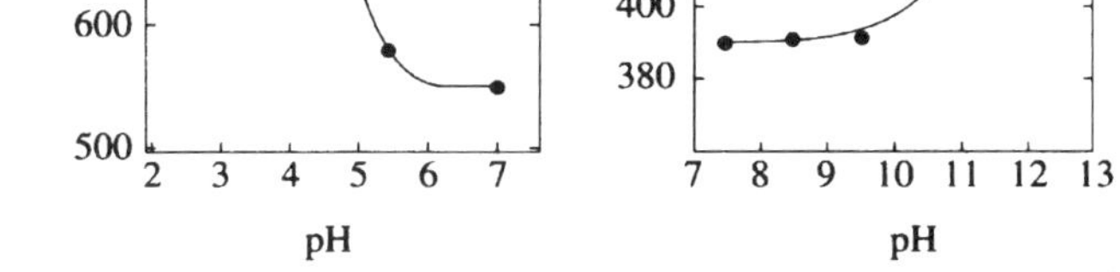

Fig. 3 Water up-take of EGT-T and EGT-N.

The Trp residue exists only in human serum albumin. It shows high compatibility with heterocyclic molecules, which show negative charge and quench its fluorescence [50,51].

The EGT-T shows absorption at around 500 nm and exhibits red color at pH 4.0, which is where the Trp residue dissociates. By adding warfarin to EGT-T, the intensity of this absorption decreased, resulting in a yellow color. The fluorescence of EGT-T diminishes by the addition of warfarin. Hence TFA-treated Trp that has undergone structural change is likely to have interacted with warfarin. Figure 4 shows the dissolution behavior of warfarin using TFA-treated microspheres. In comparison to the untreated sample, the TFA-treated sample increased the holding time of warfarin and, thus, there is the potential for drug delivery system (DDS) applications.

Aromatic nitro compounds have been used as qualitative and quantitative indices for metallic ions. It has been known that 1-nitro-2-naphthol forms a stable complex with Co ions by the cooperative effect of nitro and hydroxyl groups. We have evaluated the interaction between EGT-N and transition metal ions at pH 10.5 where the nitric acid-treated Trp starts dissociating using the absorption intensity of the UV spectrum. There are no spectral changes when metallic ions are added to the unrelated EGT. On the contrary, when the Co ions are added to EGT-N, the absorption intensity decreased and a green precipitate resulted.

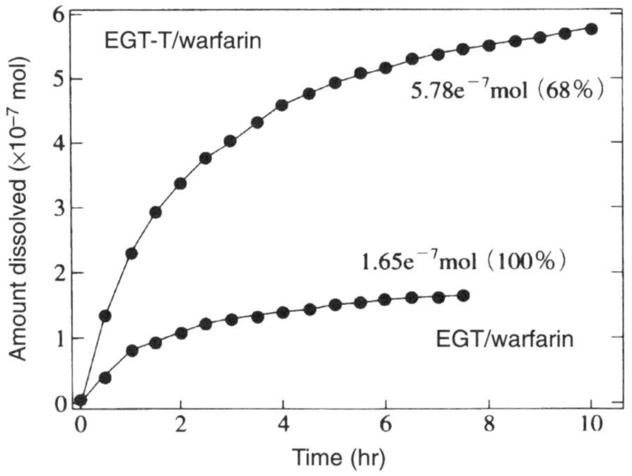

Fig. 4 Dissolution behaviour of warfarin from the peptide microspheres.

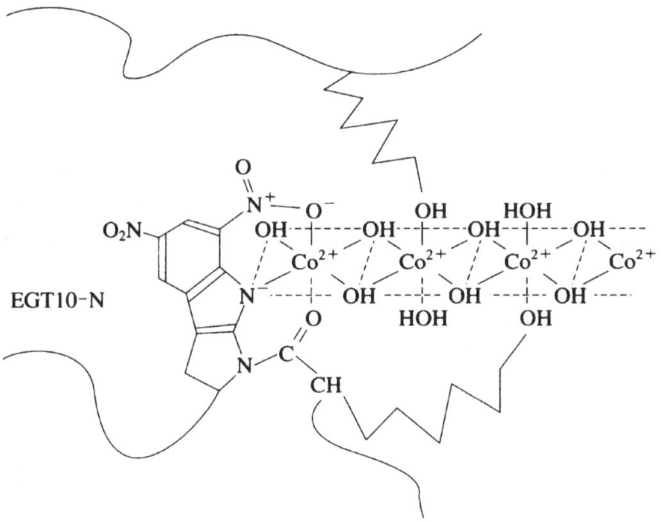

Fig. 5 A Schematic diagram of the EGT-N/Co^{2+} complex.

Judging from the intensity changes that occur only when the Co ions are added, the nitric acid-treated Trp residue selectively interacts with the Co ions. Upon analysis of this precipitate, it has been confirmed that the negatively charged Trp residue and the Co ions selectively interact as shown in Figure 5 to form a complex, thereby resulting in precipitation [52]. Accordingly, we have succeeded in preparing a gel that can monitor a Trp molecule by the color changes it undergoes.

REFERENCES

1 Osada, Y. (1987). *Adv. Polym. Sci.* **82**: 1.
2 DeRossi, D., Kajiwara, K., Osada, Y., and Yamauchi, A. (eds.) (1991). *Polymer Gels*, New York: Plenum Press.
3 Miyazaka, K. (1992). *Plastics Dictionary*, Asakura Publ.
4 Nakajima, A., Itoi, M., and Kanei, J. (1992). *Contact Lens Preparation Manual*, Nanko-do.
5 Koike, Y. (1994). *Optical Properties of Polymers*, Kobunshi Gakkai edition, Kyoritsu Publ.
6 Tanaka, T. (1978). *Phys. Rev. Lett.* **40**: 820.
7 Suzuki, A. and Tanaka, T. (1990). *Nature* **346**: 345.
8 Suzuki, A. (1993). *Adv. Polym. Sci.* **110**: 199.
9 Suzuki, A. (1994). *J. Intel. Mat. Sys. Str.* **5**: 112.
10 Suzuki, A., Ishii, T., and Maruyama, Y. (1996). *J. Appl. Phys.* **80**: 131.

11 Tanaka, T., Sun, S.T., Hirokawa, Y., Katayama, S., Kucera, J., Hirose, Y., and Amiya, T. (1987). *Nature* **325**: 796.
12 Matsuo, E.S. and Tanaka, T. (1992). *Nature* **358**: 482.
13 Suzuki, A., Yamazaki, M., and Kobiki, Y. (1996). *J. Chem. Phys.* **104**: 1751.
14 Suzuki, A., Yamazaki, M., Kobiki, Y., and Suzuki, H. (1997). *Macromolecules* **30**: 2350.
15 Kamiya, S. (1975). *Ophthalmology Tomorrow*, vol. 1, Department of Ophthalmology, Nara Medical School, pp.103–206.
16 (1986). *Biochemistry*, 7th edition, A. Ichikawa, Transl., Hirokawa Publ.
17 Nakao, H. *et al.* (1993). *Folia Ophthalmol. Japonica* **44**: 274; 1107; (1994). **45**: 614.
18 Balaz, E.A. (1961). *Molecular Morphology of the Vitreous Structure of the Eye*, G.K. Smelser, ed., New York: Academic Press, pp. 293–310.
19 Asakura, A. (1985). *J. Ophthalmol. Soc., Jpn.* **89**: 179.
20 Matsuura, T. (1994). *Nara Igaku Zasshi* **45**: 433.
21 Matsuura, T. (1994). *Japanese J. Visual Sci.* **15**: 2.
22 Matsuura, T. *et al.* (1993). *Polym. Preprints, Jpn.* **42**: 3039.
23 Yamauchi, A. (1987). *Artificial Vitreous: Preparation of Functional Polymer Gels and Their Applications*, M. Irie, ed., CMC, pp. 169–179.
24 Chirila, T.V. *et al.* (1994). *J. Biomater. Appl.* **9**: 121.
25 Yamauchi, A. *et al.* (1977). *Kobunshi Ronbunshu* **34**: 261.
26 Yamauchi, A. *et al.* (1979). *Folia Ophthalmol. Japonica* **30**: 385.
27 Hara, T. (1978). *J. Ophthalmol. Soc., Jpn.* **83**: 1478.
28 Hara, T. and Yamauchi, A. (1984). *Folia Ophthalmol. Japonica* **35**: 1340.
29 Fabin, J. and Hartmann, H. (1980). *Light Absorption of Organic Colorants*, New York: Springer-Verlag.
30 Ookawara, S., Matsuoka, K., Hirashima, T., and Kitao, S. (1992). *Functional Dyes*, Kodan Publ.
31 Nishi, H. and Kitahara, K. (1992). *The Chemistry of Dies: Sequel*, Kyoritsu Publ.
32 Gregory, P. (1991). *High-Technology Applications of Organic Colorants*, New York: Plenum.
33 Tokida, S., Matsuoka, T., Kogo, Y. and Kihara, H. (1988). *Molecular Design of Functional Dies: PPP Molecular Orbital Method and Its Application*, Maruzen Publ.
34 Ikemori, C. and Sumiya, M. (1986). *Speciality Functional Dies*, CMC.
35 Ookawara, S., Kuroki, T., and Kitao, T. (1981). *Chemistry of Functional Dies*, CMC.
36 Matsuoka, K. (1994). *Chemistry of Dies and Its Application*, Chem. Soc. Jpn.
37 Dean, P.D.G. and Watson, D.H. (1979). *J. Chromatogr.* **165**: 301.
38 Turner, A.T. (1981). *Trends Biochem. Soc.* 171.
39 Karube, I., Ishimori, Y., and Suzuki, S. (1978). *Anal. Biochem.* **86**: 100.
40 Ishihara, K., Kato, S., and Shinohara, I. (1982). *J. Appl. Polym. Sci.* **27**: 4273.
41 Nakanishi, E., Sugiyama, E., Shimizu, Y., Hibi, S., Maeda, M., and Hayashi, T. (1991). *Polym. J.* **23**: 983.
42 Nakanishi, E., Hamada, K., Sugiyama, E., Hibi, S., and Hayashi, T. (1991). *Polym. J.* **23**: 1053.
43 Nakanishi, E., Shimizu, Y., Ogura, K., Hibi, S., and Hayashi, T. (1991). *Polym. J.* **23**: 1061.
44 Hayashi, T., Nakanishi E., Iizuka, Y., Oya, M., and Iwatsuki, M. (1994). *Eur. Polym. J.* **30**: 1065.
45 Hayashi, T., Nakanishi, E., Iizuka, Y., Oya, M., and Iwatsuki, M. (1995). *Eur. Polym. J.* **31**: 453.

46 Yamanaka, H., Hino, T., Nakagawa, A., and Sakamoto A. (1993). *Chemistry of Heterocyclic Compounds*, Kodan Publ., p. 81.
47 Shigimoto, H., Nakanishi E., Kondo, N., and Hibi, S. *Kobunshi Ronbunshu* (in press).
48 Sugimoto, H., Nakanishi, E., Okamoto, S., and Hibi, S. (1996). *Proc. Soc. Polym. Sci., Jpn.* **45**: 358.
49 Suzaki, K., Sugimoto, H., Nankanishi, E., Okamoto, S., and Hibi, S. (1996). *Proc. Soc. Polym. Sci., Jpn.* **45**: 358.
50 Kurono, Y., Ozeki, Y., Yamada, H., Takeuchi, T., and Ikeda, K. (1987). *Chem. Pharm. Bull.* **35**: 734.
51 Loun, B. and Hage, D.S. (1996). **68**(7): 1218.
52 Suzaki, K., Sugimoto, H., Mizuno, S., Nakanishi, E., Okamoto, S., and Hibi, S. (1996). *Proc. Soc. Polym. Sci.* **45**: 1806.

Section 7
Energy Conversion

RYOICHI KISHI

7.1 CHEMOMECHANICAL POLYMER GELS

7.1.1 Chemomechanical Materials

Mankind has developed many energy exchange technologies. Thermal-mechanical (internal combustion, friction), electrical-mechanical (motors, electric generators) and electrical light (glow discharge, solar battery) exchange systems are just a few. Of these artificial energy exchange technologies, there is one that has not been commercialized. This system creates direct exchange of chemical energy to mechanical energy and vice versa. Called a chemomechanical system, this technology would convert chemical energy directly to mechanical energy without converting it first to other energy forms [1]. Mechanical energy conversion in organisms is achieved mostly with the chemomechanical system. For example, the organic or biochemical compound commonly referred to as adenosine triphosphate (ATP) is involved in release of energy to the muscle cells, which then function by mechanical energy. Unlike thermal systems or motors, this kind of system has very high conversion efficiency due to direct energy conversion. This conversion efficiency is much higher than artificial energy conversion systems. Compared to a conversion efficiency of 5–30% in thermal systems, the conversion efficiency of living muscle is as high as 60%. Because an engine first burns fuel and generates heat at high-energy loss and then converts it into mechanical energy, efficiency is naturally low. However, in the case of a chemomechanical system, chemical energy is rather more directly converted into mechanical energy (movement), which leads to high conversion efficiency.

Many molecules change their shapes in response to changes in their environment. For example, if a small amount of sodium hydroxide aqueous solution is added to a poly(acrylic acid) (PAA) aqueous solution, viscosity increases by several tens of times. This is due to the spreading of the polymer chains as the carboxyl groups of PAA dissociate into carboxylate and repel each other by static interaction (see Fig. 1(a)). Upon addition of hydrochloric acid to this solution, the carboxylic groups return to their nondissociated state and the polymer chains coil again (see Fig. 1(c)). Katchalsky and others first constructed the chemical-mechanical energy conversion system using the deformation of polymers in solution [2, 3]. He crosslinked PAA and formed a gel membrane to obtain mechanical energy from the deformation of the polymer. The PAA membrane went through 200 expansions and contractions in which a weight that was several hundred times heavier than the membrane was moved up an down following the addition of a sodium hydroxide solution and a hydrochloric acid solution (see Fig. 1). Contractile stress and amount of work were comparable to that experienced by muscles. Katchalsky's group attempted to manufacture an engine and a turbine using a chemomechanical system as the driving force [4, 5]. Such material, which directly converts chemical energy to mechanical energy to perform mechanical work (movement), is called chemomechanical material [6, 7]. Other examples of artificial materials that are capable of performing mechanical work using chemical energy without the need for other driving forces are not known. A chemomechanical system is defined as "a thermodynamic system which converts directly the chemical energy to mechanical energy or mechanical energy to chemical potential energy". The energy for such a system can be obtained from inexpensive seawater or even urine. It can also be obtained by using the minute difference in temperature between two liquids. It is thus interesting from both resource and energy perspectives. Further, a chemomechanical system is silent and there is no exhaust gas or waste. Hence it can be said that this is an environmentally desirable

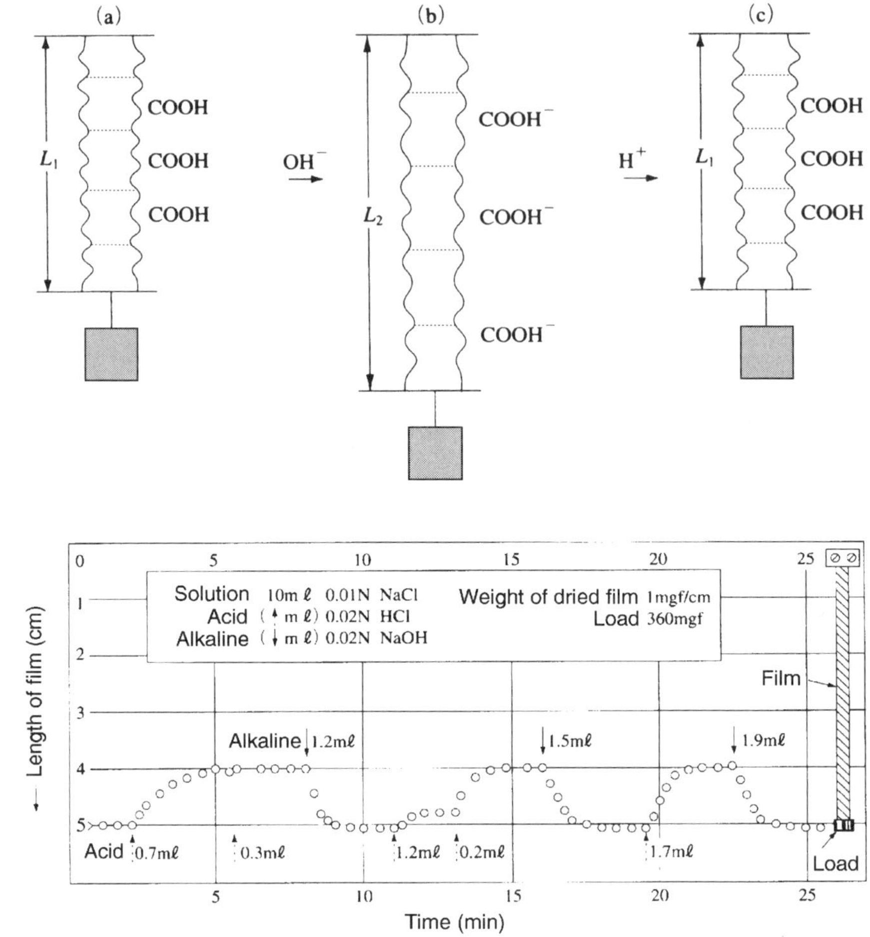

Fig. 1 Chemomechanical cycle of a crosslinked PAA film by acid-alkaline environment.

energy conversion system. If synthetic polymers are used in such a system, they are considered to be soft deformation functional materials.

There are many application possibilities, including use in artificial muscles, switches, sensors, and medical devices. The chemomechanical polymer gels introduced here are discussed based on which stimulus creates the driving force.

7.1.1.1 Chemomechanical polymer gels that function by temperature variation

7.1.1.1.1 Chemomechanical polymer gels that function by formation of interpolymer complexes Poly(methacrylic acid) (PMAA) forms an interpolymer complex with poly(ethylene glycol) (PEG) in an aqueous solution via cooperative hydrogen bonding. Nagata *et al.* noticed that this system reversibly undergoes a dissociation-aggregation cycle upon temperature changes. They developed a chemomechanical system using a temperature gradient as the driving force for the cycle. When load is applied to a crosslinked and nonsoluble PMAA membrane at 100 times its own weight and dipped into a PEG aqueous solution, the system repeated shrinking and expansion (see Fig. 2) [8–10]. This system functions based on the equilibrium between the two polymers. Thus, if the concentration of PEG in the aqueous solution increases, the system shrinks at lower temperature and vice versa. Aggregation of polymer molecules is enhanced at higher temperatures because in addition to hydrogen bonding hydrophobic bonding is also involved. The PMAA membrane, which shrinks upon addition of PEG, also can be a functional separation membrane with expansion and shrinking [11, 12]. If the PMAA membrane is fixed onto a filtering device with an O-ring on its outer perimeter and a PEG aqueous solution is added to the membrane, the pores of the membrane widen according to the mechanism shown in Fig. 3. As a result, the permeability and separation functions increase. Such an isothermal "chemical valve" will be useful to separate proteins. Figure 4 illustrates results that show PEG with molecular weight of 3000 opening a "chemical valve" following the addition of albumin aqueous solution.

When only a PMMA membrane is used, albumin adheres to the membrane surface and the solution almost stops flowing. However, in the case of a chemical valve, the solution permeates and separates with approximately 50% blocking rate. The system also showed similar high efficiency for hemoglobin and polysaccharides [13].

An interpenetrating polymer network (IPN) that forms an interpolymer complex and its expansion-shrinking responsivity has been evaluated. Bae *et al.* synthesized an IPN consisting of poly(N-acryloylpyrrolidine) (PAPy) and poly(oxyethylene) (PEO) [14]. This IPN expanded more than the gel made from PAPy or PEO alone and showed thermoresponsivity. There are also examples of IPN synthesis from poly(acrylic acid) and

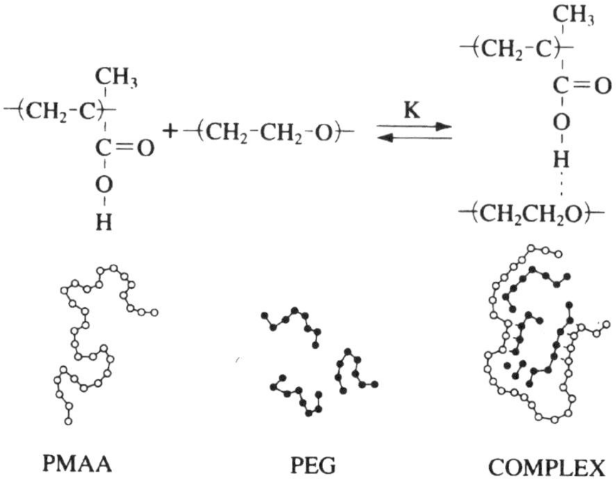

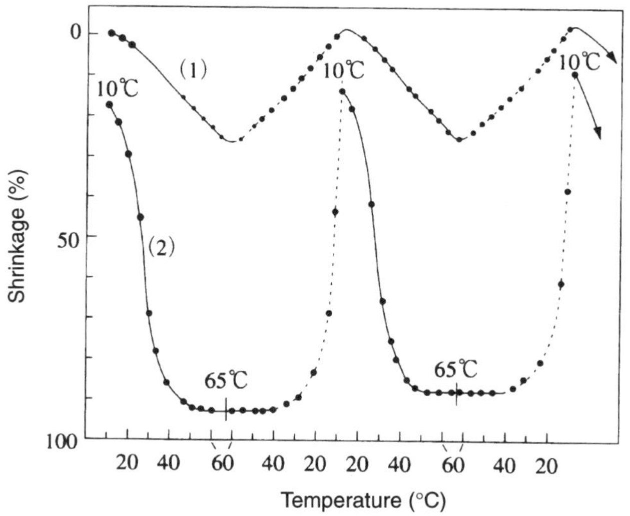

Expansion and shrinking behavior of a PMAA membrane to which a load of 490 mg is applied to the film with dry size of 10×23 mm (4.7 mgf), when the membrane is immersed into (1) pure ater and (2) 70 ml of poly(ethylene glycol)(molecular weight 2000) solution.

Fig. 2 Chemomechanical cycle of PMAA membrane accompanying temperature variation [9].

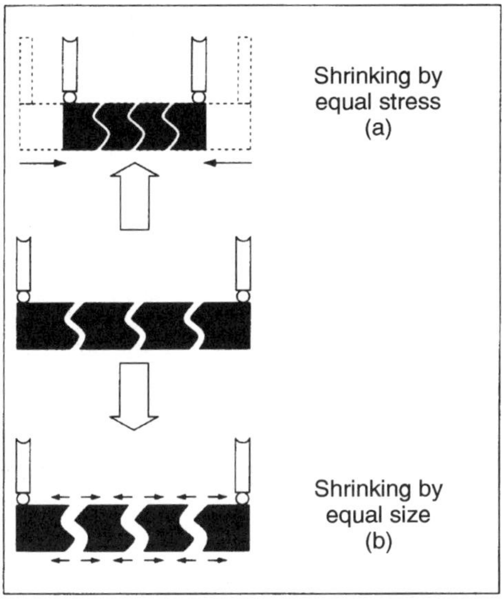

Fig. 3 Functional mechanism of a chemial valve [12].

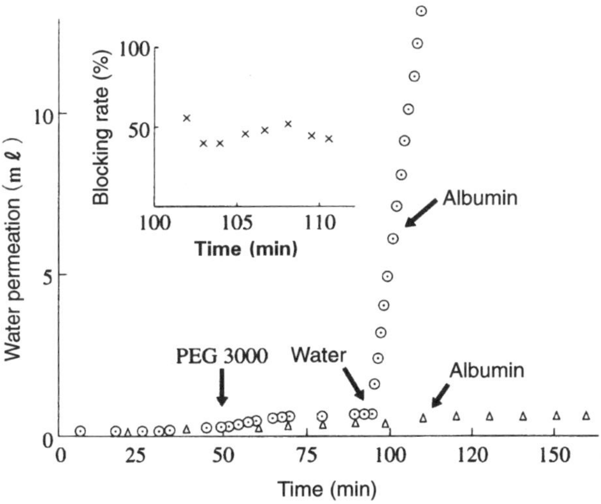

○: When 0.25% PEG (molecular weight 2000) aqueous solution is added, and Δ: when PEG is not added. Albumin (human, molecular weight 67,000), 0.025%, pressure 0.2 kgf/cm^2.
Insert: blocking of albumin by a chemomechanical membrane.

Fig. 4 Permeation control of albumin by chemical valve function of PMAA membrane.

poly(acrylamide) [15]. This IPN can control thermoresponsivity by ionizing the acrylic acid portion.

7.1.1.1.2 Chemomechanical gels that function with phase transition caused by temperature changes Various polymers exhibit reversible phase transition in aqueous solution due to temperature variations. Representative examples include poly(vinyl methyl ether) (PVME) and poly(N-isopropylacrylamide) (PNIPAAm) [16, 17]. Common features of thermoresponsive polymers are the coexistence of hydrophilic and hydrophobic portions in the same polymer chain. Increased hydrophobic interaction at an elevated temperature causes phase separation to take place. Gels obtained by crosslinking these polymers also show thermoresponsivity. The PNIPAAm gel shows the phase transition at 33°C in pure water. It swells at a temperature below the transition and vice versa (see Fig. 5) [18].

N-isopropylamide (NIPAAm) polymerizes via free radical polymerization. It readily copolymerizes with other monomers. The gel, which is synthesized by copolymerizing with a hydrophilic monomer, sodium acrylic acid (NaAA), experiences transition temperature increases following introduction of ionic groups. Its degree of swelling also increases [19]. On the other hand, if it is copolymerized with a hydrophobic monomer, the transition temperature of the gel decreases. Accordingly, the transition temperature and degree of swelling can be controlled by copolymerization. Thus, the property can be tailored based on need. Similar to PNIPAAm, N-substituted acrylamide derivative polymers have been synthesized and their phase transition behaviors are being studied [20, 21].

Polymer gels with amino acid groups or peptides in their side chains have been synthesized [22, 23]. The copolymer gel between methacryloyl-L-alanine methyl ester (MA-L-AlaOMe) and 2-hydroxypropyl methacrylate (HPMA) shows thermoresponsivity. However, as the HPMA fraction increases, the thermoresponsivity decreased. This gel not only shows thermoresponsivity but also pH responsivity. When the gel is made by irradiating γ-rays onto a sequential polypeptide, elastin, also shows thermoresponsivity [24]. These materials consist of biocompatible amino acids and peptides and therefore applications in biorelated areas are possible.

Totally new types of thermoresponsive gels have also been synthesized. The copolymer gel made of stearoyl acrylate (SA) and acrylic acid (AA) (copolymer ratio SA : AA=1 : 4) exhibits a crystalline-amorphous

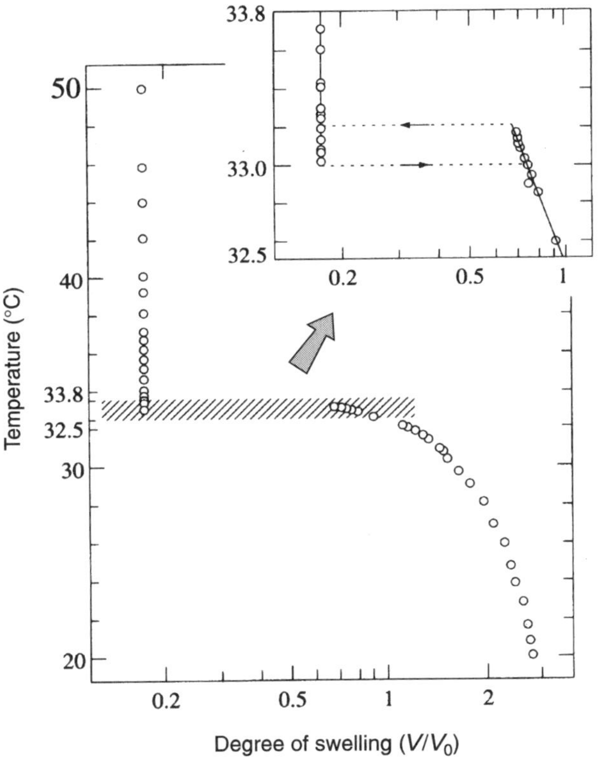

Fig. 5 Volumetric changes of a PNIPAAm gel as a function of temperature.

transition by changing temperature. The modulus changes significantly at a transition temperature of 49°C (see Fig. 6) [25]. Interestingly, this gel shows shape memory at elevated temperature [26]. This gel, which softens at 50°C, is stretched and subsequently fixed by cooling but returns to its original shape by heating it again to the transition temperature. Using this gel, gel hand and gel arm devices have been made.

7.1.1.2 Chemomechanical gels that function by solvent exchange

Polyelectrolytes literally possess dissociated groups on the polymer chain and conformation is strongly influenced by the state of the electric charge. Thus, the molecular shape changes drastically by changing pH,

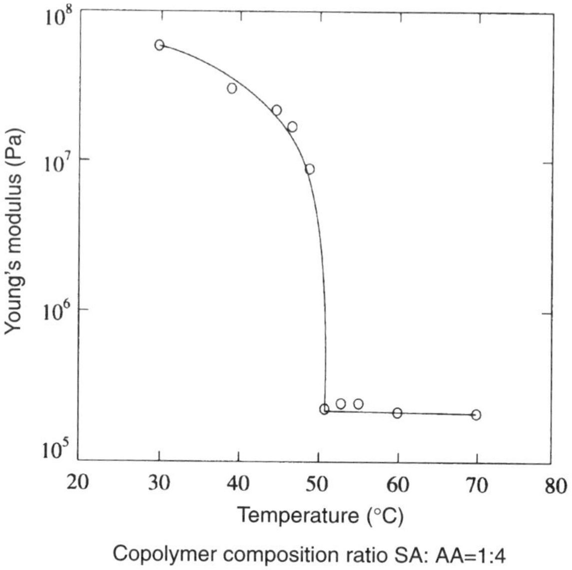

Fig. 6 The change in Young's modulus of a poly(SA-co-AA) gel as a function of temperature.

exchanging water and organic solvent, and changing salt concentration. The aforementioned chemomechanical system constructed by Katchalsky and others also functions with these stimuli serving as the trigger.

Many of the currently developed chemomechanical systems also use these stimuli. Among them, representative examples will be introduced in the following section.

Using as examples the superabsorbent polymers in paper diapers and feminine products, polyelectrolyte gels show poor mechanical properties but a high water absorption property. Thus, it is problematic for use in the mechanical exchange materials needed for chemomechanical systems. However, various gels that have excellent responsivity and strength have recently been synthesized. Suzuki repeated freezing and thawing of a mixed aqueous solution of poly(vinyl alcohol) (PVA)/PAA/polyarylamine and constructed a hydrogel composite that has excellent strength and elasticity [27]. This gel swells in water but shrinks in organic solvents such as acetone and ethanol. When the gel is synthesized a fine sponge structure is formed by controlling the freezing rate in order to improve strength, output power, and response rate. This gel-like artificial muscle had properties comparable to those in a frog's muscle. It showed output power of 0.1 W/cm^3, response time of 0.4 s, and durability of 1000 repetitions. In one example a strong, amphoteric gel was obtained by oxidizing polyacrylonitrile fibers at 200°C and then hydrolyzing the fibers [28]. This fibrous gel responded to pH and quickly stretched and shrunk (see Fig. 7). The response time was as fast as 2–3 s and the generated stress during shrinking was 20 kg/cm^2, which is comparable to the capabilities of human muscles.

In order to develop biosystem applications, it will be useful to find materials that respond in the vicinity of pH 7. Polyamine, which possesses polar and nonpolar groups in an alternating fashion, was found to repeat swelling and shrinking within a narrow pH range [29]. It was also possible to introduce a soft segment into the polymer chain. Consequently, improvement in response time was achieved.

7.1.1.3 Chemomechanical polymer gels that function by addition of a chemical substance

It is well known that chemical substances contribute to the various stimuli responses in living systems. Studies in artificial systems of chemomecha-

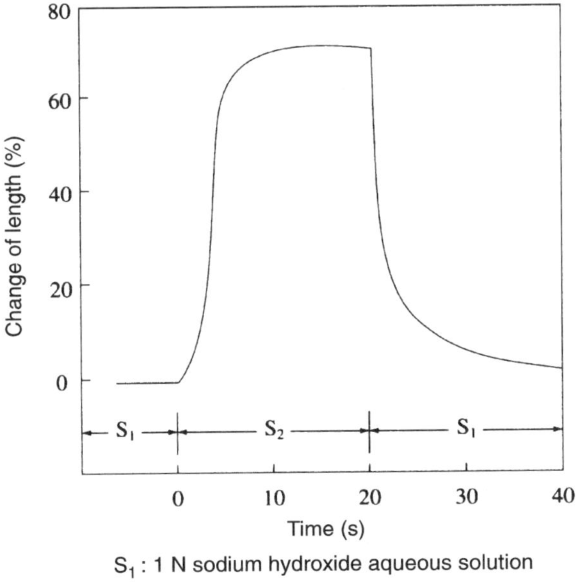

Fig. 7 Expansion-shrinking behavior of fibrous PAN gel by acid–alkaline exchange.

nical gels that are able to respond to chemical substances, have pointed the way to applications of drug delivery systems. A gel capable of responding to blood sugar concentration changes has been developed [30], a first step in the creation of an artificial pancreas. This gel is made of a polymer, which has a boric acid group on the chain. This polymer chain forms an interpolymer complex with PVA and shrinks. However, when glucose is added to this system, the formation of aggregates between boric acid and PVA is inhibited, resulting in extended polymer chains and swelling of the gel. Hence, the gel recognizes the blood sugar level and expands. If a gel contains drugs (e.g., insulin) and is in a shrunken state, the drug is not released but the gel swells in response to elevated blood sugar levels and then insulin is released.

A gel that responds to ATP has also been synthesized [31]. This gel, made of miocene crosslinked by glutaldehyde, shrinks rapidly upon the addition of ATP.

A polymer that responds only to certain ions has also been synthesized [32]. A microsphere copolymer made of NIPAAm and acryloylaminobenzo-18-crown-6 (crown ether) suddenly changes its volume at 28°C when no salt is added. However, when a potassium ion is added, the transition temperature increases by approximately 5°C (see Fig. 8). In contrast, when lithium and sodium ions are added, the transition temperature decreases and selectivity of ions has been confirmed.

If a chemomechanical system that is similar to living systems can recognize specific chemical substances and then function in response to this recognition, a feedback system can be constructed by combining multiple systems.

7.1.1.4 Chemomechanical polymer gels that function by light

In a manner similar to that of photorecording materials that are the subject of many studies, photoresponsive gels can be largely divided into two types. They are the photon mode-type, which functions by photophysical or photochemical reactions upon irradiation of light, and the heat mode-type, which functions by changing photoenergy into thermal energy.

Photoresponsive gels based on the photon mode work by cis-trans photoisomerism of photochromic molecules like azobenzene or by photoionization of triphenylmethaneloyco. Ishihara *et al.* found that poly(2-hydroxyethylmethacrylate) gel shrinks upon ultraviolet irradiation after equilibrium swelling is reached in water [33]. They further found that this gel swells again upon irradiation by visible light. Azobenzene isomerizes

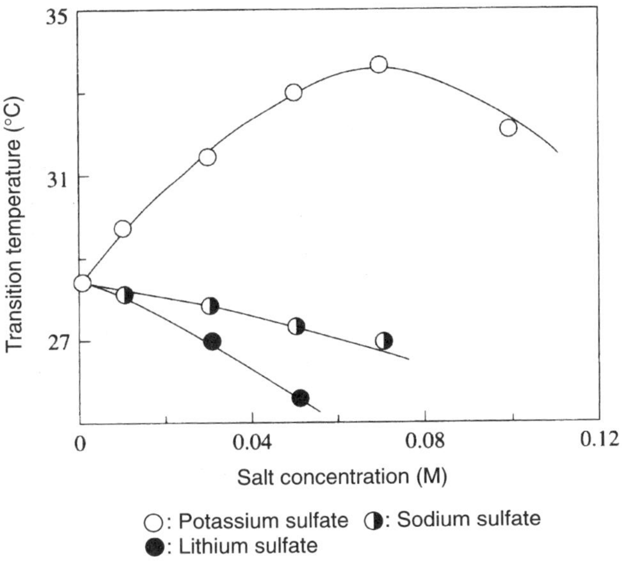

Fig. 8 Change in transition temperature of a PNIPAAm gel that possesses 12 mol% crown ether groups upon addition of salt.

to polar cis form with ultraviolet irradiation. It transforms to trans form in visible light. This gel is thought to undergo an expansion-shrinkage cycle because the hydrating water around the hydroxyl groups is influenced. Although it is not a gelled system, a monolayer of PVA to which azobenzene is introduced shows reversible area changes under ultraviolet-visible light irradiation [34].

Triphenylmethaneloyco is a photoionizable organic dye. The acrylamide (PAAm) gel that contains this unit ionizes with irradiation of light and swells by static repulsion [35, 36]. Figure 9 depicts the photovolumetric change of a PAAm gel that contains 1.9 mol% triphenylmethaneloiccyanide. When light of wavelength that is >270 nm is irradiated, the gel starts swelling. In about 2 h, the gel has increased 18-fold. If the swollen gel is left in the dark, it shrinks. However, it takes more than 20 h to return to its original size. To improve the response time of this gel, electric field response has also been studied [37]. In one example phase transition of a gel is controlled by photoionization of triphenylmethanloico [38]. The PNIPAAm gel with triphenlymethaneloicocyanide changes phase transition behavior upon irradiation with light.

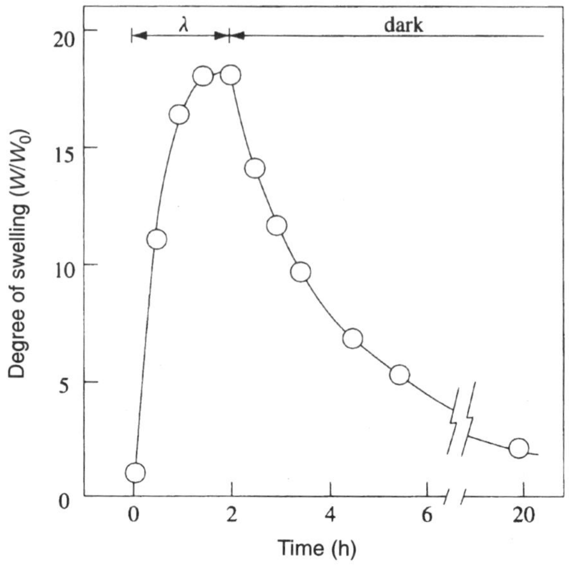

Fig. 9 Swelling behavior of PAAm gel that contains triphenylmethaneloico-cyanide upon ultraviolet irradiation [36].

A heat mode-type photoresponsive gel has also been synthesized [39]. This gel was synthesized by the copolymerization of NIPAAm/copper chlorophiline/N,N′-methylenebisacrylamide. This gel also exhibits thermoresponsivity, showing continuous volumetric changes upon temperature variation. However, it shows discontinuous volumetric transition upon irradiation using an 80 mW or higher argon ion laser. It is thought that expansion-shrinkage occurs because the thermal diffusion takes place faster than the collective diffusion of the polymer chains. According to this calculation, a gel with a diameter of 1 mm shows a response time of shorter than 5 ms.

7.1.1.5 Chemomechanical polymer gels that function by electric field

For the chemomechanical systems introduced thus far, it is necessary to install subsystems to change temperature or exchange solvents. As a result, the system becomes large. If a chemomechanical system is to be used as an actuator or transducer, electric field control is thus more favorable. This section introduces electric field responsive chemomechanical gels.

If an electrode is in contact with a polyelectrolyte hydrogel and direct current is applied, the gel shrinks by expelling water. If voltage application is halted in the middle, the shrinking process can be stopped. The gel then returns to its original size when immersed in water [40]. Whether natural or synthetic polymers, these gels with electric charges exhibit the shrinking phenomenon when direct voltage is applied. Such shrinkage of gels by electric stimulus is the result of ion transport by the electric field gradient [41]. Polyanions move to the cathode upon voltage application whereas counter ions move to the anode. Contact with the electrode then cancels the electric charge. This then eliminates the hydrating water that is expelled outside the gel, thereby resulting in shrinkage. In the case of polycations, the opposite phenomenon takes place. Figure 10 illustrates electric field shrinkage of PAA microparticles from a polyanion gel [42]. A microparticle (diameter 180 μm) shrinks quickly upon voltage application. When 6 V is applied, it takes only 50 s to shrink. The rate of

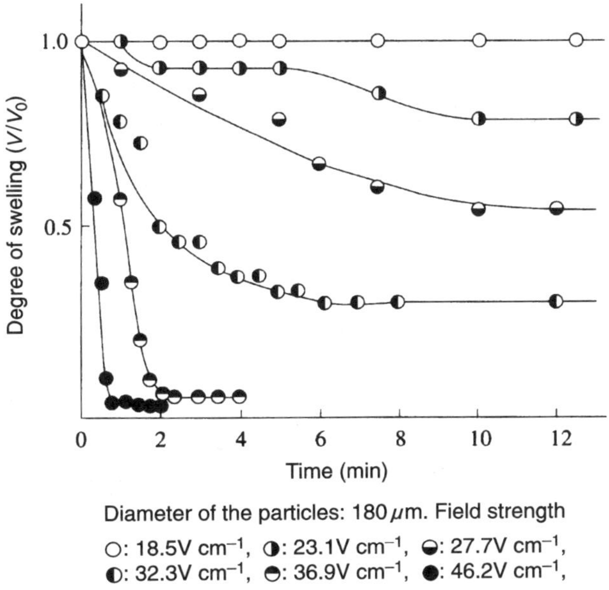

Fig. 10 Expansion-shrinkage behavior of PAA microparticles under electric field [42].

shrinking is proportional to the square of the diameter. It can be estimated that it takes only 0.23 ms to shrink a 1-μm diameter particle.

When mechanical deformation is applied to a polyelectrolyte gel such as a PAA gel, electric potential is generated (mechanoelectric phenomenon) (see Fig. 11) [43]. When the PAA gel is deformed, the polymer chain also deforms. It is thought that the stretched polymer chain enhances the dissociation of carboxylic groups, leading to generation of potential. Using this principle, it is possible to manufacture an artificial contact sensing device. An artificial contact sensing device is one in which potential is generated by pushing the gel and light is emitted by a light emitting diode (LED).

Field responsivity of a polyelectrolyte-metal composite membrane made of a perfluorosulfonic acid to which platinum is electroplated on both sides is being evaluated [44]. If a voltage is applied using the platinum layer as the electrode after the composite membrane is immersed

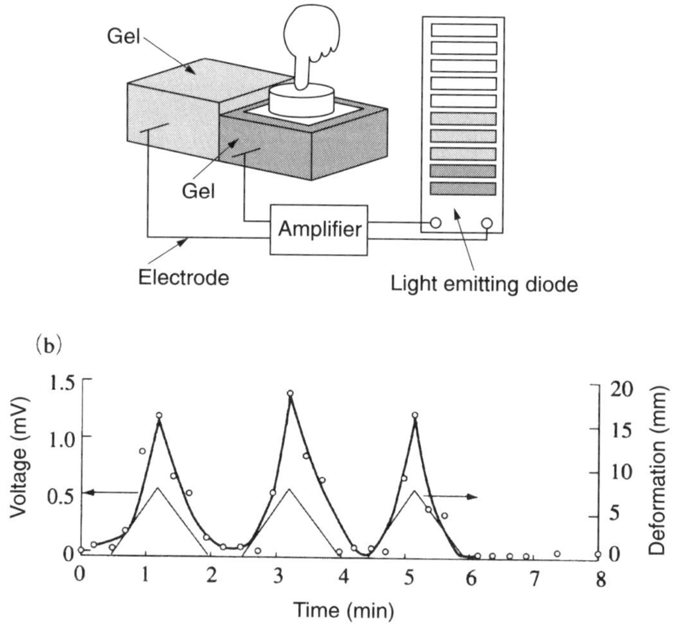

Fig. 11 The diagram of the contact-sensing device using a gel (a) and the voltage generated when force is applied to a PAA gel (b) (the size of the gel: 15 × 10 × 10 mm; degree of swelling of the gel: 16).

in a sodium hydroxide aqueous solution and the mobile counter ions are exchanged with Na^+, the membrane bends towards the anode side. Even at a voltage as small as 1 V, the membrane bends. There is no electrolysis of water or formation of bubbles observed. The response time is short. If a sample 3 mm in length is moved in air, the bending can follow the sine curve at 100 Hz.

If the water causes gas, acid, and alkaline substances to be generated during voltage application of field-responsive gels the potential for problems exists.

To avoid these phenomena, it is possible to prepare a hybrid gel with a conducting polymer (e.g., a polypyrrole). This gel can be operated with a low voltage of 0.4–1.4 V [45]. With a solution that contains an oxidation-reduction agent like hydroquinone or viologen derivatives the voltage can be reduced to 1.2 V [46].

Even nonionic gels show field responsivity. A PVS gel swollen by dimethylsulfoxide (DMSO) deforms in proportion to the voltage used even though it is nonionic. Although the amount of deformation is small, the rate of response is as fast as milliseconds (see Fig. 12) [47]. The

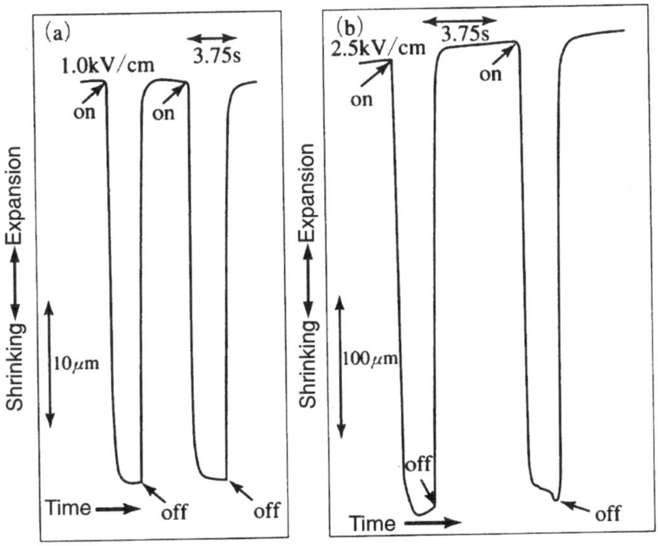

Field strength (a): 1.0 kV cm^{-1}, (b): 2.5 kV cm^{-1}

Fig. 12 Expansion–shrinking behavior of a PVA gel swollen in DMSO under voltage application [47].

deformation of polymer networks is thought to be caused by the formation of a structure by the solvent molecules in the gel under the electric field. Charge transfer complex gels with conducting polymers have also been studied [48, 49]. Poly(N,N-dimethylaminopropylacrylamide), which is swollen in dimethylformamide and further doped by an acceptor molecule like tetracyanodimethane, is reported to exhibit a conductivity of 10^{-3} S/cm. This gel not only shrinks under electric field but also shows interesting phenomena that include photoelectric exchange function, a frequency generation phenomenon, and the orientation of dopant molecules in the gel under an electric field.

7.1.1.6 Composite and organized chemomechanical polymer gels

When chemomechanical gels are used as materials or devices, the development of high performance and better handling require organized and composite materials to be manufactured. These manufacturing methods include chemical methods such as grafting, copolymerization and hybridization, and physical methods like blending, microgelation, film formation, fibrillation, and composite manufacturing.

When a stimuli-responsive gel is used for various devices, the response time is an essential property. The response time of gels is proportional to the square of the gel size [50]. Hence, the possibility of manufacturing as small a gel as possible is an important consideration. A monodisperse microsphere is such an example. The PNIPAAm, which is polymerized by precipitation polymerization, has a 0.94-μm diameter [51]. The PNIPAAm microsphere polymerized inside a micelle is 80–150 nm in diameter and its response time is excellent [52].

By utilizing the phase separation of thermoresponsive polymers, porous chemomechanical gels can be obtained. The PVME polymer has a transition around 36°C. It readily polymerizes when γ-rays irradiate its aqueous solution and a gel is formed. If the solution is phase separated by an increase in the temperature and then subsequently crosslinked, a sponge-like gel with continuous pores ranging from 100 to 200 μm can be obtained (see Fig. 13) [53]. This gel, which shows extremely fast response time in comparison to a homogeneous gel, completes expansion and shrinking in ≈60 s (see Fig. 14). It is now feasible to manufacture a fast response PVME gel [54]. When a mixed solution of sodium alginic acid (NaAlg) and PVME is spun in a spinning bath of calcium chloride aqueous solution, a fibrous composite of phase-separated NaAlg and

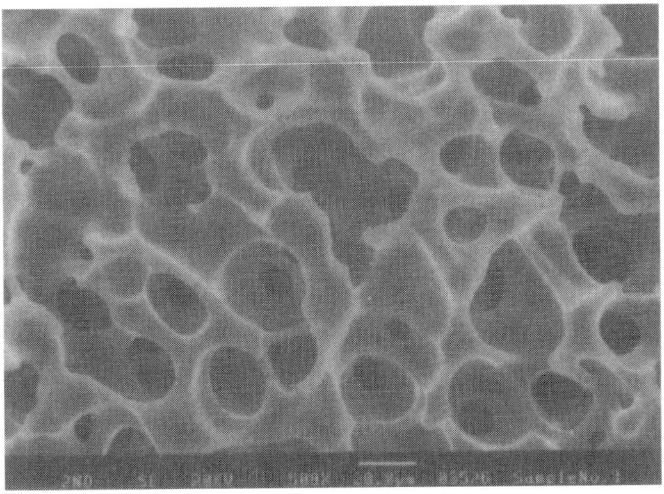

Fig. 13 SEM photomicrograph of porous PVME gel.

PVME can be obtained. If γ-rays irradiate this composite, NaAlg decomposes and PVME crosslinks to form a fibrous porous gel. A porous and fibrous gel with a swollen state diameter of 400 μm completes expansion and shrinking in less than 100 ms upon temperature change. An artificial

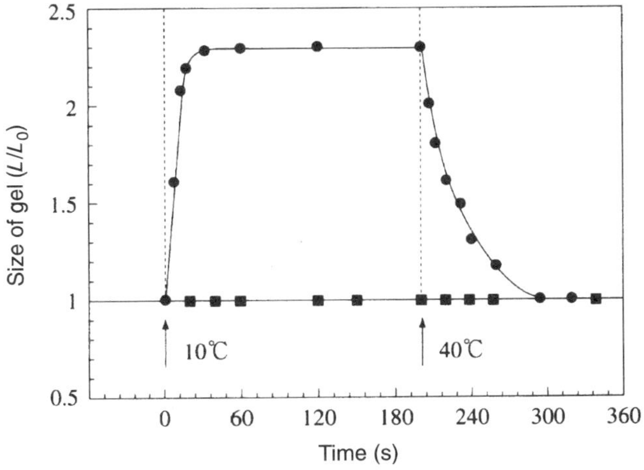

●: Porous gel ■: Homogeneous gel

Fig. 14 Expansion-shrinking behavior of PVME gel accompanying temperature changes.

muscle made by bundling 1000 fibrous gels can lift a weight of about 350 g by using hot and cold water repeatedly. By adding a small amount of carbonblack to the PVME solution and then using γ-rays, a heat mode-type photoresponsive gel can be manufactured [55]. When visible light with an intensity of $0.627\,mW/cm^2$ is irradiated onto a gel (dimensions of $22 \times 22 \times 6\,mm$) shrinking is completed in approximately 150 s.

To construct materials that resemble live tissue, composites with enzymes have been created. By adding urea to the NIPAAm-AA copolymer gel to which urease is fixed, ammonium is generated by the enzyme reaction and this causes expansion of the gel [56].

A NIPAAm gel to which concanavalin is fixed swells following addition of dextrin sulfate and shrinks when α-methyl-D-glucopyranoside is added [57]. Enzymatic reactions cause these systems to expand and contract. As well, enzyme activities also can be controlled by utilizing the expansion-shrinkage capability of the gel [58]. Control of enzyme activity is being studied using the PVME gel in which glucosidase is inclusion-fixed. Usually, the activity of free enzymes increases as the temperature increases. However, the enzyme, when inclusion-fixed in the PVME gel, will show high activity below the phase transition temperature but decreased activity above the phase transition temperature.

The materials described here are homogeneous polymer networks. However, gels with molecular level stereoregularity have also been synthesized. The gel, crosslinked in a lyotropic liquid crystal state of poly(γ-benzyl-L-gluatamate) (PBLG), is a cholesteric gel [59]. Figure 15 illustrates the polarized photomicrograph of the cholesteric structure of the liquid crystal gel. The fingerprint pattern that appears when a liquid crystal shows cholesteric structure is observed, which demonstrates that this gel maintains a liquid crystal state. Further, PBLG lyotropic liquid crystal orients in both the electric and magnetic fields and forms a nematic phase. Therefore, by crosslinking under a magnetic field, a liquid crystal gel with unidirectionally oriented PBLG molecules can be obtained. This gel is the first anisotropic material that shows anisotropic expansion-shrinkage behavior [60]. If gels with ordered structures like those of natural systems can be synthesized, construction of extremely high-performance chemo-mechanical systems will be possible one day.

A thermotropic liquid crystal gel has also been synthesized [61]. If an acrylic monomer with a cyanobiphenyl group side chain is polymerized with a crosslinking agent, a liquid crystal elastomer can be obtained. The

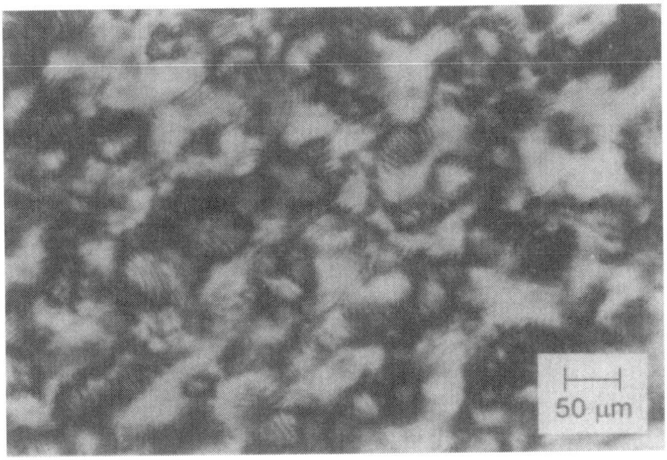

Fig. 15 Polarized light micrograph of PBLG cholesteric liquid crystal gel.

elastomer swells when immersed in a low molecular weight liquid crystal, forming a thermotropic liquid crystal gel. This liquid crystal gel deforms at a very fast response rate upon application of an electric field.

7.1.1.7 Chemomechanical polymer gels that show nonlinear response

A chemomechanical gel that shows a nonlinear response with linear stimulus has been reported. If direct voltage is applied to a polyelectrolyte gel by inserting a pair of platinum electrodes, current fluctuation with good repeatability as shown in Fig. 16 can be observed. The gel possesses the ability to convert direct current into pulsed current upon shrinking

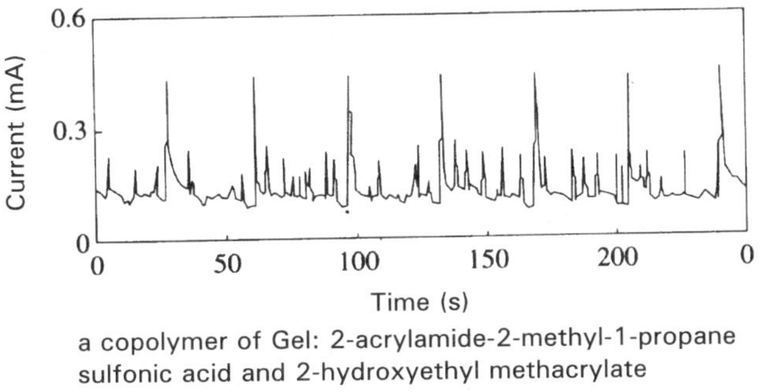

Fig. 16 Fluctuation of the current when direct voltage is applied to the gel.

[62]. The fundamental frequency transforms to a more stable waveform as a function of time.

Hirotsu studied the phase transition of gels obtained from copolymerizing NIPAAm and acrylic acid in an electric field [63]. The phase-transition temperature of this gel is approximately 39°C. However, this transition decreases to 31°C with an electric field of 10 V/cm. If the direct voltage is applied to this gel the volume shrinks while vibrating. Although the mechanism has yet to be understood, it is nonetheless a very interesting phenomenon.

Poly(L-glutamic acid) shows reversible helix-coil transition with pH changes. In an alkaline solution, the gel swells violently due to static repulsion by the carboxylic group and the shape of the membrane cannot be maintained. If a block copolymer is synthesized with hydrophobic L-lysine, a cylindrical microphase separation is observed. This material avoids macroscopic deformation. Thus, a membrane with molecular level deformation can be manufactured [64]. This membrane shows not only control of permeation by pH changes but also nonlinear responses like vibration of potential by salt concentration difference [65] and nonlinear resistance upon voltage application [66].

7.1.1.8 Conclusions

As introduced here, it is now possible to manufacture chemomechanical polymeric materials with excellent response rate and strength. In the future, gels will be combined with other materials and used as part of a system. For example, because a gel expands and shrinks in water (solvent), it is necessary to construct a system that prevents a solvent from evaporating if it is to be used in air. Further, external stimuli that include temperature change and solvent exchange will be space-intensive. Thus, in order to be included in a system, miniaturization will be necessary. If these problems can be solved, a totally new system, which moves smoothly like a living creature, may become a reality.

7.1.2 Polymer Gel Actuators

7.1.2.1 Introduction

An actuator is the collective name for a device that performs mechanical work using electrical and fluid energy. There has been interest in recent years in developing actuators with completely different and new driving forces as a result of advances in robotics and improvements in traditional electric, fluid and air-type actuators. Polymer gels have the potential to be used in new actuators. The idea of using a solution in which gels undergo expansion and shrinkage due to external stimuli such as pH, salt concentration and temperature as the actuator is surprisingly old. It was actually proposed almost fifty years ago. Due to the discovery of the volumetric phase transition phenomenon of polyelectrolyte gels by Tanaka in the late 1970s [67], polymer gels have again received attention. In addition, hydrogels have been actively studied from the biometrics point of view because their composition is similar to that of live bodies. Studies on polymer gel actuators can be largely divided into two categories: (i) for construction of biomimetic actuators to replace muscles; and (ii) artificial replication of the smooth movements of living systems. In this section, both studies will be reviewed.

7.1.2.2 Artificial muscles

An artificial muscle that uses a polymer gel is based on the expansion-shrinkage behaviors that external stimuli cause. It is thus fundamentally different from the mechanism of muscles in living bodies [68, 69]. However, it is still necessary to exhibit target values for performance power generation (0.2–1 MPa) and response rate (<10 ms) that are comparable to those found in live bodies. The former value is achieved mainly by molecular design and synthesis of polymer gels with higher modulus and larger swelling-shrinking deformation. The latter became possible with the manufacturing techniques that were developed for thin and fibrilar polymer gels. In this section, gels are described based on the type of external stimuli.

(a) Thermoresponsive type Poly(N-isopropylacrylamide) and poly-(vinyl methyl ether) (PVME) shrink in water upon increased temperature. These polymers have been known for a long time as having negative solubility [70]. It has been attracting attention because its transition approaches body temperature. In 1989, Hirasa *et al.* irradiated γ-rays to a PVME aqueous solution to crosslink and further spin this material into a gel fiber [71]. When they irradiated the γ-rays at around the transition temperature, they found that a sponge-like gel fiber could be obtained. They studied the actuator characteristics of the gel fiber by exchanging cold water at 20°C and warm water at 40°C. As a result, they found that the response rate for swelling-shrinking was greatly improved with the sponge-like structure. The PVME fiber has a diameter of 400 μm at 20°C.

Upon changing to 40°C, the fiber shrunk within 1 s with a shrinking force of 0.31 g or 24 kPa per fiber. They manufactured an artificial muscle by bundling 1000 fibers and lifted a weight of 300 g. They also manufactured a composite by forming a photocrosslinked PVA gel on one side of a PVME gel. They designed an artificial finger and attempted to grab an object in warm water with this finger [72].

(b) Solvent composition responsive type Polymer chains with ionic groups such as COONa or SO_3Na change the spreading of the chain as a result of suppression of dissociation of the ionic groups when the pH or solvent composition of the solution changes. If many polymer chains are crosslinked and gelled, the microscopic spread appears as macroscopic volumetric changes. These solvent composition, responsive-type gels undergo discontinuous change of volume as much as several tens to several hundred times at a certain solvent composition (for example, in water/acetone mixed systems, it is 50% of the acetone concentration). By utilizing this property fully, an artificial muscle can be designed. However, since the response rate is extremely slow by minute composition variation near the acetone concentration of 50%, a drastic composition change is made by exchanging water with acetone.

In the 1970s, Nambu discovered that a physically crosslinked PVA gel with fracture strength of >10 MPa and rubber elasticity could be obtained. This was achieved by repeatedly freezing and thawing an aqueous solution of poly(vinyl alcohol) with a certain degree of saponification and molecular weight [73]. This white and opaque PVA gel exhibits swelling-shrinking as shown in Fig. 1 by water-acetone exchange. In the late 1980s, Suzuki developed a new gel (water uptake of 87%) by mixing PVA, poly(acrylic acid), and poly(acrylamine) with a 1.74 : 0.24 : 0.26 ratio in order to improve the amount of deformation and response rate [74]. It was observed that a gel membrane 10-μm thick responded in 0.4 s and exhibited the stress of 0.2 MPa by exchanging water with acetone. This stress is comparable to that of a frog muscle. By utilizing a lithographic technique and manufacturing a net-like gel membrane where each fiber was 2-μm in diameter, they succeeded in achieving a response rate of approximately 100 ms.

Many researchers are studying artificial muscles by exchange of aqueous solutions of different pH. Umemoto obtained a gel fiber with many COONa groups by thermally treating a 22.5-μm wide poly(acrylonitrile) fiber at 220°C and then further saponifying it with a 1 N NaOH

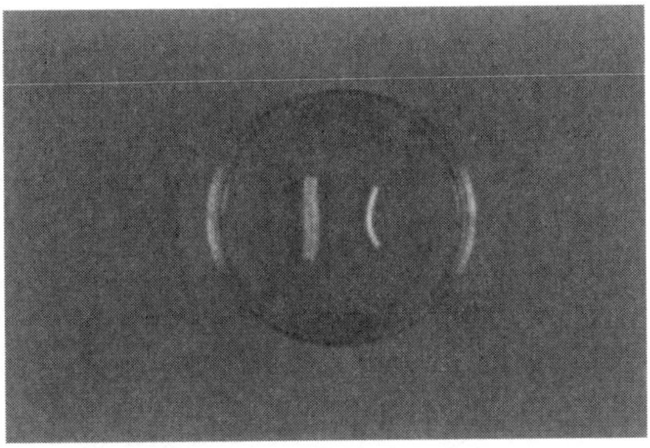

Left: in water, right: in acetone

Fig. 1 PVA-PAA gel in water or acetone.

solution. He measured the shrinkage stress of this PAN gel fiber when 1 N NaOH and 1 N HCl solutions were exchanged. The generated stress was approximately 1 MPa and the response time was 2 s (the time necessary for 50% shrinkage) [75].

(c) Electric field responsive type The aforementioned change in pH can also be induced by an electric field. For example, if direct current is applied to a sodium chloride (NaCl) aqueous solution, chlorine ions gather near the cathode and in its vicinity the solution becomes acidic; the opposite then occurs farther from the electrode. When a gel of poly(acrylic acid) (PNNNa) in an NaCl solution is near the cathode, the gel shrinks due to the acidic pH near the cathode. In 1965, Hamlen *et al.* (General Electric) reported for the first time the possibility of an artificial muscle using an electric field responsive gel [76]. A gel fiber consisting of PVA and PNNNa (length–12.8 cm; diameter–unknown) was immersed in a 1% NaCl aqueous solution. One side was connected to the cathode and a weight was hung from the other side; application of 5-V direct voltage (40 mA) followed (see Table 1). The portion in contact with the cathode started to shrink and the fiber length became 12.2 cm after 10 min. If the polarity of the field was changed, the gel swelled again and returned to its original size. They explained that shrinking and swelling was due to the pH change near the cathode. In 1972, GE further advanced the Hamlen group's work. An electric field responsive gel muscle was developed and

Table 1 Deformation of a gel by electric field (an example).

Deformation	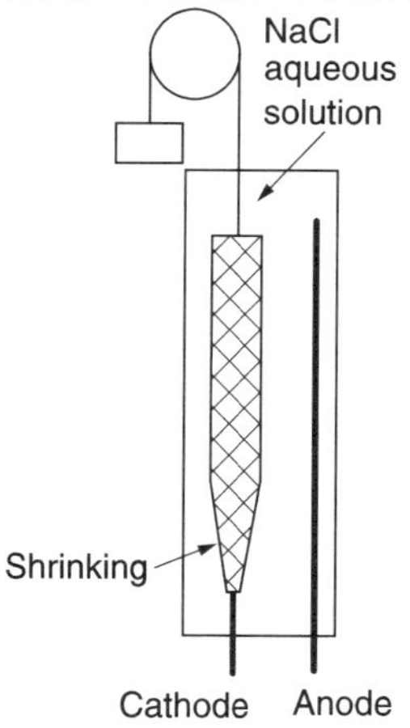	NaOH aqueous solution 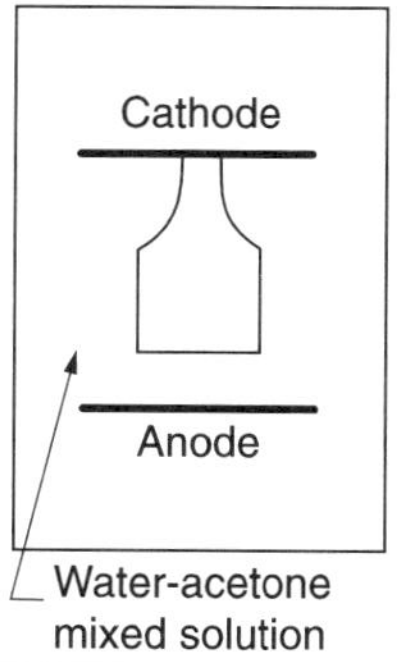	Swelling 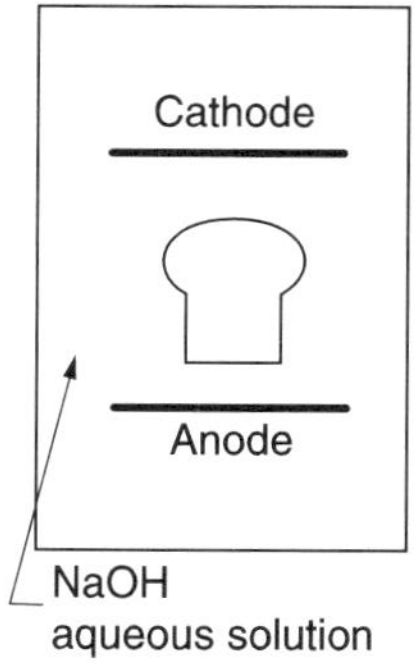	Shrinking towards the direction of the electric field 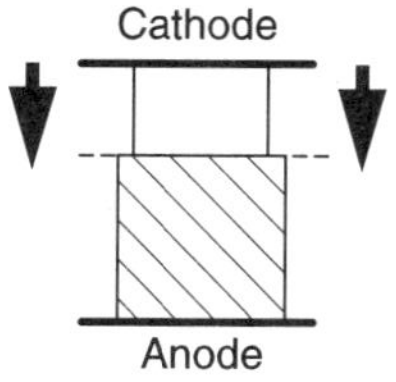
Year	1965	1982	1985	1992
Gel	PVA-PAA gel (ionized gel)	PAANa-PAAm gel (ionized gel)	PAANa-PAAm gel (ionized gel)	PVA gel-DMSO (nonionized gel)
Deformation mechanism	pH change	pH change	Change of osmotic pressure by the ion concentration difference	Structural change of DMSO

it was observed to generate a load of 10 g (response time of several minutes) [77].

Grodzisky and Shoenfeld (MIT) studied the time-dependent tensile stress that appears when the alternating electric field is applied to a collagen (a protein and amphoteric gel) membrane [78]. This collagen membrane was 45 μm in dry condition and the generated stress was 10 kPa. Shrinkage by electric field or pH has also been reported by Osada and Hasebe, and Kishi and Osada [79, 80] and by DeRossi *et al.* [81]. The generated stress is ≈10 kPa, which is one order of magnitude smaller than the aforementioned thermoresponsive type or solvent composition responsive type gels, and the response time is also slower by 1 to 2 orders of magnitude.

7.1.2.3 Biomimetic actuators

(a) Bending phenomenon with an electric field and in a robot hand An electrical signal is a common stimulus and probably the most practical one due to ease of control. In the mid-1980s polymer gels were known to shrink by direct voltage [76, 82, 83]. Since then, new electric field deformation phenomena have been observed. In 1985, the author and others discovered that a gel that is facing the cathode swells selectively when a PAANa gel square rod is placed in an aqueous solution of an electrolyte like NaOH or Na_2CO_3 without contacting the electrode. We further found that, when this local swelling occurs in a rod-like gel, the rod bends significantly towards the anode as a bimetal would (Fig. 2) [84]. This swelling deformation was later found both qualitatively and semi-quantitatively to be due to the change in osmotic pressure [85–87]. This osmotic pressure was caused by the difference in the mobile ion concentration in and out of the gel. When there is no electrolyte in contact with the gel or the concentration is low, the cathode side shrinks and the material bends towards the cathode. Thus, by changing the concentration of the electrolyte, the direction of bending can also be controlled.

Irie and Kunwatchakun irradiated the ultraviolet radiation to a poly(acrylamide) gel to which a photoresponsive group, a triphenylmethaneleuco derivative, was introduced. The ultraviolet radiation dissociated the leuco derivative. They reported that this gel showed a bending phenomenon similar to that of the PAANa gel upon application of direct voltage [88].

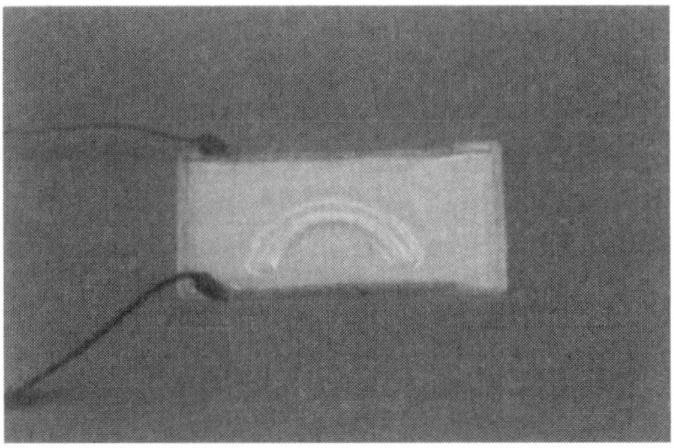

Upper side: cathode, lower side: anode, the gel is bending toward the anode

Fig. 2 Bending of the sodium salt of poly(acrylic acid) under the electric field.

The bending phenomenon under the electric field due to swelling is characterized by a large deformation in a relatively short time. On the other hand, the gels, for example a representative PAANa gel, are mechanically weak. This problem, however, was solved by preparing a composite of PVA and PAANa. Subsequently, it became possible to design an actuator to model live bodies. Figure 3 shows a robot hand with multiple fingers made of a PVA-PAA gel. The gel finger is a laminate of a PVA gel membrane in which a platinum wire (the anode) with a diameter of 50 μm is embedded, PVA-PAA gel rod (8 mm × 8 mm × 80 mm), and platinum plate (the cathode). The cathode has a plastic spacer in order not to interfere with the swelling of the PVA-PAA gel upon electric field application. The robot hand can move the fingers in 10 mM Na_2CO_3 aqueous solution upon application of an electric signal. It can catch a quail's egg without breaking it. The egg also can be released by reversing polarity [89].

Although in an aqueous solution, a biomimetic actuator using a polymer gel has been demonstrated. In fact the technology has already moved forward—it is now possible to manufacture a robot hand with gel fingers that can operate in air [90].

(b) Bending under electric field, gel fish, and gel loop In regard to the bending phenomenon under the electric field of the PVA-PAA gel, the gel responds faster as the thickness of the gel decreases. A gel with a

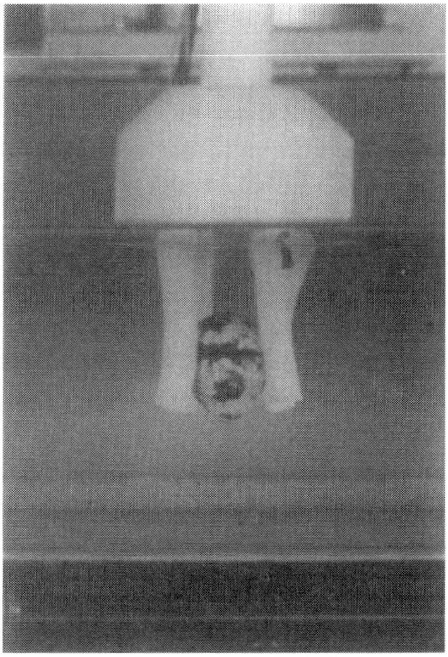

A quail's egg is retrieved from an aqueous solution of an electrolyte.

Fig. 3 A robot hand with several fingers made of PVA-PAA gel.

thickness of 1 mm responds to the alternating electric field at 1 Hz and displays swelling and shrinking behavior [91]. Utilizing this movement, a gel fish that swims via an electric signal was manufactured in 1988 (see Fig. 4). The gel fish consists of a plastic body with a PVA-PAA gel membrane tail fin 40-mm long and 0.7-mm thick. Using this tail fin's bending movement, the fish swam at 2 cm/s. In the 1990s, other materials that exhibit faster bending movement than the PVA-PAA gel were found. This was achieved by a combination of poly(2-acrylamide-2-methylpropane sulfonic acid) (PAMPS) gel and a certain surface active agent [92]. Osada *et al.* manufactured a gel loop (Fig. 5) by installing hooks at both ends of a PAMPS gel, which was then hung from a ratchet bar. An electrical signal was applied. This gel loop showed 1D movement in an aqueous solution by using bending motion.

A platinum-plated (both sides) perfluorosulfonic acid (PFSA) gel membrane (thickness, 1 mm, length 15 mm) was found to bend with a

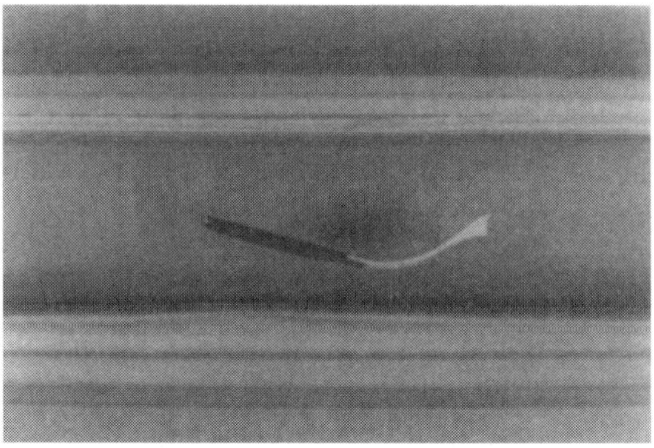

The gel fish moves forward by moving the tail fin

Fig. 4 The gel fish that possesses a PVA-PAA gel tail fin.

signal at 1 V and several Hz [93]. Among the characteristics of these systems, stable movements can be maintained for >10 million times at 1 Hz and high-speed response can be achieved if gel size is reduced. Application to a micromachine is now under evaluation because of these properties. Bending phenomena mechanisms of PVA-PAA, PAMPS, and PFSA gels do differ however.

Bending movement speeds under the electric field can be increased by decreasing the size. Artificial propulsion of a material can be achieved using this bending motion. This amounts to further progress in live organism mimetics.

(c) Bending under the electric field and with an artificial feather
The polymer gels described thus far had to possess ionic groups so as to

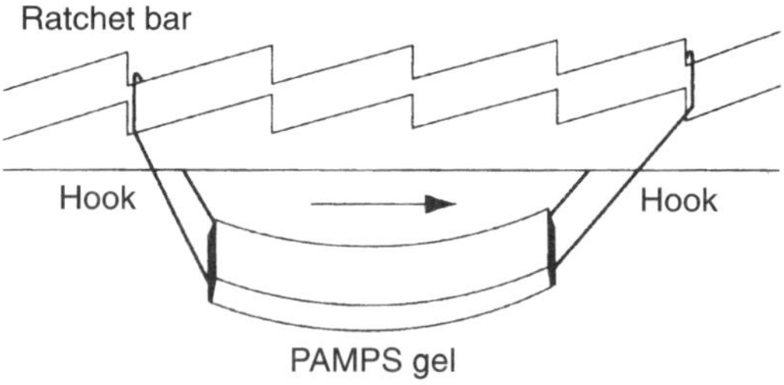

Fig. 5 A gel looper that uses a PAMPS gel.

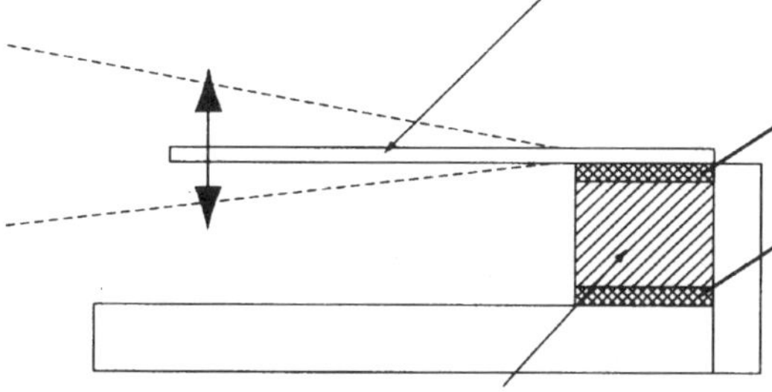

Fig. 6 Movement of a feather that uses the stretching-shrinking of a PVA gel under alternating current.

promote electric field responsivity. Thus, they were necessarily polyelectrolyte gels. However, it has now become clear that nonionic gels can also deform under the electric field [94]. Hirai *et al.* made a gel by immersing a crosslinked glutaldehyde crosslinked PVA in dimethylsulfoxide (DMSO). They found that the gel shrinks (bends) towards the direction of the electric field if the gel is sandwiched by two electrodes and a high voltage at $\approx$2 kV per 1 cm is applied. This gel is 4 mm long and 20–30-μm thick. The response time of this gel is several hundred ms and is reversible. Hence, it shows response even to the alternating electric field and bends towards the electric field. Hirai *et al.* succeeded in building a system (shown in Fig. 6) that demonstrates the flapping motion of a PVA feather. It is now hoped that it might be possible to build a driving system that is like a bird's feather. Although the mechanism of bending PVA-DMSO gel in the direction of an electric field is not well understood, Hirai *et al.* hypothesize that it is due to DMSO structural changes that result from small angle x-ray scattering [95]. Similar deformation is observed in a gel with a magnetic fluid under a magnetic field [96].

7.1.2.4 Conclusions

Various prototypes of artificial muscles and biomimetic actuators capable of transforming external energy into mechanical energy have been proposed that use polymer gels. There are still problems with durability and amount of work possible. However, it is highly likely that new actuators will be developed. Because polymer gels are similar to human skin, actuators made from such materials are soft.

7.1.3 Actuators Made of Conducting Polymers

7.1.3.1 Introduction

Hard materials like metals and ceramics traditionally have been used for actuators. However, soft materials with smooth movement are also useful depending on the purpose. In particular, light and smoothly functioning actuators are desirable as medical devices and artificial organs that have direct contact with the human body. In this section, the study of a conducting polymer, polyaniline, will be used as an example. The dependence of deformation ratio on the electrolyte solution and load will be described. Furthermore, the measurement results and the functional mechanism of excess response characteristics will also be discussed.

7.1.3.2 Electrochemical oxidation and reduction of conducting polymers

Figure 1 shows the structure of representative conducting polymers, polyacetylene, polypyrrole, and polyaniline. Conducting polymers possess readily oxidized or reduced π-electrons on the main chain. The π-electrons transfer to the electrode by oxidation and the negative ions in the electrolyte solution are doped to the conducting polymer. The conductivity drastically changes from the insulating to the metallic states. Further, the negative ions are antidoped by reduction and the polymer returns to an insulator [97, 98]. Oxidation and reduction can be performed repeatedly.

In polyaniline, the lone pair of the nitrogen atom along with the p-electrons of the benzene ring contribute to electrooxidation. Thus, it behaves slightly different from other conducting polymers [99, 100]. As shown in the cyclic voltammogram (CV) shown in Fig. 2, polyaniline in hydrochloric acid aqueous solution take three states. They are the emeraldine (ES) between two oxidation peaks, oxidative pernigranine (PS) at the higher potential side, and reductive leucoemeraldine (LS) at the lower potential side. In the oxidative process, chlorine ions are doped into polyaniline due to the LS$\rightarrow$ES process. Furthermore, in the ES$\rightarrow$PS process, hydrogen ions are produced. Although in the LS$\leftrightarrow$ES process the reaction is reversible, in the ES$\rightarrow$PS process, partial hydrolysis can take place.

7.1.3.3 Measurement techniques for electrodeformation

To measure electrodeformation there is a technique that builds a bimorph actuator and calculates the deformation ratio from the curvature [101]. As this method exaggerates even a 1% strain, it is effective for qualitative

(a) Polyacetylene

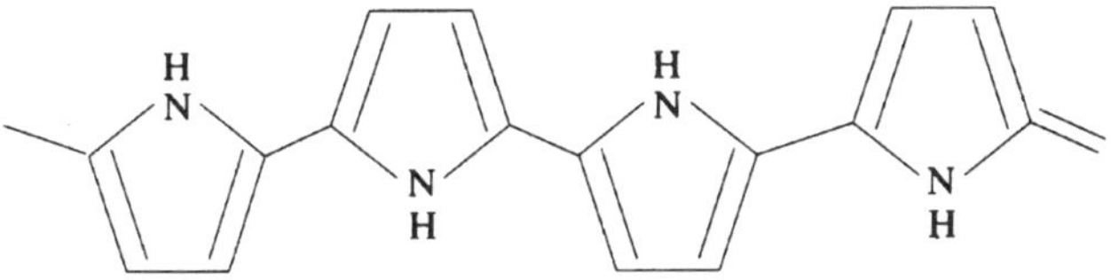

(b) Polypyrrole

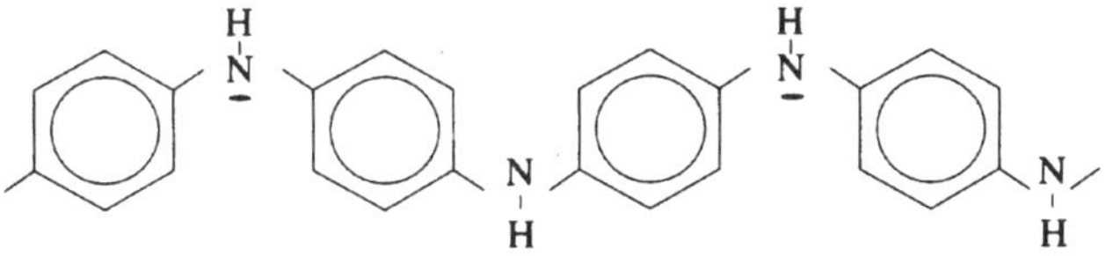

(c) Polyaniline

Basic lucoemeraldine

Fig. 1 Molecular structures of representative conducting polymers.

measurement or demonstration purposes. Other methods include one that uses chemical balance to measure deformation directly [102] and one that takes out deformation through the pinhole at the bottom of the electrochemical cell and measures the strain and dependence on the load [103, 104]. These methods are quantitative because direct measurement of strain and local dependence is possible. This section discusses mainly the results from a cell with a pinhole.

The experimental procedure used follows here. A platinum electrode is affixed to a rectangular sample (15 mm × 1 mm × 30 μm) and placed in the cell as shown in Fig. 3. A glass fiber is attached to the bottom of the sample and taken out from the pinhole at the bottom of the cell. A small plate is hung at the end of the fiber. By observing movement with a laser tensiometer, sample strain can be obtained. By putting a weight on the plate, load dependence can be measured. The output of the laser

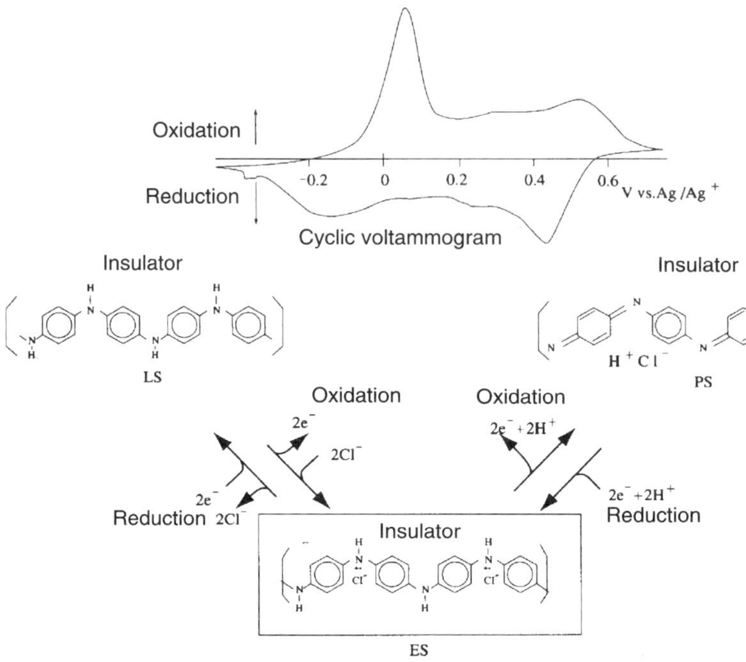

Fig. 2 Cyclic voltammogram and the corresponding structure of polyaniline with respect to the reference silver electrode in 1M/l hydrochloric acid aqueous solution.

tensiometer and CV are sent to a personal computer and data analysis is then performed.

From the CV characteristics and excess response upon step voltage of electrolytic deformation, the dependence of strain and diffusion coefficient (D) on the ionic type can be obtained. The diffusion coefficient can be obtained from the time dependence of the applied electric charge at an early period upon application of a step voltage [102],

$$f = 4D^{1/2}t^{1/2}\pi^{-1/2}d^{-1} \tag{1}$$

where f is the normalized electric charge by the saturation value and d the thickness of the membrane. The diffusion coefficient obtained from this equation is based on the model in which diffusion takes place from the surface of an infinitely large plane to the inside of the membrane of thickness d. Thus, determination of thickness in the porous sample is the problem.

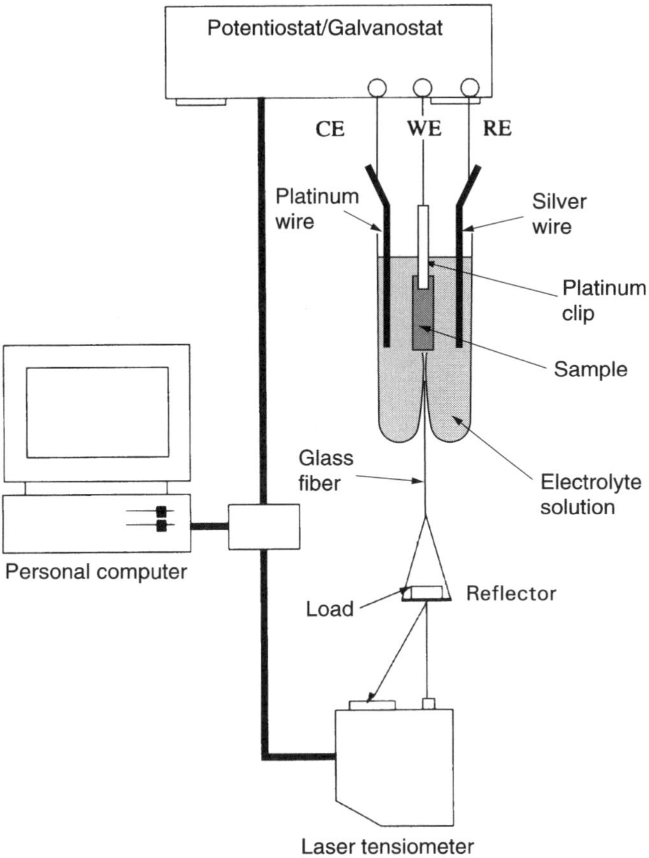

Fig. 3 Experimental apparatus for the electrodeformation.

7.1.3.4 Cyclic voltammograms and deformation behavior

Figure 4 illustrates the cyclic voltammograms (CV) of polyaniline and deformation behavior along the stretching direction. If the potential is scanned from the LS state at −0.2 V (Fig. 4 (**1**)) to the higher potential, the sample starts stretching. When the oxidation current is at maximum, stretching is also at maximum. If it is oxidized further, the stretching becomes gradual. As shown (**2**), when the potential is at +0.5 V, the sample shrinks slightly. If the potential is reversed from the PS state at +0.7 V (**3**), the sample stretches slightly with the reducing current. As shown (**4**), the sample shrinks slightly and returns to its original length < 0.2 V. The deformation behavior is similar to the results obtained by

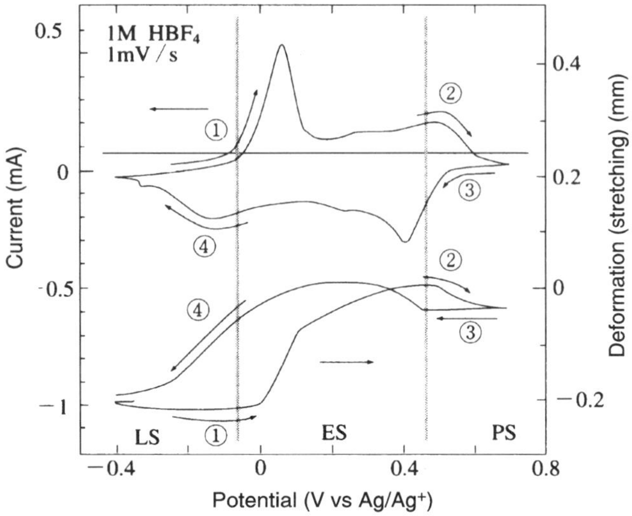

Fig. 4 The CV curves (upper curves) and deformation behavior (lower curves) of polyaniline thin membrane in 1 M/l hydrochloric acid.

microbalance when a quartz generator is used [105]. This result suggests the strong influence of in and out of the dopant for the electrodeformation.

In the LS→ES→PS oxidation process, there is some difference in deformation in the transverse direction of stretched versus unstretched film. However, they stretch homogeneously and in the opposite reduction process the samples shrink.

In the transverse direction, the sample stretches gradually (creep phenomenon) with even a slight load if oxidation-reduction is repeated.

7.1.3.5 Negative ion dependence of strain

From the electric charge, which is injected during the CV process, and the sample mass the oxidation or reduction state can be determined. The oxidation-reduction state is determined as in the insert in Fig. 5 and is expressed by y. Here, if the ES state is expressed as $y = 0, y = -0.5$ is the LS state and $y = 0.5$ is the PS state. The strain is defined as $\Delta l/l_0$ (%) with the length of the sample at the ES state l_0, and the increment of the length Δl.

Figure 5 shows the dependence of the degree of reduction in the electrodeformation of polyaniline in various electrolyte solutions. At $y = -0.2$ the sample shrank 2–3% and the strain depended on the

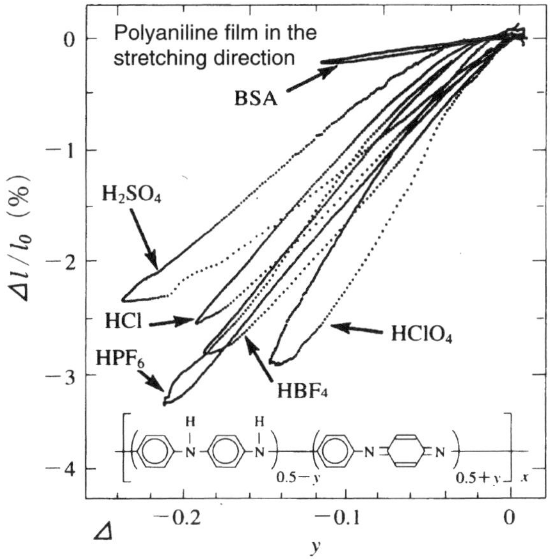

Fig. 5 Relationship between the strain along the stretching direction $\Delta l/l_0$ (%) and the degree of reduction y in various acids (the rate of voltage sweep is 1 mV/s where BSA is 0.1 mV/s).

negative ion. In particular, in benzene sulfonic acid (BSA), the sample shrinks very little. This is because the size of the negative ion, the dopant, is large and this makes it difficult to penetrate the sample. Considering the fact that the sulfate ion is divalent, in general, the heavier the ion is, the larger the deformation. Hysteresis occurs because the sample has not sufficiently reached equilibrium by the potential sweep.

The electrodeformation behavior of polyaniline was investigated in a nonaqueous electrolyte, $LiClO_4$/propylene carbonate. In this case, there is no generation of protons in the secondary oxidation. Since the doping of the negative ion occurs, larger deformation than in the aqueous system is expected. However, in reality, the strain is extremely small and the response time is much slower than in the aqueous system. These results are probably due to the significant difference in the dielectric constant of water and organic solvent and ion mobility.

7.1.3.6 Excess response of electrodeformation

Figure 6 shows the injected electric charge and strain response when the voltage is changed stepwise from 0 V. The time dependence of the

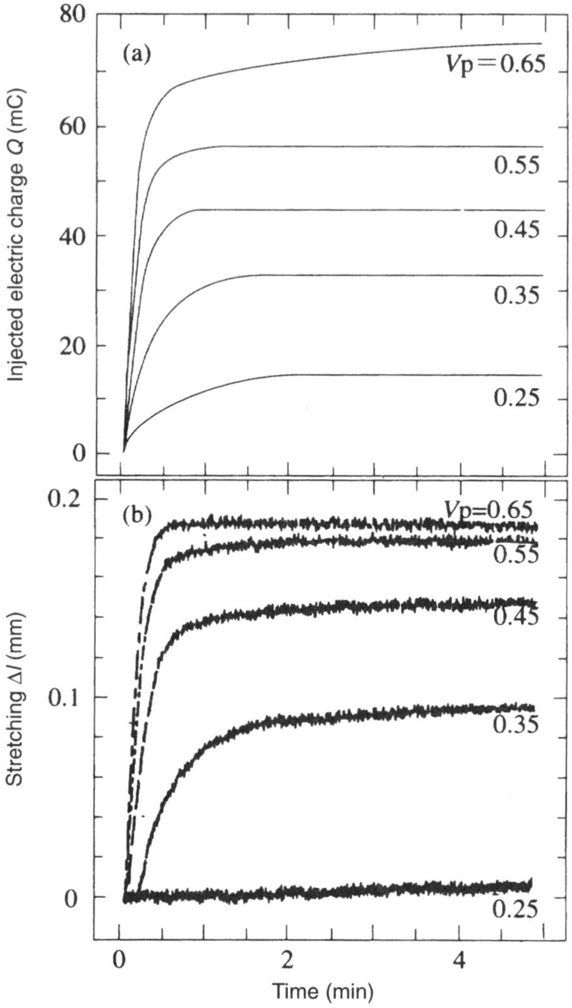

Fig. 6 The response (a) of injected electric charge (Q) and response of stretching (b) when the potential is changed stepwise from zero.

stretching shows a response similar to that of the injected electric charge. The higher the potential, the faster the deformation response. However, if the result is observed in detail, it can be seen that the response of the deformation is slightly delayed from the injected electric charge and tailed. The diffusion coefficient of various dopants can be estimated according to Eq. (1).

Figure 7 shows the relationship between the occupied volume of ions and the strain and diffusion coefficient in various acidic aqueous solutions. As is naturally expected, the larger the volume of the ion, the greater the strain and the smaller the diffusion coefficient. This result supports the contention that electrodeformation is due to the penetration of the voluminous ion. It is noteworthy that the strain approaches a finite value (approximately 1%) when ionic volume is extrapolated to zero. This strongly suggests that the electrodeformation is in part due to the static repulsion and conformation changes of the polycation.

7.1.3.7 Load dependence of strain

Figure 8 shows the degree of shrinkage against tensile load. The direction of stretching and transverse direction shows the expected anisotropy. As the load increases, the shrinking strain shows more hysteresis.

When the load is small, the shrinking strain along the transverse direction is larger. However, when the load is large, the stretching direction has higher strain. The strain does not change significantly until 1–2 MPa (2 MPa corresponds approximately to 200 gf/mm^2). Actual muscle tissue shrinks ≈30% under a force of approximately 30 gf per

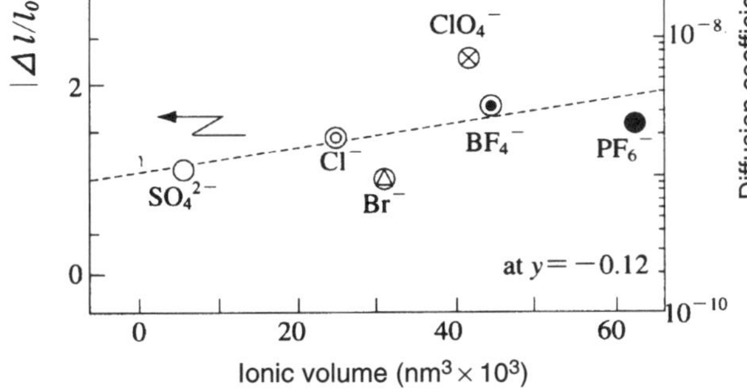

Fig. 7 The relationship between the strain and diffusion coefficient of the ions and the volume of negative ions in various acids.

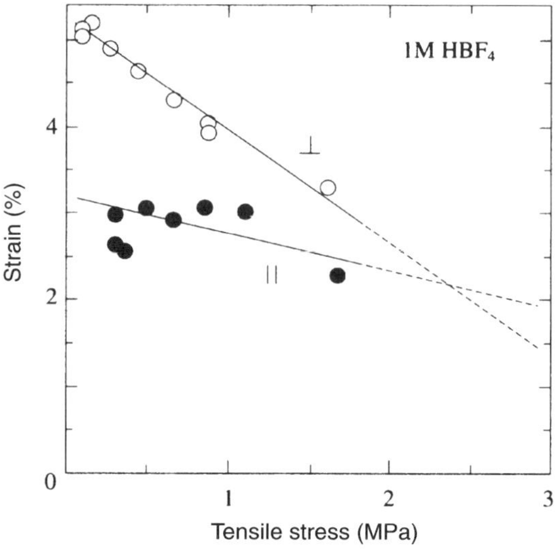

Fig. 8 Tensile load dependence and anisotropy of electrostrain of polyaniline film.

1 mm^2. Their response time is approximately 0.1 s [106]. Judging from this, the strain of conducting polymers is ≈10% of that of muscles while the generated force is one order of magnitude larger.

7.1.3.8 Deformation mechanism

First of all, in the mechanism shown in Fig. 9(a) the polymer deforms due to the insertion of counter ions by electrooxidation. In this case, the volume of the dopant and the formation of solvated ions with their supporting electrolytes lead to an increase in volume that is included in the conducting polymer. In addition to this mechanism, there are other factors that induce electrodeformation. These include a drastic increase in conductivity due to the formation of a polycation by oxidation, the rigidity of the double bonds as compared with the single bonds, and large electron-lattice interaction due to the one dimensionality aspect. In reality, all of these factors are thought to contribute to electrodeformation.

The second reason for deformation involves the change in polymer conformation as shown in Fig. 9(b). In order for the oxidation to cause the conducting polymer to exhibit high conductivity, the conjugation must be extended and the charged chain must pack closely. In other words, if the polymer is highly doped, the polymer becomes rigid and expands.

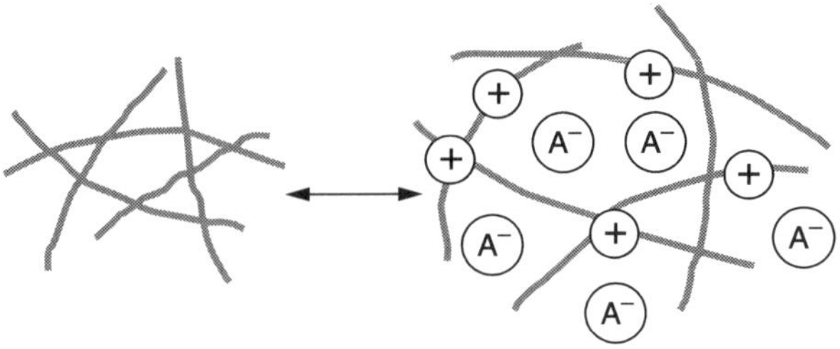

(a) Insertion of bulky ions

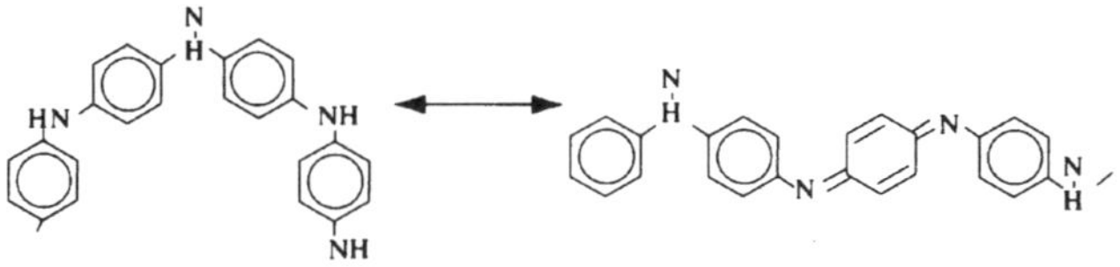

(b) Change in bonding state

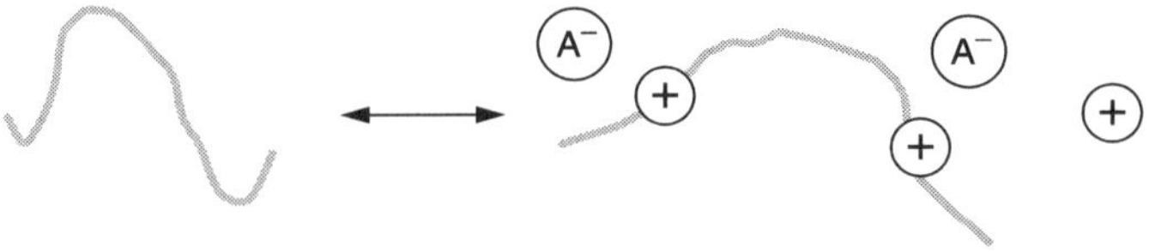

(c) Intramolecular static repulsion

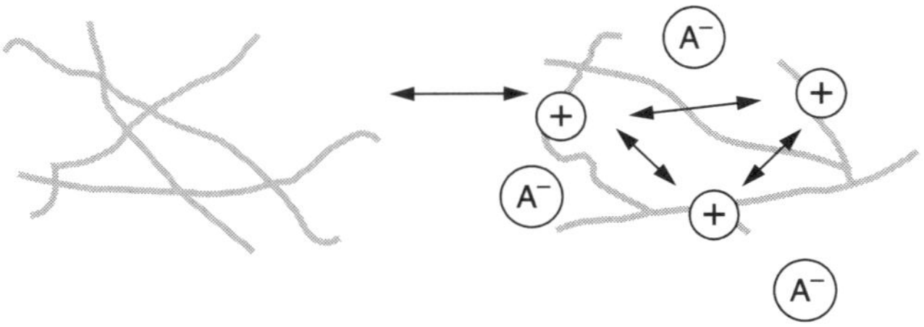

(d) Intermolecular static repulsion

Fig. 9 Mechanisms of electrodeformation.

The third mechanism of electrodeformation is the static repulsion among the same charges seen in polymer gels. This static repulsion is shielded mostly by the negative charges. The charges create inter- and intramolecular repulsive forces. Hence, as shown in Fig. 9(c) and (d), the electric charges on the same polymer chain make the chain rigid and the charges in the neighboring chains increase the intermolecular distance.

7.1.3.9 Electrodeformation of other conducting polymers

The majority of studies on electrodeformation done on substances other than polyaniline used polypyrrole [101, 107–109] and there is also a study on alkylthiophene [110]. For example, there is a report on the bending motion in an electrolyte from a biomorph-type actuator [107]. This actuator was made by electropolymerizing polypyrrole of several micrometer thickness on a platinum electrode, with the membrane affixed to one side of an adhesive tape. Pei and Inganas vapordeposited gold onto rectangular polyethylene to make an evaporated electrode on which pyrrole was then electropolymerized [101]. They made a biomorph-type actuator and experimentally obtained the approximate strain and response time of polypyrrole from the radius of the bending motion. In another study a microactuator on the order of several tens of micrometers was manufactured using a bilayer of gold and polypyrrole; it was capable of motion [108]. The strain and generated force were quantitatively studied with an Instron tester using a rectangular polypyrrole thin film. There is also an example in which the deformation behavior of a biomorph-type actuator was examined using solid and gel polythiophene [110].

7.1.3.10 Application of electrodeformation

Conducting polymers created by electrodeformation are soft, able to operate at low voltage, and show surprisingly large shrinking forces and the ability to maintain position. Many applications can be developed using these properties. As primitive examples, Baughman *et al.* proposed applications using tweezers, microbulbs, and directional guides for optical fibers [111]. When an electrodeformed actuator is used, a 3-layer structure in which a pair of conducting polymers is adhered to both sides of an electrolyte has various advantages. The first advantage is that, since the voltage is applied to a pair of films, if one side is oxidized and stretched, the other side is reduced and shrunken. Thus, the film is bent by twice the force. The second advantage is that the amount of electrolyte can be minimized because the dopant ions move from one side of the film to the other side through the electrolyte. This advantage will be useful when a small and lightweight device is to be constructed. Further, these conducting polymers function as secondary batteries.

Two types of layered actuators using polyaniline were constructed [112, 113]. One uses double-sided cellophane tape on which two polyaniline films are affixed. This system operates in an electrolyte smoothly. Another type is to adhere a polyaniline film onto a single-sided cellophane tape and surround the film with a filter paper in which an electrolyte solution is absorbed. This shell-type actuator is a self-sustaining actuator that operates in air. However, its response time is slow. The seal of this actuator is problematic and requires improvement.

Actuators that use electrodeformation require an electrolyte and can handle only small strain at this time. However, if a solid or gel electrolyte is used, a dry, self-sustaining actuator can be constructed. If saline solution can be used as the electrolyte, the actuator can be used as artificial muscles or in medical devices. Polyaniline and other conducting polymers derived with other alkyl chains dissolve into organic solvents and gel. By incorporating these gels into systems, it may be possible to improve these systems to a practical level.

7.1.4 Conclusions

Given as an example of a conducting polymer, polyaniline was described for its electrodeformation behavior. The insertion of ions is the primary cause of the deformation. Other mechanisms include static repulsion among polycations and conformational changes due to the delocalization of π-electrons. By optimizing the molecular structure and higher structure of the polymer, static repulsion and conformational changes can be utilized to generate greater strain and force.

7.2 INFORMATION CONVERSION PROPERTY

MASAYOSHI WATANABE

7.2.1 Gels and Information Conversion

One of the functions of gels, their information conversion property, will be discussed in this section. Applications of polymer gels extend into the industrial products, civil engineering and construction, agricultural, chemical, machinery and electronics, life sciences, medical and pharmaceutical, and food areas. Currently those properties that are actually used have to do with the ability of gels to absorb and hold solvents.

However, when gels are evaluated from a microscopic point of view, they provide a large amount of information and functions. Proteins and DNA, which are the basis for life, are the foremost examples of functional polymers. They possess certain stereoregularities and exhibit remarkable functions. In order to study the mechanisms of these functions in biopolymers, gels must be crosslinked to form networks. They display microscopic properties and interactions among polymers or solutes and solvents in the form of changes in degree of swelling and anomalous behavior. In fact, it has been found by recent research that a molecular folded gel (highly crosslinked resin) selectively adsorbs certain materials [114–117]. A gel made of a copolymer of various monomers exhibits various certain degrees of swelling and forms stable phases [118–120]. These findings provide great insights into the molecular recognition mechanisms of live beings as well as specific structure formation of biopolymers.

Let us consider the information conversion function of gels. The degree of swelling of gels greatly changes due to the composition of an external solution, pH, ionic type and ionic strength, chemical substance, temperature, light, and electric field. This phenomenon can be regarded as the conversion of chemical and physical stimuli information into information called volumetric change. However, futuristic attempts to use gels to construct censors and molecular switches will be explored.

7.2.2 Information Conversion by Swelling and Shrinking of Gels

A gel changes osmotic pressure by using the physicochemical changes of the external environment. As a result, the gel swells and shrinks by absorbing and expelling a solvent. Following this change in degree of swelling, not only are there volumetric changes but also changes in concentration of solute and solvent and the diffusion coefficient, and furthermore a change in polarity of the gel networks also occurs. In particular, poly(N-isopropylacrylamide) gel (PNIPAAm gel) shows an interesting property. As the temperature increases in pure water, the gel changes its volume from swelling state to shrinking state (phase transition) at $\approx 33^{\circ}C$ [121]. This phenomenon is considered to be strongly influenced by the following two effects:

1) hydrophobic interaction with the PNIPAAm gel; and
2) hydrogen bonding with solvent or polymer chains by use of electronegative O or N.

Here, we describe mainly chemical information conversion using the fact that the phase transition temperature of the PNIPAAm gel changes either by interaction with a solvent or interaction between an additive and a gel. This can be induced by the addition of a small molecular weight substance. For materials that showed large changes in phase transition behavior, the relationship between the concentration of the additive and the limiting current will be investigated using the gel treated microelectrode described in Volume I, Chapter 3, Section 2.7. Based on this relationship, converting chemical information into electrical information using the phase transition of gels is possible.

Table 1 lists the phase transition temperature of a PNIPAAm gel when various materials are added to the water in which the gel is immersed [122]. In this table, the degree of swelling change is designated as (C) or (D), dependent upon whether it is continuous or discontinuous, respectively. To date, the effect of additives on the phase transition temperature of PNIPAAm gels has been studied using both inorganic and organic salts [122–124].

The characteristic most commonly responsible for decreasing the phase transition temperature is that the compounds are ionic (inorganic salts (Nos. 1–14), sodium salts of organic acids (Nos. 7–12), and amino acids (Nos. 19–23)). The reduction is observed even with a relatively hydrophobic compound like 1-phenylurea (No. 18). On the other hand, increase in phase transition temperature was observed in urea derivatives (Nos. 14–17), which exhibit strong hydrogen bonding.

Factors affecting phase transition temperature include the changes in hydrophobic interaction caused by the structural changes of water;

Table 1 Changes in phase transition temperature of poly(N-isopropylacrylamide) in water by addition of various compounds.

No.	Compounds added	Concentration (M)	Phase transition temperature (°C)	Discontinuous (D) or continuous (C)
0	None	0	33.2	D
1	LiCl	1	26.4	D
2	NaCl	1	20.6	D
3	$MgCl_2$	1	21.0	D
4	NH_4Cl	1	25.2	D
5	NaSCN	1	37.6	C
6	Sodium 2-hydroxy benzoate	0.1	34.5	C
7	Sodium 3-hydroxy benzoate	1	31.8	D
8	Sodium 4-hydroxy benzoate	1	27.5	D
9	Monosodium maleic acid	0.25	31.5	C
10	Disodium maleic acid	0.1	29.1	C
11	Monosodium fumaric acid	0.1	30.0	D
12	Disodium fumaric acid	0.1	30.0	D
13	Urea	2	31.6	D
14	1-Methylurea	1	34.3	C
15	1-Ethylurea	1	33.3	D
16	1,1-dimethylurea	1	36.0	D
17	1,3-dimethylurea	2	37.8	D
18	1-Phenylurea	0.02	31.8	D
19	Glycine	1	21.5	C
20	Alanine	0.1	32.5	D
21	Valine	0.1	33.0	D
22	Asparagine (amide of aspartic acid)	0.1	32.3	D
23	Phenylalanine	0.04	33.3	D

changes in the hydrogen bonding between gel networks as a result of the hydrogen bonding between the gel and the compound, and the adsorption of the compound by the gel networks via hydrophobic interaction.

The addition of an ionic compound (structure-forming ion) (which showed reduction in the phase transition temperature) is thought to cause a reduction in the hydrophobicity of the gel itself. This is because these compounds participate in ionic hydration and the water is expelled from the swollen gel. This can be considered a kind of salting-out effect. The strength of this salt precipitation is organized by the Hoffmeister series. For the inorganic salt in Table 1, the Hoffmeister series for the positive ions is $Li^+ > Na^+ > NH_4^+ > Mg^{2+}$, which compares with a phase transition temperature order of $Na^+ > Mg^{2+} > NH_4^+ > Li^+$. The negative ions followed according to the Hoffmeister series. Regarding the effect of inorganic salt on the phase transition temperature of the PNIPPAm gel, the effect of the anion is stronger than that of the cation. By using the B coefficient of the Jones-Dole equation, which expresses the strength of the interaction between ion and solvent, the magnitude of the B coefficient of the anions and phase transition temperature is proportional to each other [123]. In aqueous solutions, a good relationship between the decrease in the lower critical solution temperature (LCST) of poly(ethylene) oxide and the reduction of the phase transition temperature of the PNIPAAm gel upon addition of a salt has been reported [124].

On the other hand, the addition of strongly hydrogen bonded urea derivatives weakens hydrogen bonding among gel networks as a result of the hydrogen bonding of the urea derivatives with gel or water. This further destroys the structure of water, resulting in the weakening of the hydrophobic interaction and increase of the phase transition temperature.

The 1-phenylurea, a urea derivative, is strongly hydrophobic and thus it adsorbs onto the gel networks by the hydrogen bonding with the amide bonds of the gel or hydrophobic interaction. As a result, phase transition falls by strengthening the hydrophobicity of the gel itself.

Similarly, the phenomenon of lowering the phase transition temperature is observed with tetrapentylammonium chloride [123].

As representative examples, the addition of NaCl increases the phase transition temperature, and the addition of 1,3-dimethylurea decreases it. This is shown in Fig. 1 and the swelling curves of PNIPAAm gels are displayed [122]. If the phase transition temperature shifts upon the addition of a certain compound, the degree of swelling of the gel changes at different concentrations of the compound at a constant temperature. Electrochemical measurement of the redox molecule using the gel-treated microelectrode creates the possibility that the change in the degree of swelling upon the addition of a compound can be converted to a change in the current [125, 126]. As stated in Volume 1, Chapter 3, Section 2.7, the limiting current of this measurement is proportional to the product of the diffusion coefficient and concentration of the redox molecule. Hence, change in the degree of swelling is expected to influence these behaviors.

Figure 2 shows the NaCl concentration dependence on the degree of swelling of a PNIPAAm gel at 25°C [122]. This gel shrinks as the NaCl concentration increases. The transition from the swelling phase to the shrinking phase is observed around a NaCl concentration of 0.7 M. Therefore, measurement of the limiting current could be done with the

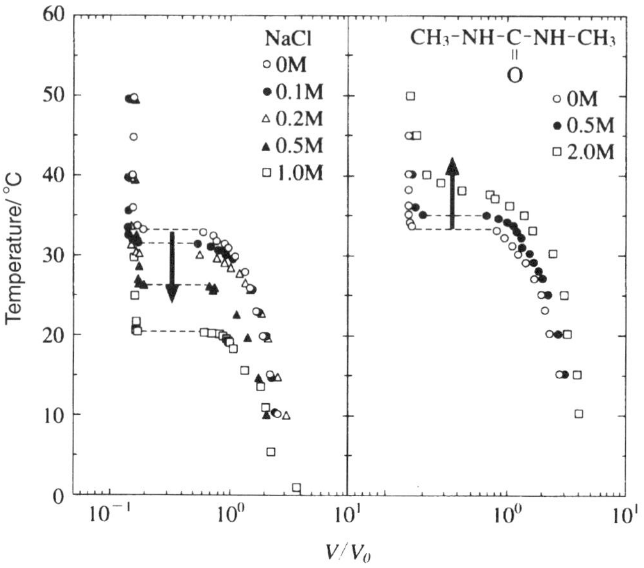

Fig. 1 Effect of the addition of NaCl and 1,3-dimethylurea on the temperature dependence of the degree of swelling of poly(N-isopropylacrylamide) gels.

gel-treated microelectrode. As the NaCl concentration increases change in the limiting current is observed. There is no current response observed on addition of 0.8 M NaCl. This limiting current is also shown in Fig. 2. The reduction of the degree of swelling of the gel and the limiting current correspond well. In particular, a sudden reduction is observed around 0.7 M where the gel shows the phase transition.

Reduction in the phase transition temperature of PNIPAAm gels can be done by using NaCl. Similar measurements were made—this time at 35°C—with 1,3-dimethylurea, the substance used to increase the phase transition temperature of a gel. Figure 3 shows the 1,3-dimethylurea concentration/degree of swelling of the PNIPAAm gel [122]. The gel swells as the concentration of 1,3-dimethylurea increases. The transition from the shrinking phase to the swelling phase is observed at around 0.5 M. Similarly, the limiting current as a result of electrochemical measurements is shown in Fig. 3. When the gel swelled, the limiting current increased. However, when the concentration is around 0.5 M (at which the gel exhibits transition) the limiting current showed a minimum. This unusual phenomenon can be explained by the overlapped effect of

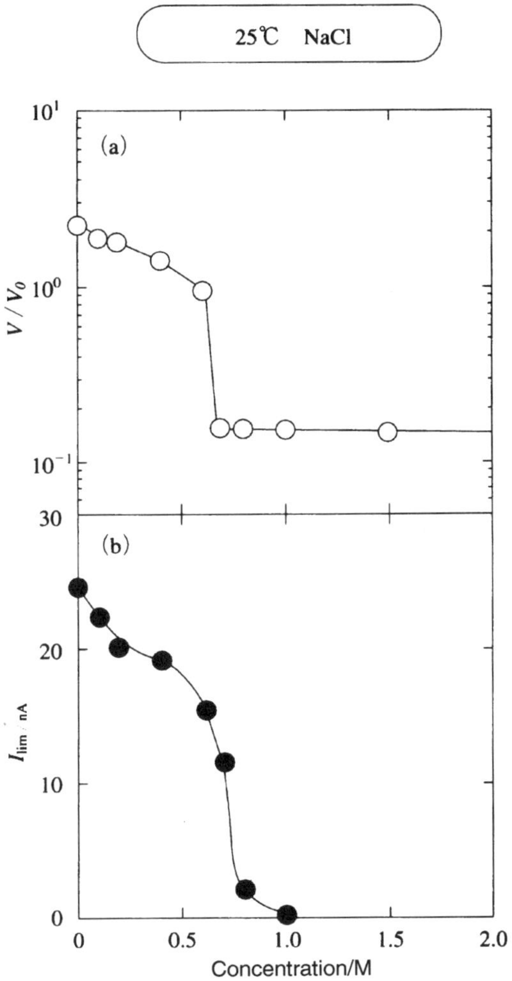

Fig. 2 (a) NaCl concentration dependence of the degree of swelling of a poly(N-isopropylacrylamide) gel; (b) NaCl concentration dependence of the limiting current of 5 mM $[Ru(NH_3)_6]Cl_3$ detected by the poly(N-isopropylacrylamide) gel-treated microelectrode.

increased redox molecules in the shrinking phase and the effect of greater diffusion coefficient in the swelling phase [122].

These results show that the degree of swelling of PNIPAAm gel changes with a change in concentration of the additive. Accordingly, it is confirmed that this change is reflected on the limited current that is

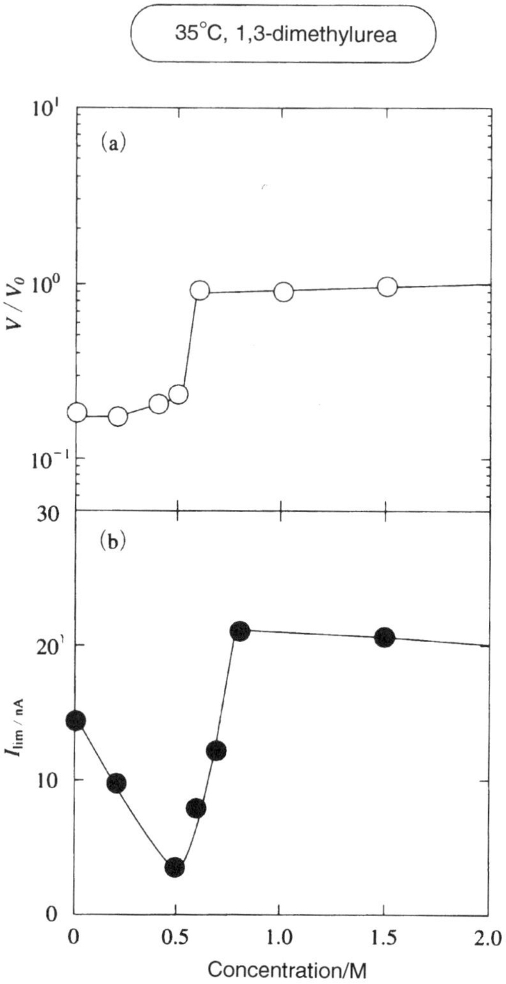

Fig. 3 (a) 1,3-Dimethylurea concentration dependence of the degree of swelling of a poly(N-isopropylacrylamide) gel; (b) 1,3-dimethylurea concentration dependence of the limiting current of 5 mM $[Ru(NH_3)_6]Cl_3$ detected by the poly(N-isopropylacrylamide) gel treated microelectrode.

obtained by the gel-treated microelectrode. Hence, the conversion of chemical information into an electrical signal has been demonstrated.

Next, we introduce an example for which the phase transition temperature is electrochemically controlled [127]. The PNIPAAm gel is electrically inactive. However, by copolymerizing with an electrically active biphenylferrocene, it will be an electrically active gel. The oxidation-reduction potential of the ferrocene site in the gel is approximately 200 mV against the NaCl-saturated calomel electrode (SSCE). Therefore, the ferrocene site in the gel is a hydrophobic ferrocene state below this potential and a hydrophilic (cationic) ferrocenium state above this potential. (See Fig. 4.) The degree of swelling at the reduced state changed significantly in the range from 15–30°C. This gel had the characteristic behaviors of PNIPAAm-type gels—low-temperature swelling and high-temperature shrinking. This gel was fixed on one side of an evaporated gold electrode on a quartz frequency generator. The impedance of the quartz frequency generator was measured as the potential of this electrode was changed. The equivalent circuitry of the quartz frequency generator is shown in Fig. 5 (inset). In this equivalent circuitry the R_1 value reflects gel thickness and viscoelastic properties of the gel on the electrode and its value increases when the loss of energy generated increases. Figure 5 shows the R_1 value when the electrode potential was alternated between +480 mV and −50 mV (based on the SSCE) [127].

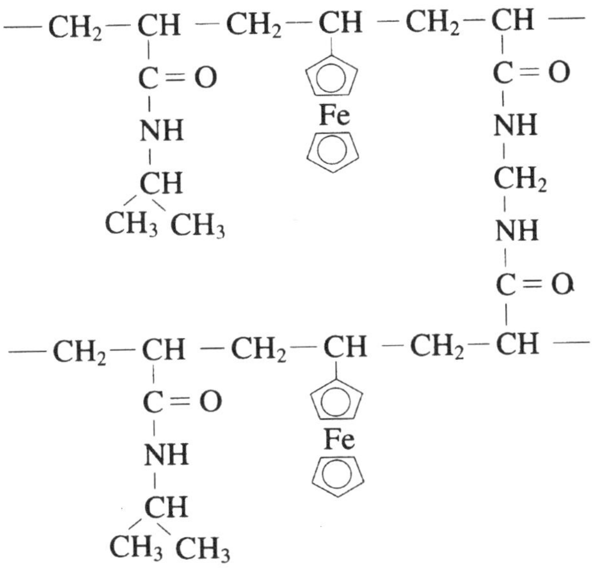

Fig. 4 Structure of the copolymer gel made of vinylferrocene and N-isopropylacrylamide.

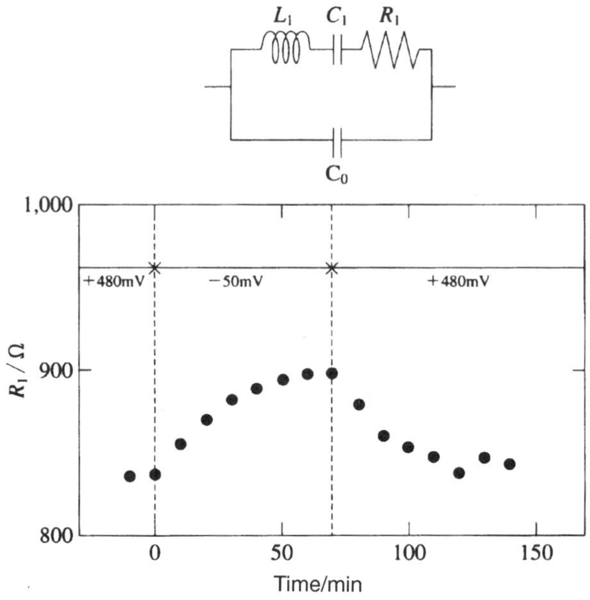

Fig. 5 Applied voltage dependence of the equivalent circuitry parameter R_1 measured by a gel-treated quartz frequency generator.

The R_1 value increased when the gel changed into a reduced state (−50 mV vs SSCE) and it decreased when the gel changed into an oxidized state (+480 mV vs SSCE). It was found that this process is reversible. At the temperature used in Fig. 5, the gel in the reduced state is in a shrunken state. When this gel is electrochemically oxidized, the degree of swelling at the same temperature increases. This change in state, caused electromechanically, is thought to cause the change in the R_1 value.

7.2.3 Information Conversion Using Specific Adsorption of Gels

The molecular imprinting technique is a method to synthesize highly specific adsorption characteristics (molecular recognition ability) [114–117]. This technique makes use of a specific molecule as a male mold (mold molecule, or guest molecule). A monomer, which can specifically interact with the male mold molecule through interactions such as hydrogen bonding, is then polymerized with a crosslinking agent. The male mold molecule is extracted from the highly crosslinked polymer and a supplemental female mold is formed (polymer host, mold polymer) (see Fig. 6(a)). This host synthesis method is in contrast to the multistepwise controlled synthesis that is based on molecular modeling.

This molecular imprinting technique has the potential to provide a molecular recognition molecule simply, quickly, stably, inexpensively, and in a large quantity. The usefulness of the molecular imprinting technique is being evaluated in various application studies. It is not only important to organic chemistry to be able to synthesize such molecular recognition molecules but also for the development of separation agents, analytical agents, medical drugs, catalysts, and censors.

These studies can be classified into two major categories. The first category involves research on selective adsorption agents, which happens

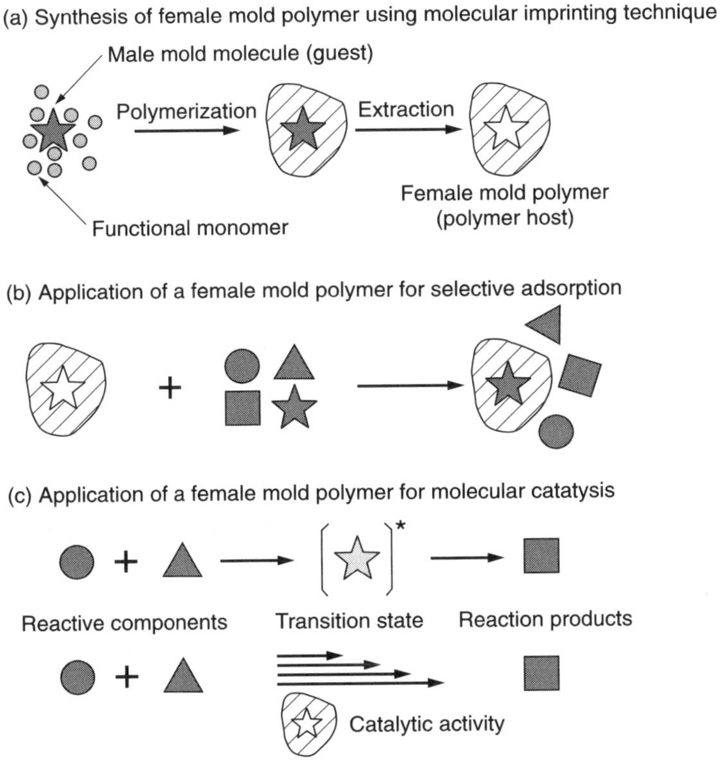

Fig. 6 Molecular imprinting technique as the synthetic method using a molecular recognizing host.

because the mold polymer (female mold molecule) is expected to show specific bonding only for the male mold molecule (see Fig. 6(b)). In this category, there are application studies on artificial antigens for analytical purposes [128], separation agents for asymmetric molecules [129], and separation agents for molecules that are similar [130]. The other category involves research on molecular catalysts [131, 132]. If the transition state of a reaction (intermediate) is used as the male mold molecule, the polymer host may bind with the reaction and exhibit catalytic activity.

Mechanisms of molecular recognition capability (selective adsorptivity) of the host molecule, which is obtained by the molecular imprinting technique, have been studied. It is recognized that the multipoint spatial interaction (represented by hydrogen bonding) between the host and guest is important. The molecular imprinting technique is then viewed as similar to the antigen/enzyme process [133]. This not only indicates the technological importance of the molecular imprinting technique as a new host synthetic method but also leads to fundamental knowledge of the recognition mechanism of biomolecules [133].

This section presents application examples in which the molecular imprinting technique is used in the construction of information conversion elements, including sensors [134–136]. The polymer obtained by the molecular imprinting technique is usually highly crosslinked and thus is different from a gel that is lightly crosslinked in order to show swelling and shrinking responses. Nonetheless, it is meaningful to introduce examples of approaches in the construction of an information conversion element because there now is a study that combines both fields [137].

A polymer membrane obtained by the molecular imprinting technique is used as the sensing layer of an electric effect-type capacitor [134]. The molecular recognition information is detected as the change in the capacitor (C)-voltage (V) curve. Figure 7 depicts the element used in the study. A polymer layer, which is prepared by using a guest molecule, phenylanilinanilide (PAA), is placed on a p-type silicon wafer where an oxidized layer from the evaporated Al layer is on one side.

On top of the polymer is a thin platinum layer with an Al electrode. The measuring solution is in contact with the polymer membrane from the platinum/Al electrode side. Preparation of the female mold polymer is achieved by copolymerizing methacrylic acid and ethyleneglycol dimethacrylate under PAA after introducing the methacrylate group on oxidized silicon using a silane coupling agent. After extracting the guest molecule

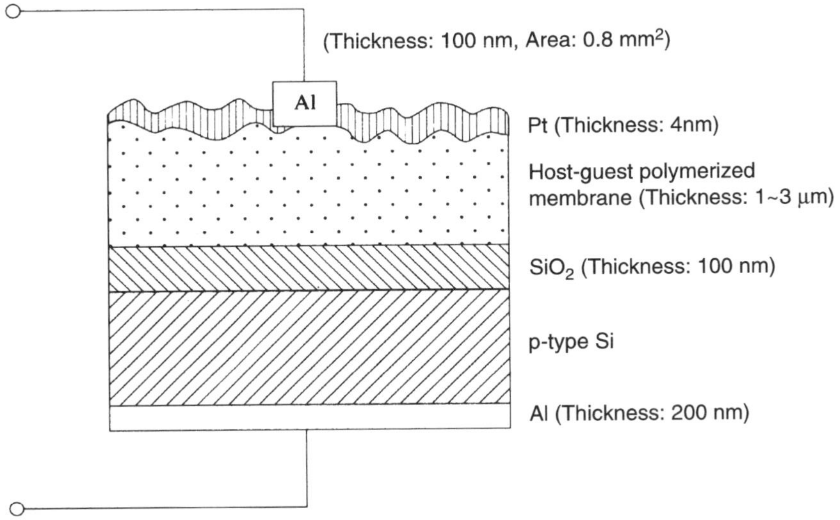

Fig. 7 Structure of electric field effect type capacitor using the host–guest polymerized membrane.

by a mixed solvent of acetic acid and methanol, the aforementioned upper electrode is attached. Figure 8 illustrates the voltage-capacitance curves of this electric field effect-type capacitor. In all cases, blind tests were performed using ethanol including 5% pure water followed by taking measurements after adding the desired compound. If PAA, which is the guest molecule, is added, capacity greatly decreases regardless of the applied voltage as shown in Fig. 8(a). It is thought to be induced by the change in the degree of swelling from the adsorption of the guest molecules onto the molecular imprinting polymerized membrane. When tylosinanilide (TA), which is an analog of guest molecule PAA, is added, similar reduction in capacitance is observed as shown in Fig. 8(b). There was no selectivity between PAA and TA. On the other hand, when phenylalaninol (PA) is added, the reduction in capacitance is small; in particular, there was no reduction of capacitance observed in the vicinity of the applied voltage at 0 V. This is probably because the PA molecule did not adsorb onto the imprinted polymerized membrane.

There is also a report on the application of a polymer obtained by the molecular imprinting technique for amperometric sensor [135]. Traditional biosensors that use biopolymers such as enzymes, antigens, or

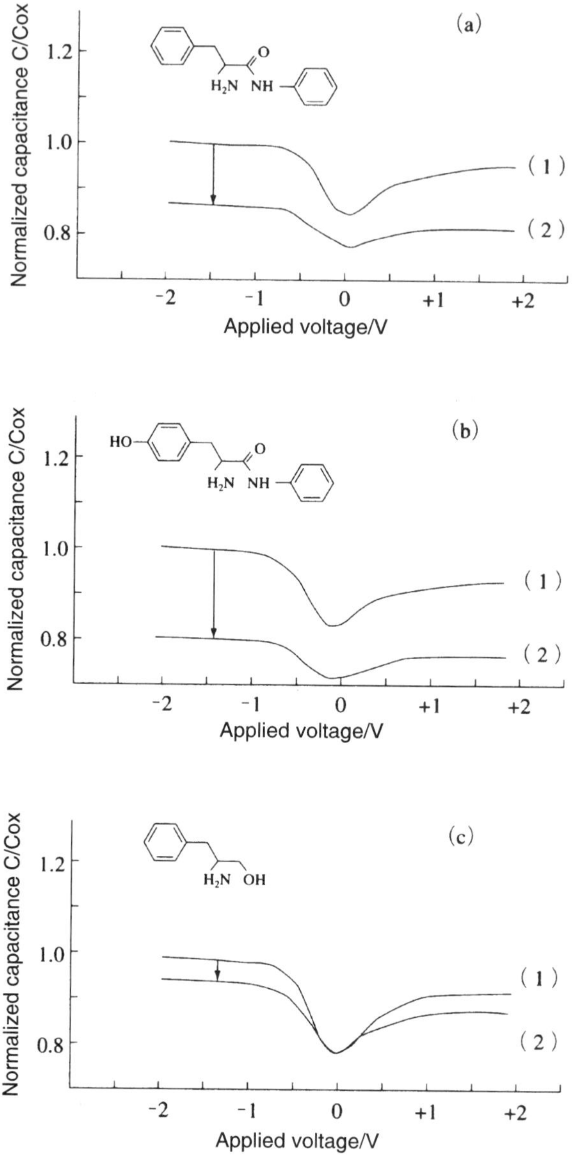

(a) Phenylalanine anilide (PAA), (b) tyrocine anilide (TA), and (c) phenylalaninol (PA)

Fig. 8 Applied voltage dependence of a field effect capacitor made of template polymerized film by phenylalanineanilide in $EtOH_{wt}$ (1) and when 25 mM of the compounds shown in the figure is added.

antibodies have problems including long-term stability, loss of irreversible inactivity at high temperatures or in severe chemical environments, and application environment limits regarding aqueous solutions. Therefore, if a synthetic polymer made by molecular imprinting can be used instead of biopolymers, this would be extremely valuable from the engineering point of view. A morphine sensor will be described here. A platinum electrode is immersed into an agarose solution to which a resin obtained by molecular imprinting using morphine as the guest molecule is dispersed. Epichlorohidrin was used to crosslink the resin. Accordingly, a platinum electrode (M-MIP) on which a molecular imprinting resin-fixed agarose gel was fixed at a 0.5-mm thickness was obtained. For comparison, an electrode on which a molecular imprinting resin made from an L-phenlyalanine anilide is fixed was also prepared. The sensor characteristics were measured electrochemically using these electrodes as the acting electrodes and the Ag/AgCl electrode as the reference electrode. Figure 9 shows the chemical structures of morphine and codeine. Morphine is oxidized at 0.5 V whereas codeine will not be oxidized at this voltage. Figure 10 shows the oxidation current at 0.5 V as a function of time when morphine and codeine are added [135]. When morphine was added at t_m, the current increased initially and reached a steady current i_m in about 2 h. This steady current appears as a result of constant morphine concentration at the electrode surface by achieving the steady state between the sorption of morphine in the molecular imprinting resin and the morphine oxidatively consumed in the platinum electrode. At this point, if codeine is added, the current increased again and then decreased, showing a maximum. This current peak is due to the partial replacement of morphine in the molecular imprinting resin by electrochemically inactive codeine whose chemical

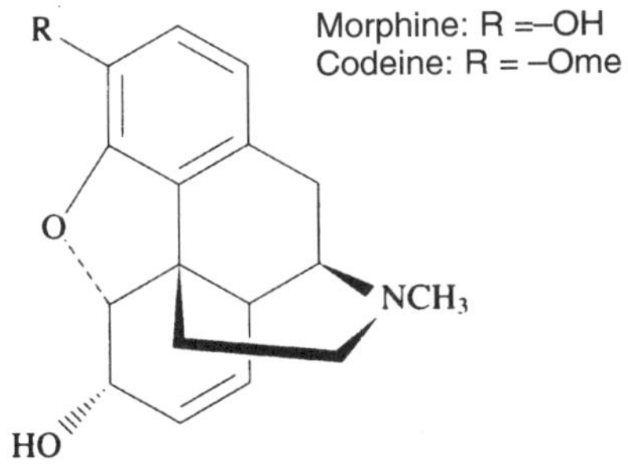

Fig. 9 Chemical structures of morphine and codeine.

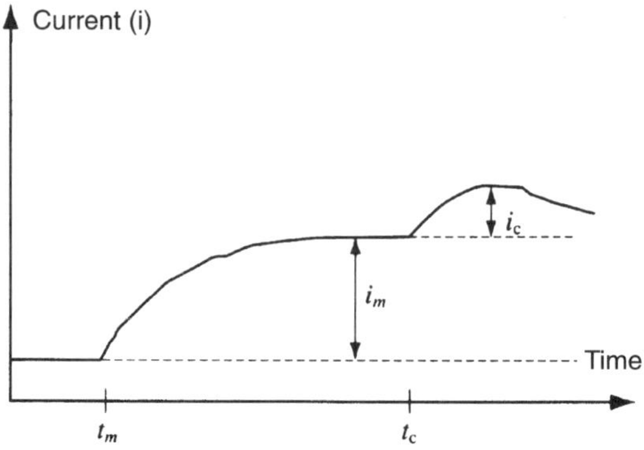

t_m, the time when morphine was added; t_r, the time when codeine was added; i_m, steady current by the addition of morphine; and i_c, current peak by the addition of codeine.

Fig. 10 Time-dependent change of the current at the electrode on which a resin made by molecular imprinting using morphine is fixed.

structure is extremely similar to that of morphine. Figure 11 shows the i_c values as a function of morphine concentration on an M-MIP electrode, an O-MIP electrode, and a platinum electrode (Pt-Ag) on which only agarose is fixed [135]. As expected, the increase in the peak current value of the

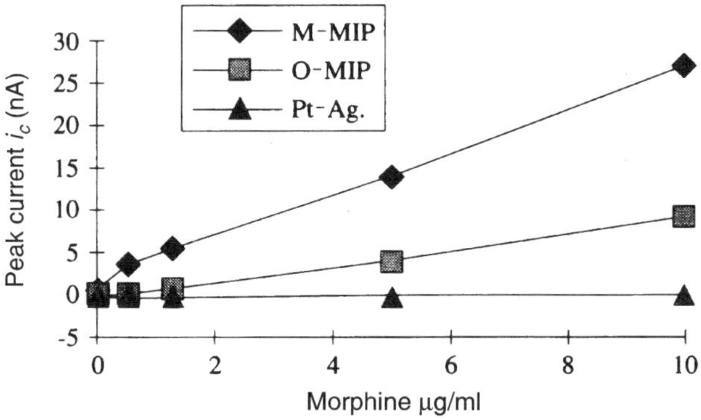

M-MIP, an electrode on which a molecular imprinting resin made by morphine is fixed; O-MIP, an electrode on which a molecular imprinting resin made by L-phenylalanine anilide is fixed; Pt-Ag, an electrode on which only agarose is fixed.

Fig. 11 Relationship between the morphine concentration and the peak current by the addition of codeine.

O-MIP electrode, the reference sensor, was lower than the M-MIP electrode value. With the Pt-Ag electrode, after codeine was added the peak current was not observed. Consequently, it has been shown that quantification of morphine is possible using the M-MIP electrode. The advantage of this sensor is that it can quantitatively determine morphine in the presence of various materials by using codeine as the competitively adsorbing molecule. Although experiments were repeated using other similar compounds, only codeine influenced the desorption of morphine. Furthermore, there is great potential under application conditions in which ordinary biosensors cannot be used because this created sensor is durable at high temperatures, and in the presence of strong acids and bases, and organic solvents and heavy metals (Ag^{+}, Hg^{2+}).

In addition to sensors that can detect capacity and current, optical sensors are also being evaluated (see Fig. 12) [136]. Using dancil-L-phenylalanine, a fluorescing amino derivative, as a guest molecule, molecular imprinting resins made from methacrylic acid or 2-vinylpyridine that is crosslinked by ethylene glycol were synthesized [136]. An optical sensor was manufactured as shown in Fig. 13 using the powdered form of this resin. An optical fiber for light source (350 nm) and an optical fiber for fluorescence detection (485 nm) are both fixed onto this resin. Upon immersing this sensor into a solution of the L-form and of the D-form of dancilphenylalanine, it was found that the molecularly imprinted L-form exhibited stronger fluorescence. This is due to the selective adsorption of the L-form by the molecular imprinting resin (see Fig. 14).

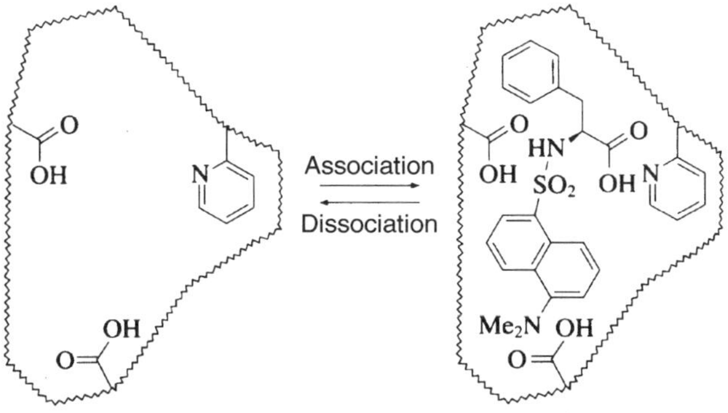

Fig. 12 A schematic diagram of complex formation and dissociation of dancylalanine with a molecular imprinting resin made from dancylphenylalanine.

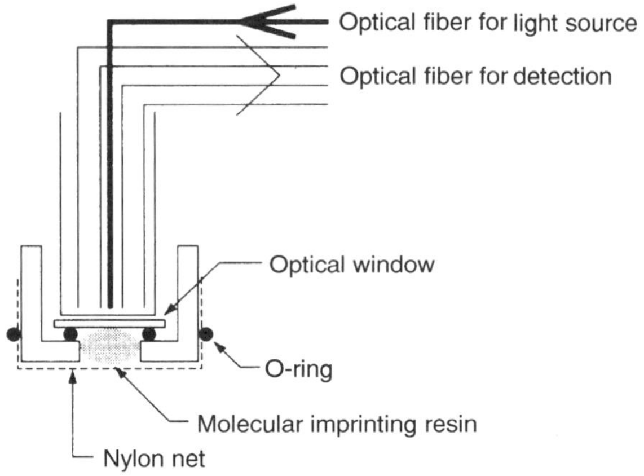

Fig. 13 Structure of an optical fiber-type sensor.

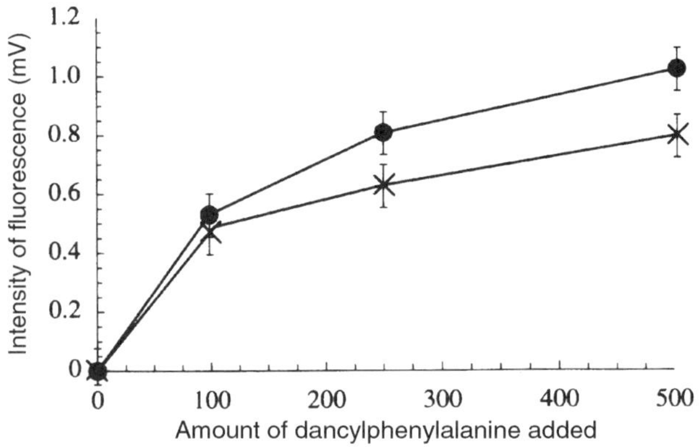

Fig. 14 Enantiomer concentration dependence of optical fiber-type molecular recognition sensor using a molecular imprinting resin made from dancil-L-phenylalanine (•, L-form; ×, D-form).

REFERENCES

1 Osada, Y. (1987). *Konbunshi*. **36**: 353. (This system was proposed by Katchalsky *et al*. and was named a mechanochemical system. This terminology is more appropriately "chemomechanical system" or "chemomechanical reaction" from its concept.)
2 Kuhn, W., Hargitay, B., Katchalsky, A., and Eeizenberg, H. (1950). *Nature* **165**: 514.
3 Osada, Y. (1984). *Sekkei Seizu* **19**: 289.
4 Steinberg, I.Z., Oplatka, H., and Katchalsky, A. (1966). *Nature* **210**: 568.
5 Sussman, M.V. and Katchalsky, A. (1970). *Science* **167**: 45.

6 Committee on Polymer Laboratory, Society of Polymer Science, Japan. (1974). *Functional Polymers*, Chapter 9, Kyoritsu Publ.
7 Study Group on Polymer Complexes. (1983). *Polymer Assembly*, Chapter 9, Gakkai Publ. Center.
8 Osada, Y. and Saito, Y. (1975). *Makromol. Chem.* **176**: 2761.
9 Osada, Y. and Saito, Y. (1976). *J. Chem. Soc. Jpn.* 171.
10 Osada, Y. (1980). *J. Polym. Sci. Polym. Chem. Ed.* **18**: 281.
11 Osada, Y. and Takeuchi, Y. (1981). *J. Polym. Sci., Polym. Lett., Ed.* **19**: 303.
12 Osada, Y. and Koike, M. (1983). *J. Chem. Soc. Jpn.* 812.
13 Osada, Y. and Takeuchi, Y. (1983). *Polym. J.* **15**: 279.
14 Bae, Y.H., Okanao, T., and Kim, S.W. (1988). *Makromol. Chem. Rapid Commun.* **9**: 185.
15 Ilman, F., Tanaka, T., and Kokufuta, E. (1991). *Nature* **349**: 400.
16 Horne, R.A., Almeida, J.P., Day, A.F., and Yu, N. (1971). *J. Colloid Interface Sci.* **35**: 77.
17 Schild, H.G. (1992). *Prog. Polym. Sci.* **17**: 163.
18 Hirokawa, Y. and Tanaka, T. (1984). *J. Chem. Phys.* **81**: 6379.
19 Hirotsu, S., Hirokawa, Y., and Tanaka, T. (1987). *J. Chem. Phys.* **87**: 1392.
20 Itoh, S. (1989). *Kobunshi Ronbunshu* **46**: 437.
21 Itoh, S. (1990). *Kobunshi Ronbunshu.* **47**: 467.
22 Yoshida, M., Yang, J.-S., Kumakura, M., Hagiwara, M., and Katakai, R. (1991). *Eur. Polym. J.* **27**: 997.
23 Yoshida, M., Suzuki, Y., Tamada, M., Kumakura, M., and Katakai, R. (1991). *Eur. Polym. J.* **27**: 493.
24 Urry, D.W., Harris, R.D., and Prasad, K.U. (1988). *J. Am. Chem. Soc.* **110**: 3303.
25 Matsuda, A., Sato, J., Yasunaga, H., and Osada, Y. (1994). *Macromolecules* **27**: 7695.
26 Osada, Y. and Matsuda, A. (1995). *Nature* **376**: 219.
27 Suzuki, M. (1989). *Kobunshi Ronbunshu* **46**: 603.
28 Umemoto, S., Okui, N., and Sakai, T. (1991), in *Polymer Gels–Fundamentals and Biomedical Applications*, D. DeRossi *et al.*, eds., New York, Plenum Press, p. 257.
29 Oguri, H., Kitano, S., Nabeshima, Y., Tsuruta, T., Kataoka, K., and Sakurai, Y. (1990). *Proc. Soc. Polym. Sci.* **39**: 628.
30 Kitano, S., Koyama, Y., Kataoka, K., Okano, T., and Sakurai, Y. (1992). *J. Controlled Release* **19**: 162.
31 Kawahara, J., Oomori, K., Hattori, S., and Kawamura, M. (1991). *Proc. Soc. Polym. Sci., Jpn.*, **40**: 629.
32 Irie, M. (1993). *Advances in Polymer Science, Responsive Gels: Volume Transition* II, Berlin: Springer-Verlag, pp. 110, 49.
33 Ishihara, K., Hamada, N., Kato, S., and Shinohara, I. (1989). *J. Polym. Sci. Chem. Ed.* **22**: 121.
34 Seki, T. and Tamaki, T. (1993). *Chem. Lett.* 1739.
35 Irie, M. and Kunwatchakum, D. (1985). *Makromol. Chem. Rapid Commun.* **5**: 829.
36 Irie, M. and Kunwatchakum, D. (1986). *Macromolecules* **19**: 2476.
37 Irie, M. (1986). *Macromolecules* **19**: 2890.
38 Mamada, A., Tanaka, T., Kunwatchakun, D., and Irie, M. (1990). *Macromolecules* **23**: 1517.
39 Suzuki, S. and Tanaka, T. (1990). *Nature* **346**: 345.
40 Osada, Y. and Hasebe, M. (1985). *Chem. Lett.* 1285.
41 Kishi, R., Hasebe, M., Hara, M., and Osada, Y. (1990). *Polym. Adv. Technol.* **1**: 19.

42 Kishi, R. and Osada, Y. (1989). *J. Chem. Soc. Faraday Trans.*, 1, **85**: 655.
43 Sawahata, K., Gong, J.P., and Osada, Y. (1995). *Macromol. Rapid Commun.* **16**: 713.
44 Oguro, K. (1995). *Zairyo* **44**: 681.
45 Shinohara, K. and Aizawa, M. (1989). *Kobunshi Ronbunshu* **46**: 703.
46 Umemoto, S., Okui, T., and Sakai, T. (1990). *Proc. Soc. Polym. Sci., Jpn.* **39**: 624.
47 Hirai, T., Nemoto, H., Hirai, M., and Hayashi, S. (1994). *J. Appl. Polym. Sci.* **53**: 79.
48 Ohnishi, S. and Osada, Y. (1991). *Macromolecules* **24**: 3020.
49 Ohnishi, S. and Osada, Y. (1991). *Macromolecules*. **24**: 6588.
50 Tanaka, T. and Filimore, D.J. (1979). *J. Chem. Phys.* **70**: 1214.
51 Kawaguchi, H., Fujimoto, K., and Mizuhara, Y. (1992). *Colloid Polym. Sci.* **270**: 53.
52 Fujiwara, S., Ito, S., and Hirasa, K. (1991). *Report of the Institute for Fiber and Polymer Materials* **167**: 67, 75.
53 Hirasa, K. (1986). *Kobunshi* **35**: 1100.
54 Hirasa, K., Morishita, K., Onomura, R., Ichijo, H., and Yamauchi, A. (1989). *Kobunshi Ronbunshu* **46**: 661.
55 Kishi, R., Ichijo, H., and Hirasa, O. (1993). *J. Intel. Mater. Struc.* **4**: 533.
56 Kokufuta, E. and Tanaka, T. (1991). *Macromolecules* **24**: 1605.
57 Kokufuta, E., Zhang, Y.-Q., and Tanaka, T. (1991). *Nature* **351**: 302.
58 Kokufuta, E., Ogane, O., Ichijo, H., Watanabe, S., and Hirasa, O. (1992). *J. Chem. Soc. Chem. Commun.* 416.
59 Kishi, R., Shishido, M., and Tazuke, S. (1990). *Macromolecules* **23**: 3779.
60 Kishi, R., Sisido, M., and Tazuke, S. (1990). *Macromolecules* **23**: 3858.
61 Kishi, R., Suzuki, Y., Ichijo, H., and Hirasa, O. (1994). *Chem. Lett.* 2257.
62 Umezawa, K. and Osada, Y. (1987). *Chem. Lett.* 1795.
63 Hirotsu, S. (1985). *Jpn. J. Appl. Phys.* **24** (Supple. 24-2, Proc. 6th Int. Meet. Ferroelectro.): 396.
64 Minoura, N., Aiba, S., and Fujiwara, Y. (1986). *J. Appl. Polym. Sci.* **31**: 1935.
65 Minoura, N., Aiba, S., and Fujiwara, Y. (1986). *Kobunshi Ronbunshu* **43**: 803.
66 Minoura, N., Aiba, S., and Fujiwara, Y. (1993). *J. Am. Chem. Soc.* **115**: 5902.
67 Tanaka, T. (1978). *Phys. Rev. Lett.* **40**: 820.
68 Yamamoto, K. and Maruyama, K. (1986). *Muscles*, Kagaku Dojin.
69 Tomita, T. and Sugi, H. (1986). *Physiology of Muscles*, New Physiology Science Series. vol. 4, Igaku Shoin.
70 Hirawa, O. (1986). *Kobunshi* **35**: 1100.
71 Hirasa, K., Morishita, K., Ono, R., Ichijo, H., and Yamauchi, A. (1989). *Kobunshi Ronbunshu* **46**: 661.
72 Kishi, R., Hirasa, K., and Ichijo, H. (1995). *Zairyo Kagaku* **32**: 69.
73 Mambu, A. (1982). *Tokkyo Kaiho* 130543.
74 Suzuki, M. (1989). *Kobunshi Ronbunshu* **46**: 603.
75 For example, Umemoto, S. (1991). *Polymer Gels*, New York: Plenum, p. 257.
76 Hamlen, R.P., Kent, C.E., and Shafer, S.N. (1965). *Nature* **206**: 1150.
77 Fragala, A., Enos, J., LaConti, A., and Boyack, J. (1972). *Electrochimica Acta* **17**: 1507.
78 Grodzisky, A.J. and Shoenfeld, N.A. (1977). *Polymer* **18**: 435.
79 Osada, Y. and Hasebe, M. (1985). *Chem. Lett.* 1285.
80 Kishi, R. and Osada, Y. (1989). *J. Chem. Soc. Faraday Trans.* **85**: 655.
81 DeRossi, D.E., Chiarelli, P., Buzzigoli, G., and Domenici, C. (1986). *Trans. Am. Soc. Artif. Intern. Organs* **32**: 157.
82 Tanaka, T., Nishio, I., Sun, S.T., and Ueno, S. (1982). *Science* **218**: 467.
83 Hirotsu, S. (1985). *J. Appl. Phys., Jpn.* **24**: 396.

84 Shiga, T. and Kurauchi, N. (1985). *Polym. Prepr., Jpn.* **34**: 508.
85 Shiga, T. and Kurauchi, T. (1990). *J. Appl. Polym. Sci.* **39**: 2305.
86 Hirose, Y., Giannetti, G., Marquardt, J., and Tanaka, T. (1992). *J. Phys. Soc., Jpn.* **61**: 4085.
87 Doi, M., Matsumoto, M., and Hirose, Y. (1992). *Macromolecules* **25**: 5504.
88 Irie, M. and Kunwatchakun, D. (1986). *Macromolecules* **19**: 2476.
89 Shiga, T., Hirose, Y., Okada, A., and Kurauchi, N. (1989). *Kobunshi Ronbunshu* **46**: 709.
90 Shiga, T., Hirose, Y., Okada, A., and Kurauchi, T. (1993). *J. Intell. Mater. Struct. Systems* **4**: 553.
91 Shiga, T., Hirose, Y., Okada, A., and Kurauchi, T. (1993). *J. Appl. Polym. Sci.* **47**: 113.
92 Osada, Y., Okuzaki, H., and Hori, H. (1992). *Nature* **355**: 242.
93 For example, Oguro, K. and Yasuzumi, K. (1995). *Zairyo Kagaku* **32**: 70.
94 Hirai, T., Nemoto, H., Hirai, M., and Hayashi, S. (1994). *J. Appl. Polym. Sci.* **53**: 79.
95 Hirai, M., Hirai, T., Sukumoda, A., Nemoto, H., Amemiya, Y., Kobayashi, K., and Ueki, T. (1995). *J. Chem. Soc. Faraday Trans.* **91**: 473.
96 Hirai, T. (1995). *Zairyo Kagaku* **32**: 59.
97 Kaneto, K., Hayashi, S., Ura, S., and Yoshino, K. (1985). *J. Phys. Soc. Jpn.* **54**: 1146.
98 Kaneto, K. (1988). *Kobunshi* **37**: 526.
99 Shacklette, L.W., Wolf, J.F., Gould, S., and Baughman, R.H. (1988). *J. Chem. Phys.* **88**: 3955.
100 Huang, W.S. and MacDiarmid, A.G. (1993). *Polymer* **34**: 1833.
101 Pei, Q. and Inganas, O. (1993). *Synthetic Metals* **55–57**: 3718.
102 Herod, T.H. and Schlenoff, J.B. (1993). *Chem. Mater.* **5**: 951.
103 Kaneto, K., Kaneto, M., and Takashima, W. (1995). *J. Appl. Phys., Jpn.* **34**: L837.
104 Takashima, W., Fukui, M., Kaneko, M., and Kaneto, K. (1995). *J. Appl. Phys., Jpn.* **34**: 3786.
105 Daifuku, H., Kawagoe, T., Yamamoto, N., Ohsaka, T., and Oyama, N. (1989). *J. Electroanal. Chem.* **274**: 313.
106 Alexander, R.M. (1992). *Living Organisms and Movement*, Tokyo: Nikkei Science, p. 13.
107 Otero, T.F., Rodriguez, J., Angulo, E., and Santamaria, C. (1993). *Synthetic Metals* **55–57**: 3713.
108 Smela, E., Inagas, O., and Lundstrom, I. (1995). *Science* **268**: 1735.
109 Gandhi, M.R., Murray, P., Spinks, G.M., and Wallace, G.G. (1995). *Synthetic Metals* **73**: 247.
110 Chen, X. and Inganas, O. (1995). *Synthetic Metals* **74**: 159.
111 Baughman, R.H., Shacklette, L.W., Elsenbaumer, R.L., Plichta, E.J., and Becht, C. (1991), in *Electrochemical Actuators Based on Conducting Polymers*, P.I. Lazarev, ed., The Netherlands: Kluwer Academic Publishers, pp. 267–289.
112 Kaneto, K. (1994). *Kobunshi* **43**: 856.
113 Kaneto, K. (1994). *Fiber and Industry* **50**: 628.
114 Shea, K.J. (1994). *Trends Polym. Sci.* **2**: 166.
115 Mosbach, K. (1994). *Trends Biochem. Sci.* **19**: 9.
116 Wulff, G. (1993). *Trends Biotechnol.* **11**: 85.
117 Matsui, J. and Takeuchi, S. (1995). **48**: 1259.
118 Annaka, M. and Tanaka, T. (1992). *Nature* **355**: 430.
119 Tanaka, T. and Yasunaka, M. (1993). *Kagaku* **63**: 124.

120 Tanaka, T. (1995). *Kobunshi* **44**: 8.
121 Hirotsu, S., Hirokawa, Y., and Tanaka, T. (1987). *J. Chem. Phys.* **87**: 1392.
122 Nakayama, D., Akahoshi, T., Sasaki, K., and Watanabe, M. (1995). *Polym. Prepr., Jpn.* **44**: 1618.
123 Inomata, H., Goto, S., Otake, K., and Saito, S. (1992). *Langmuir* **8**: 687.
124 Suzuki, A. (1993). *Adv. Polym. Sci.* **110**: 199.
125 Watanabe, M., Tadenuma, Y., Ban, M., Sanui, K., and Ogata, N. (1993). *J. Intell. Mat. Syst. & Struct.* **4**: 216.
126 Watanabe, M. (1995). *Hyomen Gijyutsu* **46**: 324.
127 Tatsuma, T., Takada, K., Matsui, H., and Oyama, N. (1994). *Macromolecules* **27**: 6687.
128 Vlatakis, G., Andersson, L.I., Muller, R., and Mosbach, K. (1994). *Nature* **361**: 645.
129 Sellegren, B., Lepisto, M., and Bosbach, K. (1988). *J. Am. Chem. Soc.* **110**: 5853.
130 Dabulis, K. and Klibanov, A.M. (1992). *Biotech. Bioeng.* **39**: 176.
131 Robinson, D.K. and Mosbach, K. (1989). *J. Chem. Soc., Chem. Commun.* **1989**: 969.
132 Beach, J.V. and Shea, K.J. (1994). *J. Am. Chem. Soc.* **116**: 379.
133 Schultz, P.G., Lerner, R.A., and Benkovic, S.J. (1990). *Chem. Eng. News*, May 28, 26.
134 Hedborg, E., Winquist, F., Lundstrom, I., Andersson, L.I., and Mosbach, K. (1993). *Sensors and Actuators A* **37–38**: 796.
135 Kriz, D. and Mosbach, K. (1995). *Anal. Chim. Acta* **300**: 71.
136 Kriz, D., Ramstrom, O., Svensson, A., and Mosbach, K. (1995). *Anal. Chem.* **67**: 2142.
137 Akahoshi, T., Nakayama, D., and Watanabe, M. (1996). *Polym. Prepr., Jpn.* **45**: 1830.

Section 8
Electrical and Magnetic Properties

YOSHIHITO OSADA

8.1 ELECTRICAL PROPERTIES

8.1.1 Introduction

After Galvani accidentally discovered in the late eighteenth century that he could cause a frog's muscle to contract (twitch) by touching a nerve with a pair of scissors during an electrical storm, there has been tremendous interest in exploring the electrical properties of living creatures. Although this electrical response of muscles is known to be caused by a form of ionic transfer, there has been very little quantitative analysis. Likewise, study of the electrical properties of gels will help add to information on the ionic transfer mechanism of live organisms. It is expected that new functions will be discovered for polymeric gels.

The electrical conductivity of polyelectrolyte solutions has been extensively studied but it is a new area with a very short history. Only recently has it been possible to achieve electrical conductivity that is both accurate and reproducible because the electrical properties of a gel remain dependent upon the preparation methods. It is also due to the experimental difficulties encountered in electrolysis and the electrode polarization effect.

In solid state physics the alternative quadruple method is the primary one chosen for electrical resistivity measurements. This method is excellent because it can measure true resistance and exclude both thermally induced electricity and contact resistance between an electrode and a sample. Unfortunately, this method cannot be used for electrical resistance measurements of gels because gels are structurally heterogeneous and it is not known how far the effect of electrode polarization will extend. The voltage terminal detects polarization resistance. Accordingly, the resistivity value depends significantly on the position of the voltage terminal.

Based on these experimental considerations, the authors succeeded in making accurate measurements of the electrical conductivity of gels. Here the electrical conductivity and other electrical properties of poly(2-acrylamide-2-methylpropane sulfonic acid) (PAMPS) will be introduced.

Resistance, an intrinsic physical quantity of a material, can be used to learn about material properties. A general background on resistivity is provided by the resistivity of various materials at 25°C as given in Table 1. The resistivity of polymer gels is similar to that of silicon and germanium. For example, a PAMPS gel (a strong electrolyte) has $\approx$100 Ωcm, and poly(acrylic acid) gel, a weak electrolyte, has $\approx$5 kΩcm at room temperature. A nonelectrolyte poly(ethylene oxide) (PEO) has $\approx$10 kΩcm.

8.1.2 Electrical Conductivity

8.1.2.1 Measurement techniques

Electrical conductivity measurements of gels are done with the alternating current dipolar method (the variable frequency method) using an LCZ meter. For the same reason as in polyelectrolyte solutions, alternating current is used in order to avoid electrolysis and the polarization effect of solvent molecule orientation.

To eliminate the polarization effect that is due to solvent molecule orientation it is necessary to use black-covered platinum as the electrode.

For an electrical conductivity measurement, a cubed material, approximately 1 cm^3, was used. The gel was sandwiched between electrode plates. The sample was placed in a hot bath to maintain a constant temperature. It is at this point in the experiment that the time dependence of the conductivity should be monitored to ensure thermal equilibrium.

Table 1 Resistivity of various materials (Ωcm).

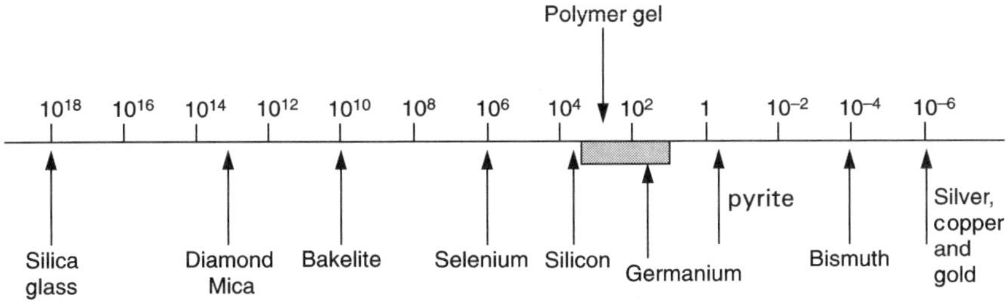

Upon achieving equilibrium temperature, 1 V is applied (electric field strength of 10^{-2}V/m) and conductivity is measured by changing the frequency from 50 Hz to 100 Hz. The result of the measurement is shown in Fig. 1. As can be seen in the figure, there is increased resistivity in the low-frequency region <200 Hz due to electrode polarization. From 200 Hz to 1 kHz, the resistivity is proportional to frequency with $-\frac{1}{2}$ power. By approximating this region with the least squares method and then extrapolating to a higher frequency region, true resistivity can be obtained. This method is the same as that used for polyelectrolytes and the frequency dependence is also the same as the $f^{1/2}$ dependence of electrolyte polarization [1].

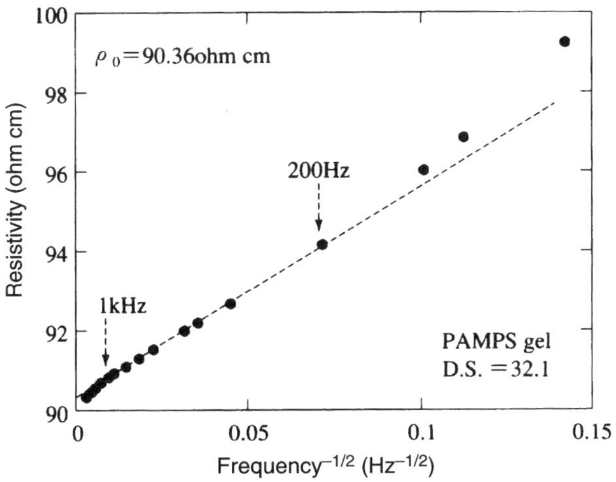

Fig. 1 Frequency characteristics of electrical resistance of PAMPS gel (the straight line indicates $-\frac{1}{2}$ dependence on the frequency).

8.1.2.2 Electrical conductivity of PAMPS systems

In general, the magnitude of electrical conductivity of polyelectrolyte gels depends on carrier concentration (that is, the crosslink density or degree of swelling) and the type of carrier ions in addition to physical factors such as temperature. Therefore, as is similar to the situation with electrolyte sulutions, we use the value of conductivity σ divided by the carrier concentration c, that is, the molar conductivity Λ, where the unit of L is $\mathrm{S\,cm^2\,mol^{-1}}$, σ is S/cm and c is mol/l.

A PAMPS gel is an anionic one that has a sulfonic group as the side chain and proton as a carrier. Figure 2 shows the relationship between resistivity R of a PAMPS gel and the distance between electrodes l. A straight line is obtained, indicating that the polyelectrolyte gel is an insulator that follows the following equation,

$$R = \sigma \frac{l}{S}$$

Thus, in addition to the aforementioned variable frequency method the variable electrode distance method [2] also can be used. From the slope ρ of the $R - l/S$ graph, accurate electrical conductivity can be determined. Using this method, impedance due to the electrode polarization can be completely eliminated.

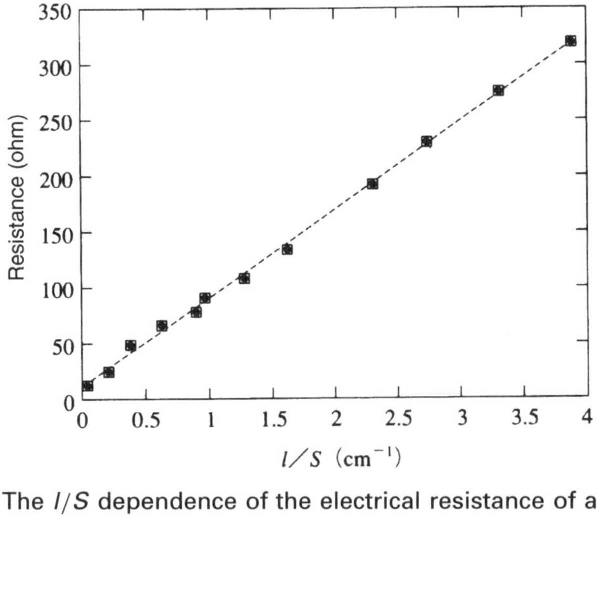

Fig. 2 The l/S dependence of the electrical resistance of a PAMPS gel.

Figure 3 shows the concentration dependence of the electrical conductivity of 2-acrylamide-2-methylpropane sulfonic acid (AMPS), its polymer solution, and the gel at 25°C [3].

As seen in a strong electrolyte solution, the Kohlraush relationship holds in the monomer solution:

$$\Lambda = \Lambda_0 - kc^{1/2}$$

where Λ ($S\,cm^2\,mol^{-1}$) is the molar conductivity, Λ_0 is the equivalent conductivity at infinite dilution, k is a constant, and c (mol/l) is the solution concentration. On the other hand, the polymer and the gel do not follow this rule. The higher the concentration, the greater the molar conductivity. This phenomenon is explained as due to the reduction of friction coefficient by the coiled polymer chains [4]. The molar conductivity of the polymer and the gel is approximately equal. Considering that the ions of the gel are fixed and that the ions contribute to the electrical conductivity of the polymer, this result indicates that the mobility of the counter ions in the gel is greater than in the polymer.

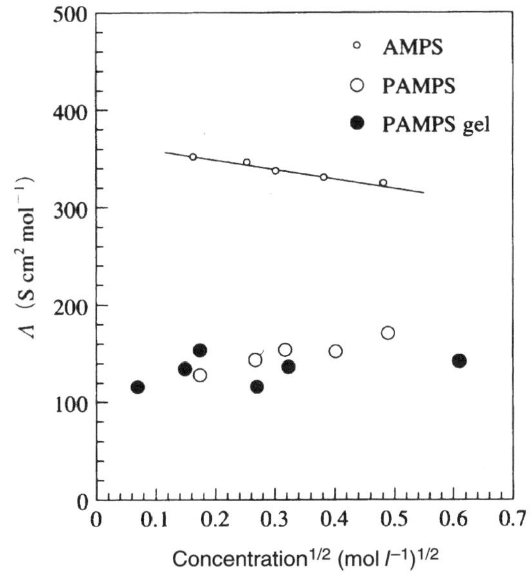

Fig. 3 Concentration dependence of molar conductivity of AMPS, PAMPS and a PAMPS gel.

Figure 4 shows the temperature dependence of the molar conductivity of AMPS, PAMPS, and a PAMPS gel. In general, unlike electron conductance of metals, carriers of ionic polymer gels are ions. Hence, electrical resistance decreases as temperature increases. This trend shows an Arrhenius-type temperature dependence as seen in semiconductors,

$$\sigma = \sigma_0 \exp\left(-\frac{E}{kT}\right)$$

where σ ($\mathrm{S\,cm^2\,mol^{-1}}$) is the electrical conductivity, k (J/K) is the Boltzmann constant, T(K) is absolute temperature, and E(J) is the activation energy.

Table 2 lists the molar conductivity values at 25°C and the activation energy obtained from the temperature dependence of a PAMPS solution and a PAMPS gel. The PAMPS gel that contains mobile protons has much greater molar conductivity than the gel with alkali metal salts [3]. This is due to the proton conductance mechanism in which protons transfer through hydrogen bonds.

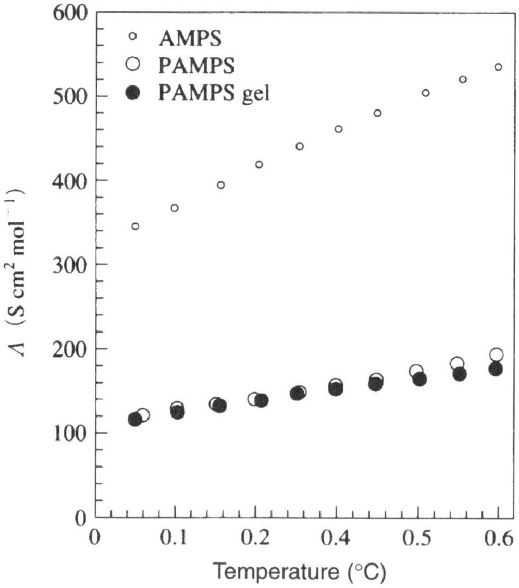

Fig. 4 Temperature dependence of molar conductivity of AMPS, PAMPS, and a PAMPS gel.

Table 2 The molar conductivity and activation energies of PAMPS and a PAMPS gel.

	Molar conductivity ($S\,cm^2\,mol^{-1}$)	Activation energy (kJ/mol K)
Polymer solution	119	9.5
Polymer gel	123	7.5

8.1.3 Dielectric Relaxation

The dielectric relaxation phenomenon is a memory phenomenon (as in viscoelasticity). These kinds of phenomena are generally called residual effect ones. Dielectric relaxation of solids, polymeric materials, liquids, and gases are described in detail in the monographs given here [5, 6].

Reports on dielectric relaxation of gels have begun to appear in recent years. These studies deal with sol-gel transition mechanisms and the static electrical field to which counter ions are exposed. Examples of such reports include studies of the sol-gel transition mechanism done by measuring direct current conductivity and the modulus of the gel obtained from a toluene solution of polydiacetylene [7–9]. Another example involves finding strong relaxation below the gelation of κ-carrageenan, a polysaccharide electrolyte [10]. This showed a relaxation strength of 1000 that cannot be explained by the ordinary dielectric relaxation phenomenon of dipoles. This is likely caused by heterogeneity, which is due to the higher-order structure of the gel.

There is a report on the colombic force field of a polyelectrolyte gel based on the analysis of dielectric relaxation spectra. High electron density of a polymer ion forms an extremely strong coulombic field in its vicinity (see Fig. 5) [11]. This distribution diagram is obtained by the numerical calculation based on the Poisson-Boltzmann equation. An ionic polymer gel possesses a static potential well. The counter ions that dissociated from the polymer ions then gather around them and form a restricted phase. Unlike free ions, these restricted counter ions show dielectricity. From dielectric relaxation spectra, the insight on the coulombic force field around the polymer ions and microscopic morphology of the gel can be obtained [12–15].

In addition, unfrozen and free water also have been studied with the dielectric relaxation method [16].

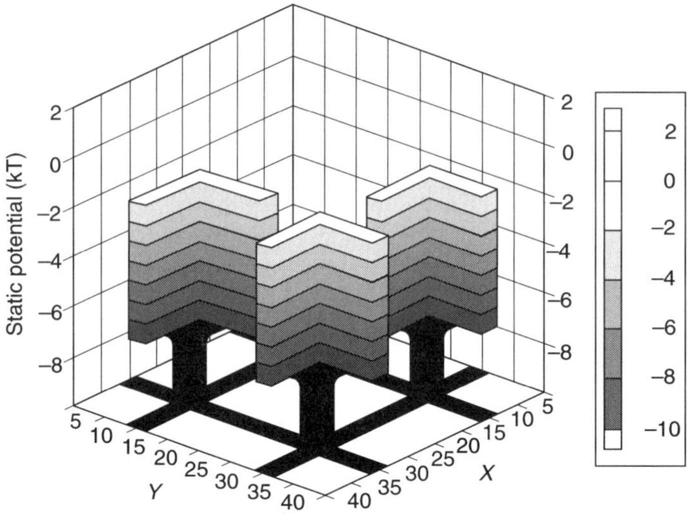

Fig. 5 Static potential distrubtion of ionic polymer networks obtained by 3D numerical calculation. The calculation used a stacking model approximation of a 2D mesh.

8.1.4 Electrical Shrinking and Electrical Permeation Phenomena

If an electrode contacts a polyelectrolyte gel that contains water and a direct current of $\approx 10\,\mathrm{V}$ is applied, the gel spews out water and shrinks. Such shrinkage of gels by electric stimulation is generally seen for all natural and synthetic polymers as long as they are ionic polymer gels.

This electrical shrinkage of ionic networks is found to be due to electric permeation of water. For an anionic polymer gel, the polymer ions try to move towards the anode while their counter ions move towards the cathode. However, because the polymer ions are fixed they can barely move, while their small molecular weight counter ions move to the cathode (electrophoresis). At the same time, the hydrating water of the ions in the gel also moves to the cathode (electropermeation). The electrical charges of the counter ions that reached the electrode are eliminated by electrochemical reactions and the hydrating water is expelled at the cathode, resulting in gel shrinkage. The opposite will occur for a cationic polymer gel.

Assuming that this phenomenon is similar to the electropermeation of water in a capillary, the author and others have proposed a 1D capillary model [17]. In this model, the distribution [11] of mobile ions was calculated by the Poisson-Boltzmann equation. Further, electrophoresis of mobile ions in an electric field and the movement of the accompanying solvent were calculated by the Navier-Stokes equation. Polymer chains parallel to the external electric field were considered in this model but those that were perpendicular to the field were ignored. This model clarified that the shrinking rate of the gel was proportional to the electrical field as well as reciprocally proportional to the viscosity of the gel. Furthermore, the shrinking efficiency, the amount of shrinkage per unit coulomb, is independent of the applied field but reciprocally proportional to the charge density of the gel.

8.1.5 Appearance of Vibration

The aforementioned movement of ions in a gel is an anomalous phenomenon that accompanies vibration [18, 19]. If a pair of platinum electrodes are inserted into a gel and direct current is applied, vibration with a good reproducibility can be observed. Gels have the ability to convert direct current into pulsed vibrations.

In systems that have rhythm or vibration, sometimes many rhythms synchronize and then a collective, concerted single rhythm can be observed. This phenomenon is called entrainment. The electrical vibration of gels exhibits entrainment. If a small alternating voltage as well as direct voltage is applied to a gel, the vibration frequency of the gel changes and eventually it is entrained to the phase of the alternating current wave.

It is interesting to note the similarity between this phenomenon and electrical aspects of the excitement of nervous systems, vibration of the Belousov-Zhabotinsky reaction, and the vibration of the potential between oil and water.

8.1.6 Piezoelectricity

When a polyelectrolyte gel is deformed by an external force, not only the elastic energy of the polymer chains but also its static energy changes considerably. Hence, the mechanical properties of a gel (e.g., Young's modulus) receive a significant contribution from static interaction as well as entropic elasticity.

In fact, if a weak electrolyte gel such as a poly(acrylic acid) is deformed, macroscopic dielectric polarization is observed. This results from the stretching of polymer chains by the deformation and resultant automatic acceleration of ionization. Utilizing this phenomenon, it is possible to develop a piezoelement that converts deformational stress into electrical energy. A pressure sensor whose diode emits light when the gel is pressed and an artificial contact sensing device also have been proposed [20].

8.1.7 Conversion

It has been discovered that if a PAMPS gel, an anionic gel, and a poly(dimethylaminopropylacrylamide) (PDMAPAA) gel (a cationic gel) are connected and a voltage is applied through the interface, a current will pass from the cationic gel to the anionic gel but almost no current will pass in the opposite direction [21]. This is similar to the nonlinear characteristics of a semiconductor diode that does not obey Ohm's law. The current ratio of the two directions can be ≈ 25 even if this process is repeated 40 times.

8.1.8 Photoinduced Electricity

If tetracyanobenzene (TCNB) is doped into a PDMAPAA gel in an N,N′-dimethylformamide (DMF) solution, a photoinduced electricity effect, which is often seen at a semiconductor interface, can be observed. The photobattery efficiency of approximately $10^{-2} \sim 10^{-1}\%$ can be obtained [22]. Conversion efficiency is different from a solution by the factor of $10^3 \sim 10^6$. The lifetime of the excited species does not differ in either gel or solution. The fluorescence intensity increases as crosslink density increases. Thus, a carrier's movement speed is thought to be important for the large photoelectric conversion efficiency of gels.

8.2 ELECTROVISCOUS FLUIDS

KIYOHITO OYAMA

8.2.1 Review

8.2.1.1 Introduction

The study of electroviscous fluids was done by Duff [23] and Quinke [24] at the end of the nineteenth century. They studied the viscosity changes of simple dielectric fluids following electric field application. However, due to the insignificant increase in viscosity, serious application studies were not conducted. Herzog, Kudar, and Paersch [25] further studied simple fluids in the 1930s, focusing on the mechanism of viscosity increase. Winslow reported in 1947 that the viscosity of suspended particles drastically increased. Since then, studies on electroviscous fluids have attracted a great deal of attention. Recent extensive studies have been conducted on electrorheological fluids (ER fluid) from the standpoint of materials, mechanisms, and applications.

Electroviscous fluid is a fluid that can drastically change its viscoelasticity by applying an external electrical field, with this effect called the ER effect. Electrorheology materials can include fluids but also ER rubbers, ER gels, and other forms [26, 27].

A typical ER fluid is a low-viscosity one in the absence of an electric field. However, by applying an electric field and shear as shown in Fig. 1, it assumes a solid-like behavior. Upon returning to the nonelectric field state, the fluid recovers its original viscosity. In rare instances there are ER fluids for which the viscosity with an electric field is lower than without the field. A noteworthy property of ER fluids is the ability to control freely the viscosity by electric field. This indicates that *voltage*, an electrical signal, can be directly converted into a *mechanical signal* (see Fig. 2). Hence, ER fluid functions like a *transistor*.

8.2.1.2 Materials for ER fluids

Many materials have been studied as potential ER fluids. When these materials are systematically analyzed, it is possible to predict what kinds of materials exhibit the ER effect and which conditions are necessary to show the ER effect. Figure 3 shows the classification. The ER fluids can be largely divided into suspended particle systems and homogeneous systems. However, this classification is not based on an analysis of

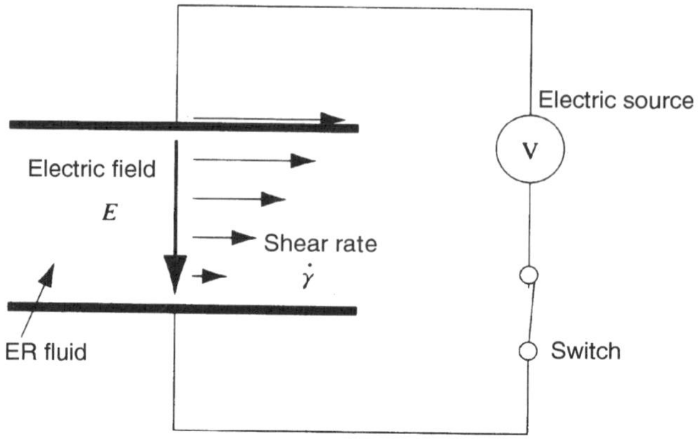

Fig. 1 Experimental configuration of ER fluid.

mechanisms. The mechanisms of many ER fluids are under study. Upon elucidation of the mechanism, further classification might be necessary.

8.2.1.1.1 Suspended particle systems

Suspended particle ER fluids [28, 29] are a suspension of polar micro-particles with diameters ranging from 1 to 100 μm in an insulating fluid. In early ER fluids, a small amount of water was included because the ER effect was enhanced when a very small amount of water was present [28, 29]. However, the stability of the system as a function of temperature or

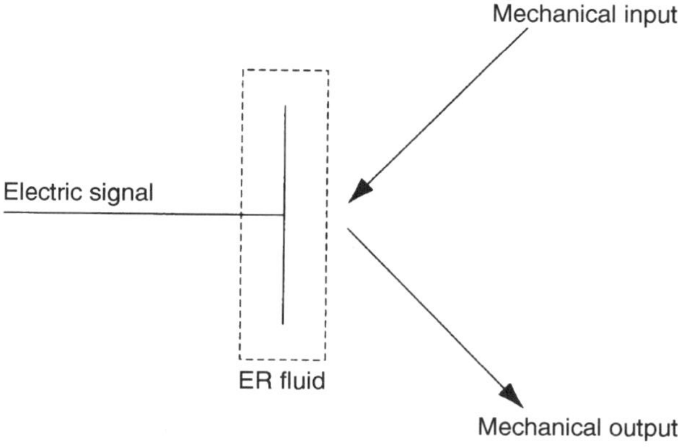

Fig. 2 Operation of ER fluid.

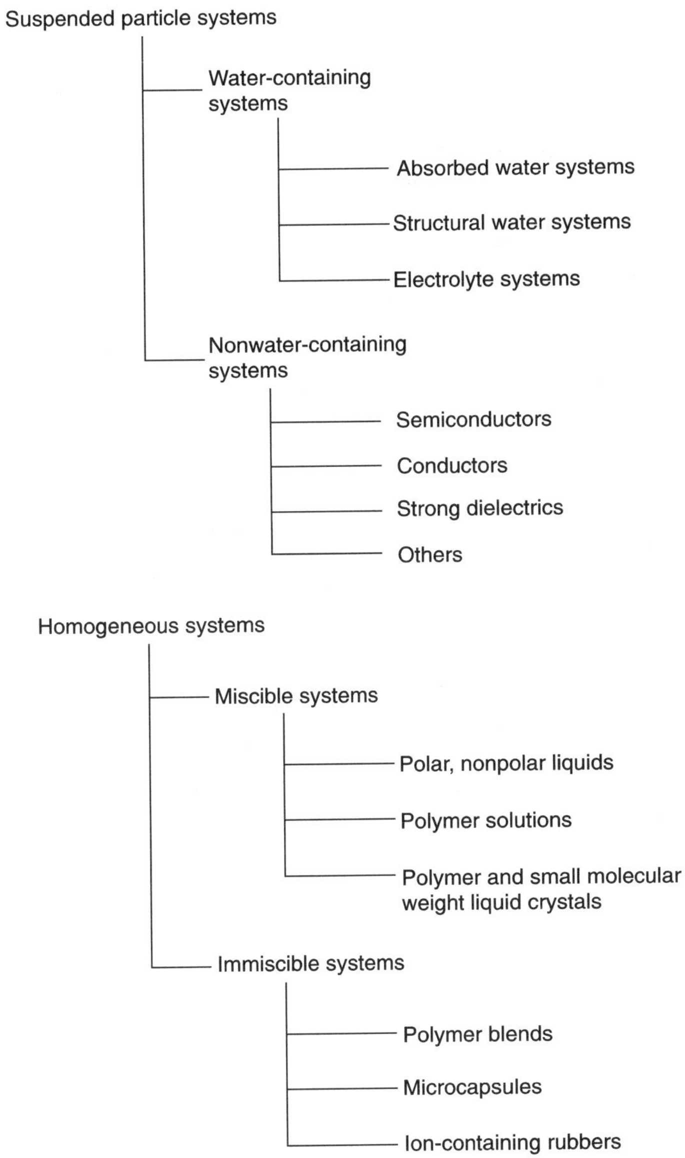

Fig. 3 Classification of ER fluids.

time is poor because of the presence of water. Hence, a majority of recent studies involved nonwater-containing ER fluids [30–32]. The classification of the ER fluids shown in Fig. 3 is based on the types of particles. It indicates that the detailed ER effect mechanism depends upon the particle/dispersant system.

(i) Flow Patterns If the shear stress of a suspended particle type ER fluid is measured during electric field application, the flow pattern generally will be like the ones shown in Fig. 4 [33]. When there is no electric field, it behaves as a Newtonian fluid in which shear stress t is proportional to shear rate γ. When an electric field is applied, shear stress increases for the corresponding shear rate and behaves as a Bingham fluid. This is because the ER effect is due to *the changes in apparent viscosity* caused by the changes in induced shear stress. It is believed that the true viscosity of an ER fluid itself will not change by the electric field,

$$\tau = \tau_y + \eta\gamma \qquad \text{(Bingham equation)}$$

where τ_y is called yield stress. When the external force exceeds this stress τ_y under the electric field, flow can occur for the first time. This τ_y is quite often proportional to the square of the electric field strength.

(ii) Mechanism of the ER Effect Appearance From the flow behaviour shown in Fig. 4 and the observation of the suspended particles under electric field application, the following mechanism is proposed. Among

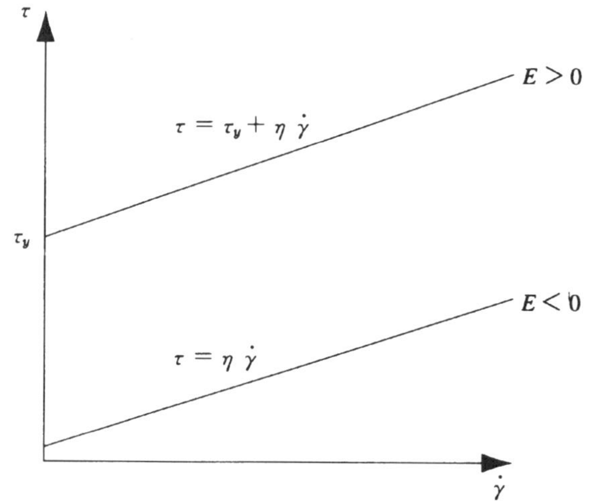

Fig. 4 Flow behavior of a suspended particle system.

the common items reported in many reports of suspended particles is the behaviour of microparticles in the dispersant. These particles are randomly dispersed when there is no electric field. However, under electric field application, clusters are formed between the electrodes. The particles connect like a chain and the electrodes, shearing planes, are physically connected (see Fig. 5).

The clusters that grew perpendicularly to the flow direction become resistant to flow and behave as though the viscosity of the fluid increased. However, the controversial issue among various theories is how these particles form clusters and move. Even the formation of polarization, the

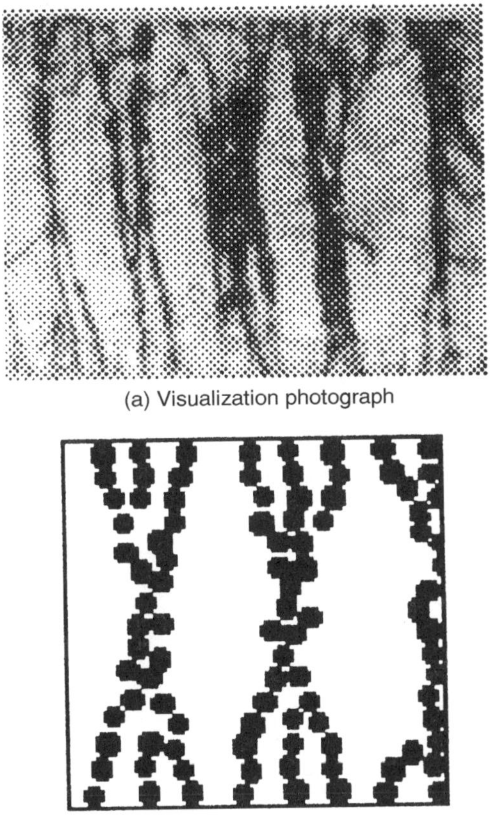

(a) Visualization photograph

(b) Model of the clusters

Fig. 5 Clusters of suspended particle system under applied electric field (titanium-coated iron particles/silicone oil system).

first stage of the cluster formation, varies depending on ER fluids. To date, various theories as described in the following have been proposed (see Fig. 6).

1. Attractive forces are generated among particles by the bulk conduction within the suspended particles and generate polarization. This theory is applicable to the ER fluids that contain materials with high free electrons or ions [34, 35].
2. Static attractive forces are generated by the surface conduction and polarization takes place near the surface.
3. Dipoles within strong dielectric molecules that cause self-polarization undergo polarization reversal and the entire particle polarizes [36].
4. When there is no electric field, adsorbed ions around suspended particles form a relatively homogeneous electrical double layer.

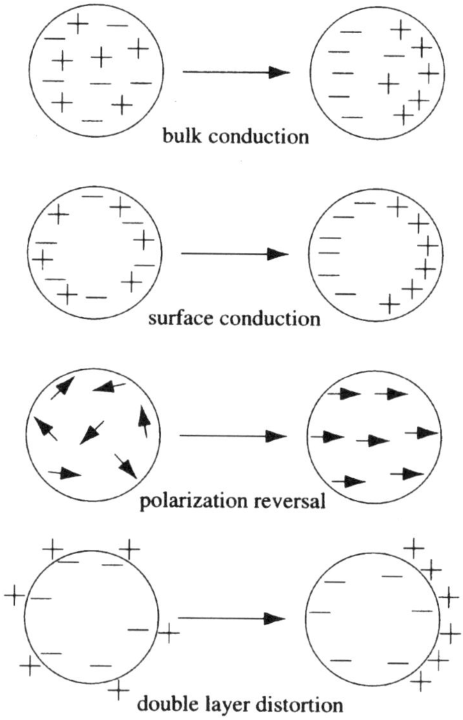

Fig. 6 Polarization mechanisms of particles.

When an electric field is applied, localization of ionic distribution takes place and electrical dipoles generate static attractive forces. This electric double layer distortion theory [37–39] is supported by the experimental observation that the ER effect is drastically affected by the addition of water or a surfactant, or the difference in the electric conductivity of the suspended particles and the dispersant.

Hence it is necessary to consider not only polarization of particles but also the relationship with the dispersant.

If the conductivity of particles and dispersant varies, the state of polarization and charge transfer among particles changes. Here, an example of negative ER effect will be shown (see Fig. 7). This can be prepared by using particles with lower conductivity and dispersant with higher conductivity than ordinary ER fluid. Boissy *et al.* [40] reported on a suspended system in which 30% PMMA particles were mixed with two kinds of mineral oil that had a higher conductivity than the particles. They observed the highest shear stress in the absence of an electric field as shown in case (a). As the electric field strength increases from (b) $E = 0.5\,\mathrm{kV/mm}$ to (f) $E = 3.0\,\mathrm{kV/mm}$, shear stress decreased. It is interesting even from the application point of view to have discovered the negative ER effect by the control of conductivity.

Theoretical studies have been reported on the behaviour of particles upon polarization using a model equation of state. For simplification, it is assumed that the radius of suspended particles is constant and the density

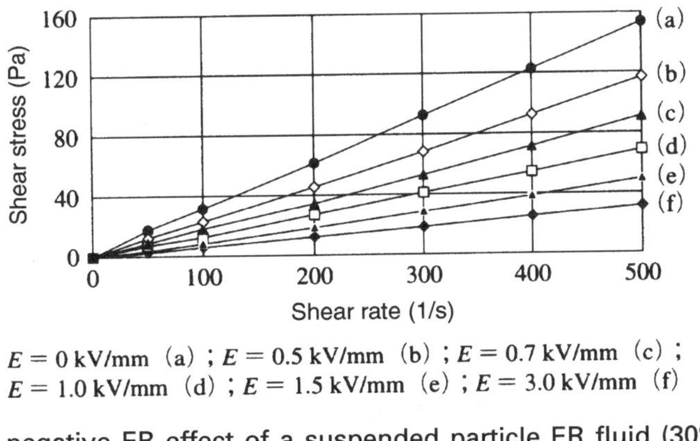

Fig. 7 A negative ER effect of a suspended particle ER fluid (30% PMMA/T-ELF TF50).

of the suspended particle and the dispersant is the same. This allows the balance of mechanical forces in a quasistatic manner. Simulations indicated that such a system behaves as a Bingham fluid [41, 42].

8.2.1.1.2 Homogeneous systems Homogeneous systems have been studied ever since they were labeled electroviscous fluids [23–25]. However, the majority of the systems studied were those of low molecular weight liquid crystals, polymeric liquid crystals, and polymer solutions.

(i) Flow Patterns Figure 8 provides a qualitative description of the flow behavior of various liquid crystals [43, 44] as examples of homogeneous systems. Unlike suspended particle systems, they do not show yield stress and the viscosity increases with electric field application by several factors. They are nonNewtonian regardless of the presence of an electric field and the degree of nonlinearity depends on the type of liquid crystal. In addition, homogeneous ER fluids have a different shear rate regime depending on the material.

(ii) Mechanism of the ER Effect Appearance The mechanism and characteristics of the ER effect of homogeneous systems differ in comparison with suspended particle systems. Among low molecular weight liquids, increase in apparent viscosity with electric field application has been observed with glycerine, turpentine oil, ethyl alcohol, water,

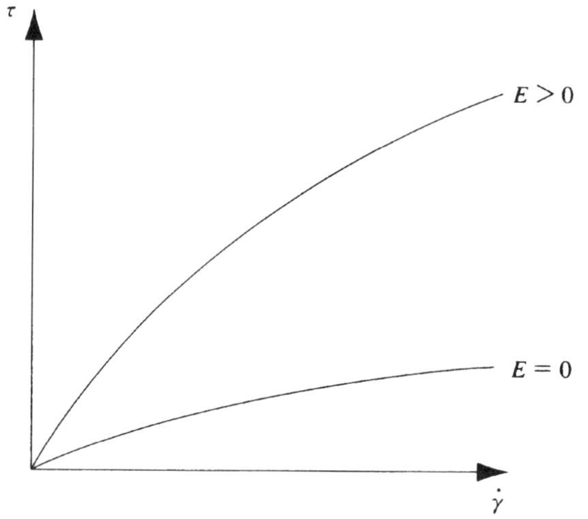

Fig. 8 Flow behavior of homogeneous systems (liquid crystals).

ether, and benzene. However, even in a nonpolar solvent, there is no assurance that such an effect is absent. Hence, various theories, including cluster formation by polar molecules, secondary flow by ionic conduction [45], and increase in apparent viscosity by the interference of the electric charges formed near an electrode [46] have been proposed. However, these theories do not consider polarization of the fluid and attempt instead an explanation only as relates to the ion conduction of fluid impurities. Unfortunately, these theories can explain only a small ER effect and are not applicable to the ER fluids that show significant ER effect (such as liquid crystal and polymer solution systems). This significant ER effect is possibly related to physical gel formation caused by orientation or electric field or changes in phase-separated structures.

(iii) Examples Because the mechanisms of the ER effect of homogeneous systems are poorly understood, actual examples are used to demonstrate it. Among homogeneous ER fluids, polymer liquid crystal ER fluids show a superior ER effect to ordinary suspended particle ER fluids. According to Inoue [47], liquid crystals having mesogens on both ends of a flexible chain had a higher ER effect than for a one-end chain (see Fig. 9). This is because the ER effect depends not only on the orientation of liquid crystal domains but also on the formation of physical gels by mesogens that act as crosslink points. Because the one-sided mesogens cannot form crosslinking, the ER effect will be small. This is a good example of the importance of macroscopic strength in the ER effect.

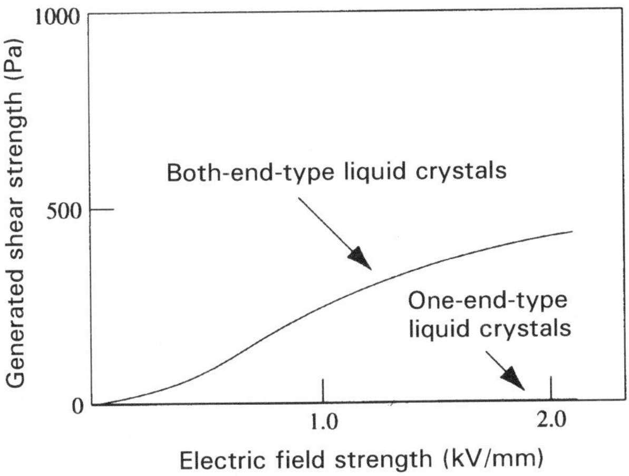

Fig. 9 The ER effects of both-end and one-end-type liquid crystals.

Among homogeneous ER fluids, there are systems that show an anomalous ER effect. Strong dielectric, low molecular weight liquid crystals [48] and polar polymer solutions [49, 50] exhibit a negative ER effect, in which the viscosity and dynamic mechanical properties decrease.

A mixed liquid crystal that was developed for a strong dielectric liquid crystal display is an example of a low molecular weight liquid crystals [48]. This system exhibits a negative ER effect from direct current to alternating current of 10^{-1} Hz. Above this frequency, however, it shows a positive ER effect (see Fig. 10). From this result, the origin of the negative ER effect is seen as movement of the domain with long relaxation times.

An example of polymer solutions is P(VDF/TrFE)/DMF solution [49] (Fig. 11). Poly(vinylidene-co-trifluoroethylene) (P(VDF/TrFE)) is known as a strong dielectric polymer. It possesses a permanent dipole moment in the main chain. The electric field frequency dependence of the relative viscosity was measured by a capillary rheometer. The result showed a positive ER effect at 10 Hz whereas it showed a negative ER effect at 1 kHz.

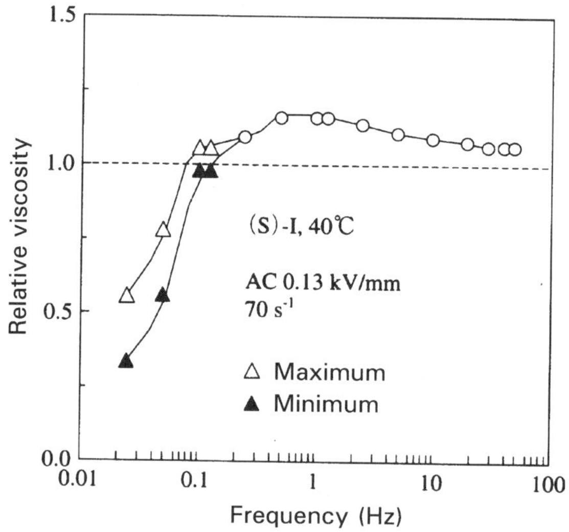

Fig. 10 Applied electric field frequency dependence of relative viscosity for a strong dielectric liquid crystal (normalized by the viscosity under no electric field) [48].

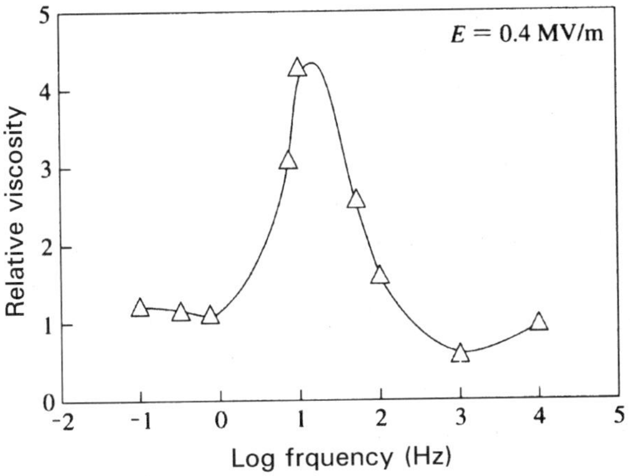

Fig. 11 Applied field dependence of relative viscosity for strong dielectric poly(VDF-TrFE)/DMS (normalized by the viscosity under no electric field) [49].

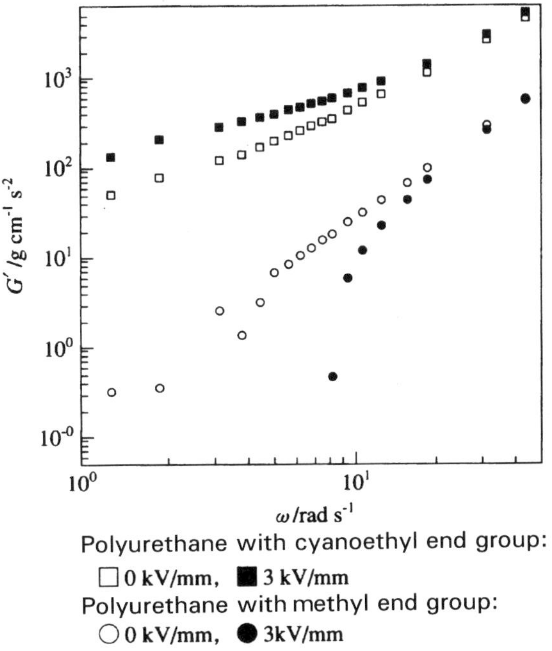

Fig. 12 Positive and negative ER effects that appear as angular frequency dependence of storage modulus G′ [50].

These two examples provide interesting results in which the ER effect changes as a function of frequency. The melt and solution of polyurethane show both positive and negative ER effects by changing the molecular structure [50] (see Fig. 12). The positive or negative ER effect that occurs depends upon whether the terminal group is cyanoethyl or methyl. If a terminal group with a large permanent dipole exists, urethane with negative ER effect assumes positive ER effect status.

8.2.2 Water Systems

YOSHINOBU ASAKO

8.2.2.1 Introduction

The phenomenon of the viscosity of a polar liquid (such as glycerin) increasing slightly when an external electric field is applied (ER effect) was observed as early as the end of the nineteenth century [51]. In 1947, Winslow applied an electric field on a suspension of limestone particles in mineral oil to refine the mineral oil. He then discovered that the viscosity of the suspension solution increased drastically [52, 53]. The electrorheological effect (ER effect) of a particle-suspension liquid is also called the Winslow effect. Winslow further found significant effects on silica and starch and thus thought that the ER effect originates from the moisture contained on the particles.

Since the discovery by Winslow, many ER fluids have been proposed. The majority of these systems contain a small amount of water [54, 55]. This is because water-containing particle-suspension liquids exhibit very large shear stress when an external electric field is applied [56, 57]. Dispersates for water-containing systems use particles that contain an ionically dissociated group (see Table 1). Inorganic gels such as silica [58, 59] and alumina [60], and crosslinked metallic salts of poly(acrylic acid) [61, 62] also show the ER effect.

Table 1 Water-containing particle-suspension ER fluids.

	Dispersate particles	Literature
Inorganic materials	Silica	[58, 59]
	Alumina	[60]
	Aluminum silicate	[60]
	Mica	[63]
	Inorganic ion exchange materials	[64]
Organic materials	Ion exchange resins	[56, 57]
	Crosslinked metallic salts of (metha)acrylic acid	[61, 62]
	Copolymers with AMPS[a] as the main component	[65]
	Copolymers with amine-containing unsaturated monomers	[66]
	Copolymers from aziridine compounds	[67]
	Amphoteric gels	[68]
	Water-containing sulfonated copolymers	[69, 70]
	Structure of water-containing sulfonated copolymers	[70, 71]

[a] 2-Acrylamide-2-methyl(propane sulfonic acid).

In general, those particles that possess dissociated groups exhibit the hygroscopic property. When such particles are used, both the water that is adsorbed on the surface and water that is added and then held by the particles will influence the ER effect significantly. Hence, those ER fluids that contain particles with ionically dissociated groups are characterized as water-containing systems. Characteristics of such systems are often discussed in relation to the contained water. In this subsection, characteristics and representative examples of water-containing systems will be described.

8.2.2.2 Examples of water-containing particle-suspension ER fluids

8.2.2.2.1 Dispersed silica-type

Klass and Martinek conducted systematic studies on the ER effect of a silica-suspension mineral oil and studied the role of the moisture contained in the silica [58, 59]. The following is a summary of the results.

1. The induced shear stress was approximately proportional to the square of the applied electric field.
2. The induced shear stress was hardly affected by the distance between the electrodes.
3. As the concentration of the dispersate (silica) increased, induced shear stress increased. When the concentration of the dispersate was increased and the electric field intensified, the liquid solidified.
4. The effect of direct and alternating electric field on the induced shear stress was hardly recognized. However, at >100 Hz, the induced shear stress decreased as the frequency increased.
5. When the SiOH density on the silica particle surface increased, the induced shear stress increased.

8.2.2.2.2 Ion exchange resin dispersion type

Sugimoto studied the ER effect of water-containing ion exchange resin-suspension fluid and evaluated various factors that influence the ER effect [56, 57]. A basic anionic ion exchange resin to which amino or ammonium group is introduced was used as the dispersate. For measurement of induced shear stress, a double tube rotational viscometer that is able to apply an electric field was used. An alternating electric field is applied. The results are summarized in the following.

1. For ion exchange resin-type ER fluids, a large shear stress was induced even when the resin had low water content.
2. The higher the water content, the larger the induced shear stress. The induced shear stress saturated above certain electric field strengths. The current density was extremely unstable and continued to increase (see Fig 1).
3. Although the fixed ion is the same, if the counter ion is different, the ER effect varied significantly.
4. When the fixed ion was ammonium, larger shear stress was induced than in the amino group.
5. The larger the ion exchange capacity, the greater the induced shear stress.

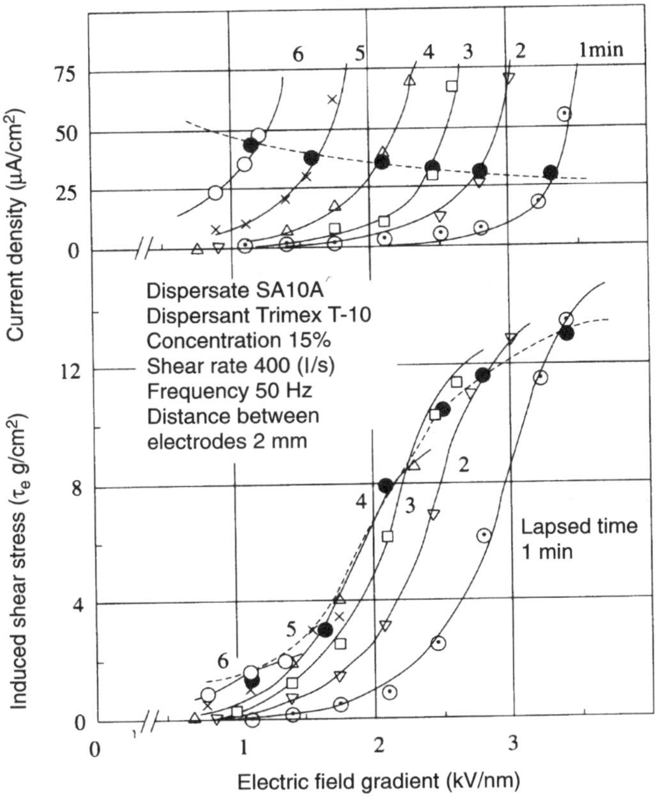

Fig. 1 Influence of electric field gradient in SA10A [51].

6. When crushed resin was used instead of a spherical resin, unstable current flowed at low electric field strength and the induced shear stress was also extremely small.
7. Depending on the type of dispersant used, a large difference in the induced shear stress was observed.
8. When a direct electric field was used, electrophoresis of suspended particles took place and the longer the application time of the electric field the greater the induced shear stress.
9. Finally, the higher the frequency of the alternating electric field, the lower the induced shear stress.

Sugimoto explained the ER effect shown by the ion exchange resin-suspension fluid by using the electric double layer theory. In the electric double layer theory a particle suspended in an electrically insulating oil has an electric double layer due to dissociation of the ionic groups by the adsorbed water on the surface. The dissociated ions are moved by the external electric field and the electric double layer deforms. Shear is applied to the polarization, resistance is created when particles try to move each other, and, consequently, viscosity increases.

8.2.2.2.3 Crosslinked metallic salt of (metha)acrylic acid type

Stangroom investigated the ER effect of suspended solid particles that had acidic groups and thus contained water. A crosslinked metallic salt of (metha)acrylic acid was the particle chosen. He reported that the solid particle-suspension fluid with acidic groups show a larger ER effect than particles that do not contain acidic groups. Because the appearance of the ER effect requires water, the solid particle containing an acid group must be rather hydrophilic. The acid group can be either dissociated or in a salt form. However, he reported on a salt that combined a monovalent salt (for example, lithium or guanidium) and a trivalent salt (for example, chromium or aluminum).

Standgroom explained the appearance of the ER effect of water-containing suspended particle-type fluid by using the water-assisted adhesion theory. This theory is defined as follows. When an electric field is applied to an ER fluid, the suspended particles form chain-like aggregates by dipole-dipole interaction. When the aggregates are formed, the electric field between contacting particles is stronger than within the particle. Thus, water with a high dielectric constant leaks out to the particle-particle gap and forms a water bridge. Because the strengths of the aggregates becomes stronger with this water, induced shear stress increases.

8.2.2.2.4 Water-containing sulfonated polymer suspension type

Asako *et al.* [68–71] studied the relationship between induced shear stress and composition and structure of suspended particles using a sulfonated polymer as a model. For this study, sulfonated poly(styrene-co-divinyl benzene) (SSD) in which sulfonic acid is almost homogeneously distributed was used. They also used non-sulfonated SSD as the core with sulfonated SSD (SSDH) as the surface layer (see Fig. 2). The water content in the dispersed phase was determined by the Karl-Fisher titration technique. The SSD composition is similar to that of a commercially available strongly acidic ion exchange resin. An ER fluid was prepared by dispersing 30% by weight of this dipsersate into polydimethylsiloxane (viscosity 20 cSt). A double tube rotational viscometer with electric field capability was used for the study. A 50-Hz alternating electric field was applied. The obtained results are summarized as follows.

1. As the water content in SSD increased, the induced shear stress and the current density that passes then increased.
2. The water content of SSD, which produces 30 $\mu A/cm^2$ upon application of 4 kV/mm, is $3 \pm 0.4\%$ by weight. This relationship was almost constant regardless of the degree of sulfonation and average particle size.
3. On the other hand, water content was 3% by weight when the thickness of the SSDH shell was >3.7 μm. When the thickness was reduced to 1.4 μm (SSDH4), the water content reduced to 1.5% by weight while for 0.7 μm (SSDH5), it was 0.7% by weight.
4. When the average diameter was kept constant at 25 μm, the higher the degree of sulfonation, the greater the induced shear stress (see Fig. 3).
5. When the degree of sulfonation of SSD is almost constant at 80%, the larger the average diameter, the greater the induced shear stress increased.
6. For SSDH, when the shell was more than 3.7-μm thick, the induced shear stress was the same. For a 0.7-μm sample (SSDH5), the stress decreased. From this result, the inner part of the particle was also found to contribute to the polarization of suspended particles under an electric field application (see Fig. 4).

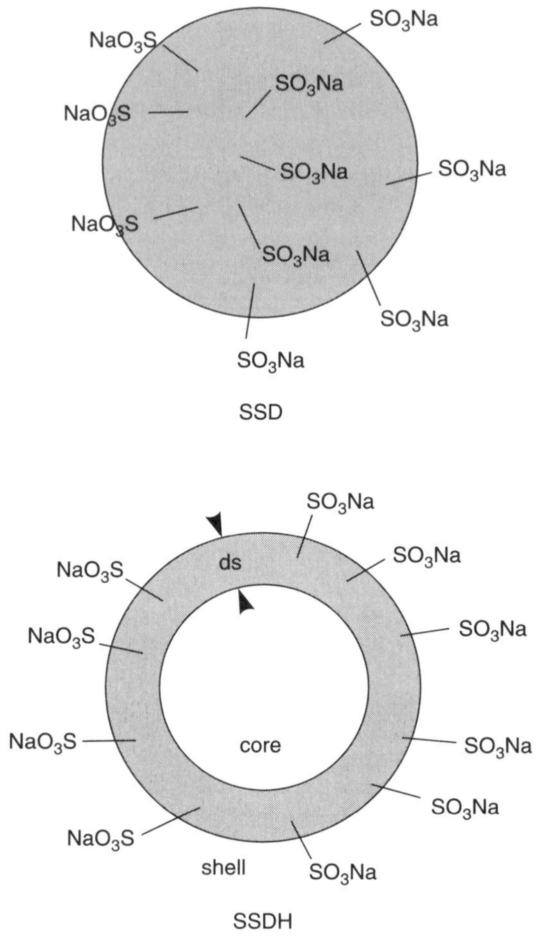

Fig. 2 A conceptual diagram of SSD and SSDH [70].

8.2.2.2.5 Problems for practical application of water-containing suspended particle ER fluids

In order for ER fluids to be used in various devices, it is necessary that: (i) the induced shear stress be large (stress properties); (ii) the current density be small (current properties); (iii) when an electric field is continuously or intermittently applied the induced shear stress or current density does not change as a function of time (electrical permanence); (iv) viscosity without electric field application be small (flow properties); and (v) the stability of the suspended particles be high (suspension stability) [72].

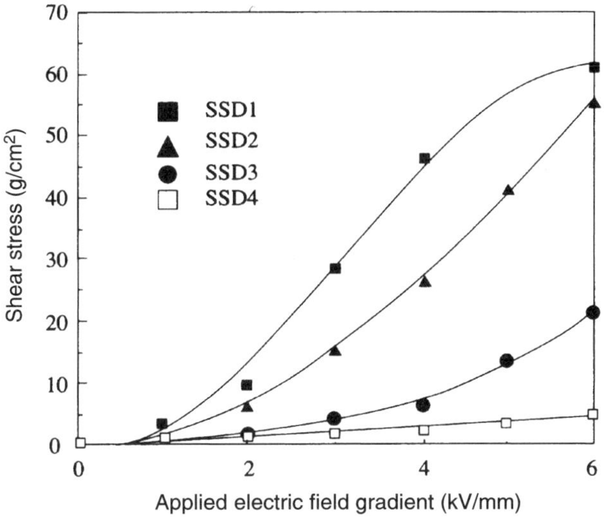

SSD1 (the degree of sulfonation: 89%; water content: 3.0 wt%)
SSD2 (43%; 3.2 wt%), SSD3 (31%; 3.2 wt%),
SSD4 (12%; 2.6 wt%).
The average diameters of SSD1-4 were all 25 μm. The degree of sulfonation was calculated based on the elemental analysis. The water content was determined by the Karl-Fisher titration method.

Fig. 3 Stress characteristics of an ER fluid when SSD is used as a dispersate (influence of the degree of sulfonation) [69, 70].

As already mentioned, water-containing suspended particle ER fluids have superior stress characteristics. However, current properties and electrical permanence are less than desirable. It was thought that water causes these undesirable properties [73].

8.2.2.2.6 Structural water-containing sulfonated polymer suspen-ded type

Asako *et al.* studied the effect of water on the stress properties, current properties, and electric permanence of an ER fluid in which a sulfonated polymer was used for the paticles in suspension. They used a sulfonated group polymer rather than one with an aromatic group. The polymer used was sulfonated poly(styrene-co-divinylbenzene) (NSP: average particle diameter = 5 μm, the degree of sulfonation = 124%; and SD: average particle size = 5 μm, the degree of sulfonation = 88%), which has the

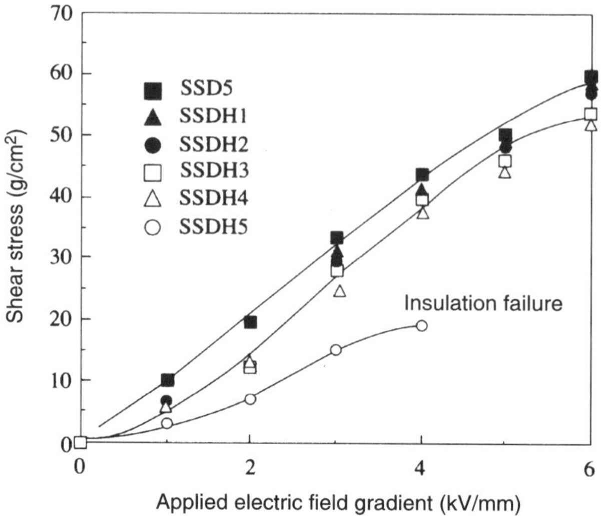

SSDH1 (the degree of sulfonation: 69%; thickness of the shell: 7.3 μm; water content: 3.3 wt%), SSDH2 (59%; 5.8 μm; 3.0 wt%), SSDH3 (42%; 3.7 μm; 3.2 wt%), SSDH4 (18%; 1.4 μm; 1.5 wt%), SSDH5 (8%; 0.7 μm; 0.7 wt%), SSD5 (82%; -; 3.3 wt%).
The average diameter of SSDH1-5 and SSD5 were all 50 μm.
The thickness of the shell was calculated from the electron microscopic observation of the distribution of sulfonic group and the degree of sulfonation.

Fig. 4 Stress characteristics of an ER fluid when SSDH is used as the dispersate [69, 70].

same composition as the commercially available strongly acidic, cationic ion exchange resin. The ER fluid was prepared by suspending 30% by weight of sulfonated polymer into silicone oil (viscosity 50 cSt). Induced shear stress was measured with a double tube rotational rheometer with electric field application capability. A 60-Hz alternating electric field was used. The water content was measured by the Karl-Fisher titration method. The results obtained are summarized in the following.

1. Even after heating at 150°C for 24 h, NSP and SSD had 1.5 and 0.8% by weight of water remaining (see Fig. 5). Moisture could not be eliminated by further heating. This residual water is extremely strongly restricted and is considered to be structural like water.

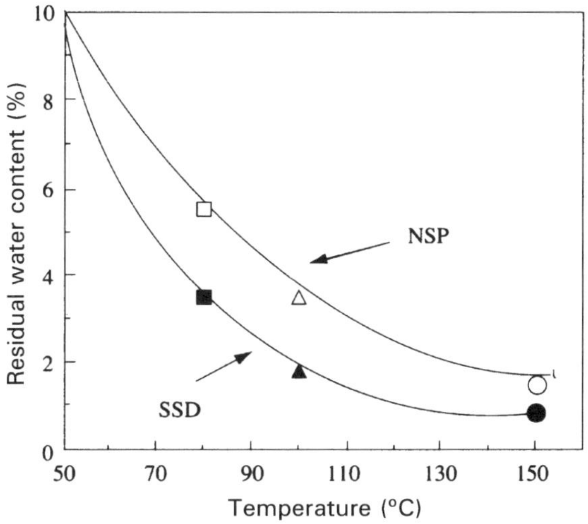

Fig. 5 Residual water contents of SSD and NSP [70, 71]

2. The lower the water content, the lower the induced shear stress of both SSD and NSP containing fluids (see for SSD: Fig. 6; NSP: Fig. 7). The current density also decreased.

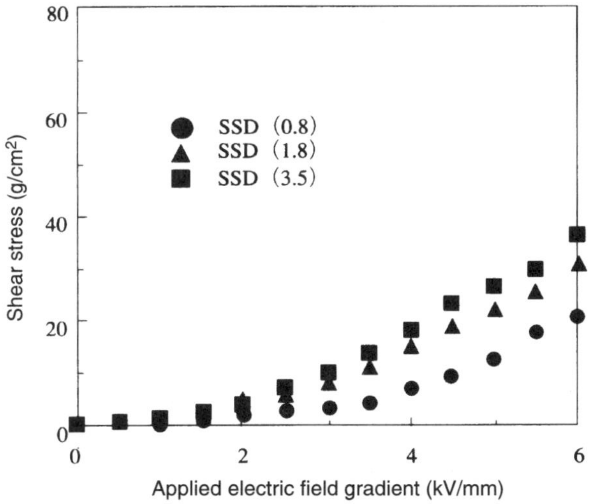

Fig. 6 Stress properties of the ER fluid with SSD as the suspended phase [70, 71].

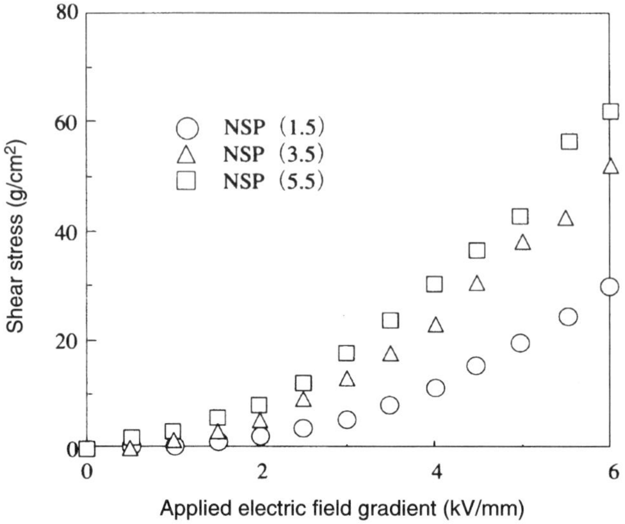

Fig. 7 Stress properties of the ER fluid with NSP as the suspended phase [70,71].

3. Induced shear stress and time-dependent stability of current density of an SSD-containing fluid were investigated. As the water content in SSD was reduced, the electric permanence improved (see for stress properties: Fig. 8; current properties: Fig. 9). For the SSD that contains only structural water, the induced shear stress and current density were nearly constant up to 20 h.
4. For NSP the electric permanence improved as the water content decreased (see for stress properties: Fig. 10; current properties: Fig. 11). For NSP (1.5) with only structural water, the induced shear stress and current density were nearly constant.

Until now, water-containing suspended particle type ER fluids were said to lack electric permanence and were impractical. However, NSP or SSD with only structural water showed remarkable improvement. Furthermore, the ER fluid that used NSP showed excellent stress and current properties in addition to electric permanence.

The ER effects of NSP and SSD that possess only structural water can be explained by the dielectric polarization theory [74], which is also used to explain the ER effect of nonaqueous fluids.

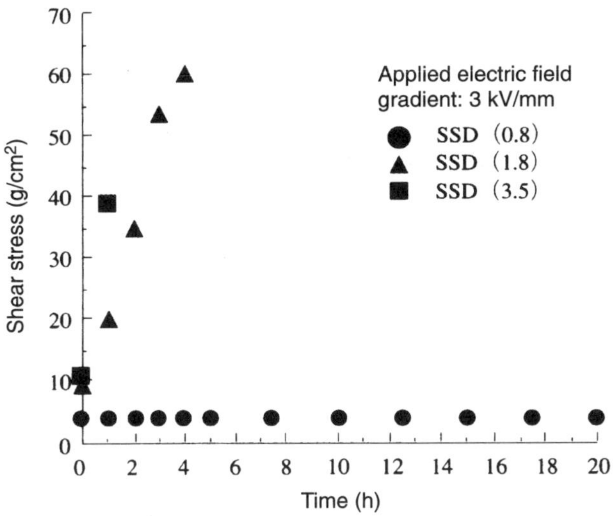

Fig. 8 Time-dependent stability of stress properties of the ER fluid with NSP as the suspended phase [70, 71].

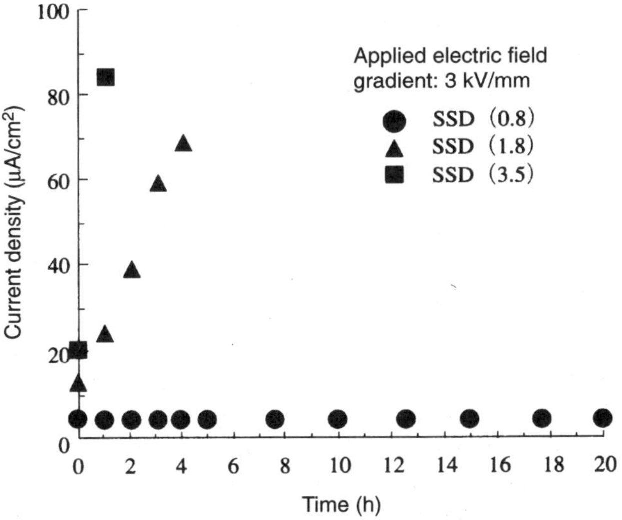

Fig. 9 Time-dependent stability of the current properties of the ER fluid with SSD as the suspended phase [70, 71].

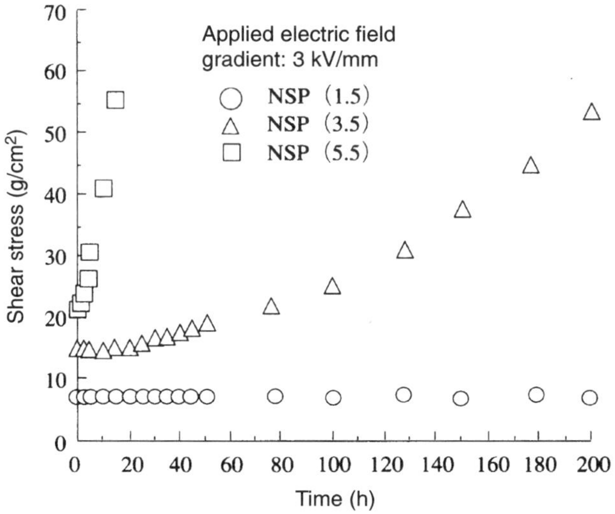

Fig. 10 Time-dependent stability of the stress properties of the ER fluid with NSP as the suspended phase [70, 71].

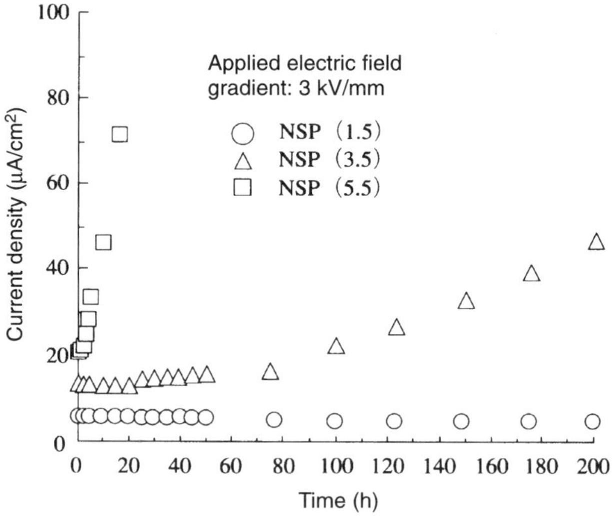

Fig. 11 Time-dependent stability of the current properties of the ER fluid with NSP as the suspended phase [70, 71].

8.2.2.3 Conclusions

Water-containing suspended particle ER fluids show large induced shear stress. The suspension composed mostly of hydrophilic particles that contain dissociated ions. The relationship between the properties of ER fluids and the water that is contained in those particles has been discussed herein and descriptive examples have been provided. If water is purposely added to suspended particles, the ER effect strengthens. However, electric permanence becomes poor and they become less practical to use. On the other hand, if only structural water is included, ER fluids with excellent electric permanence can be obtained.

To date there has been limited commercialization of devices that use the ER effect. However, ER fluids worthy of commercialization have been developed and this will accelerate further development of devices. For example, an ER cutting machine has been developed [75, 76]. This machine incorporates a variable rodless cylinder that functions by using an ER fluid with a sulfonated polymer. This machine cuts brittle ceramics, using ER fluid to control the cutting speed very accurately. It is used to manufacture a catalyst for automobile exhaust gas.

8.2.3 Hydrophobic Type†

TSUBASU SAITO

Electrorheological (ER) fluids change from fluids to solids continuously, reversibly, and instantaneously upon application of external electric field. By applying these properties of ER fluids, it is possible to develop revolutionary machines and to improve traditional instrumentation significantly. There are many application potentials, including an automotive device that controls shocks depending on road conditions, a crutch that operates smoothly and responds rapidly, a vibration dumping device for manufacturing machines and semiconductors, a robot that operates as smoothly as a human being, home appliances, and construction and civil engineering uses.

During the late 1970s, Stangroom developed stable, water-containing polymer microparticle suspended ER fluids. Since then, active research and development have been ongoing and a great deal of knowledge on the mechanisms of the ER effect has been obtained. Development of further stable and higher performance water-containing fluids accelerated the research on device application, resulting in the filing of many patents since the 1980s. Moreover, due to developments in applied research, the properties required for ER fluids have been determined. However, these early ER fluids lacked time and temperature stability, which made actual device application difficult. Recently developed ER fluids have overcome these problems and have shown great potential for application.

In this subsection, a nonwater-containing ER fluid that is considered as possessing the best potential for actual application will be described. Nonwater-containing type is a fluid that suspends polarizable microparticles without the help of water in an electrically insulating oil. In the 1980s, patents by Block *et al.* began to attract researchers' attention. The invention by Block *et al.* is based on the discovery of poly(athene or quinone) microparticles, an organic semiconductor. Since then, various fluids have been developed by many companies. Among them, polyurethane microparticles (Bayer), carbonaceous microparticles (Bridgestone), conducting microparticles coated with surface-insulating film (Asahi

---

†For the reader: the fourteen references at the end of this section (8.2.3) were uncited in the Japanese original text and thus were likely intended as general bibliographic information rather than specific in-text references [77, 90].

Chemicals), and organic-inorganic hybrid microparticles (Fujikura Chemicals) are representative.

Figure 1 shows an organic-inorganic hybrid microparticle as an example. This powder is synthesized by suspending in water inorganic microparticles and monomer droplets that contain a polymerization initiator. This system uses suspension polymerization. When polymer microparticles are formed upon polymerization, the inorganic microparticles adhere to the polymer particle surface and form a surface coating. Among various combinations, an acrylic polymer and a special titanium dioxide microparticle showed the best properties.

In order for ER fluids to be practical, they must satisfy the initial requirements for an individual device. The ER effect must also be stable and reliable. In the following, a carbonaceous microparticles suspended ER fluid, which is currently moving towards actual application, will be described as an example. Figures 2 and 3 compare temperature dependence and stability of the carbonaceous materials with water-containing systems. From these results, it can be seen that the carbonaceous systems have overcome the shortcomings found in water-containing systems. Figures 4 and 5 show the electric field strength dependence of apparent viscosity and current density whereas Fig. 6 illustrates the temperature dependence of apparent viscosity. Figure 7 shows the shear rate dependence of shear stress as a function of the electric field strength. Figures 8, 9, and 10 are the shear rate dependence of current density, the temperature dependence of response rate, and permanence of apparent viscosity and current density, respectively.

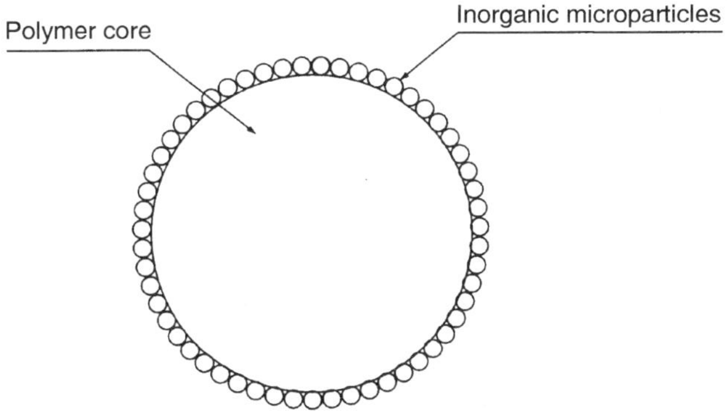

Fig. 1 Structure of organic–inorganic hybrid microparticles.

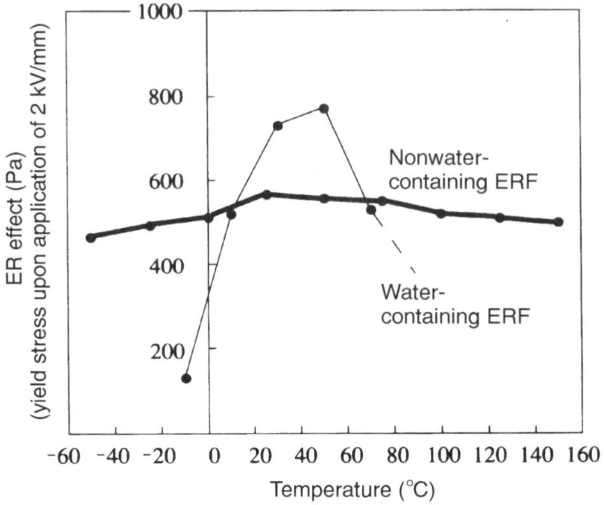

Fig. 2 Comparison of the temperature dependence of a carbonaceous ER fluid with a water-containing system.

From these results, carbonaceous fluids were found to show a large ER effect, consume a small amount of electricity, be able to be used in widely varying temperature and shear rate regions, exhibit good response,

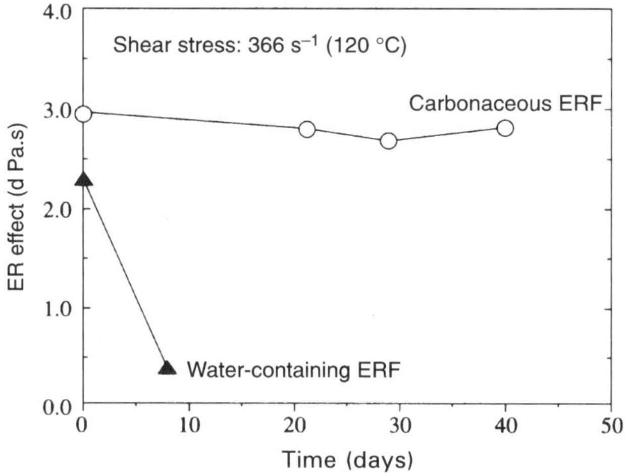

Fig. 3 Stability of carbonaceous ER fluid in comparison with water-containing particle system.

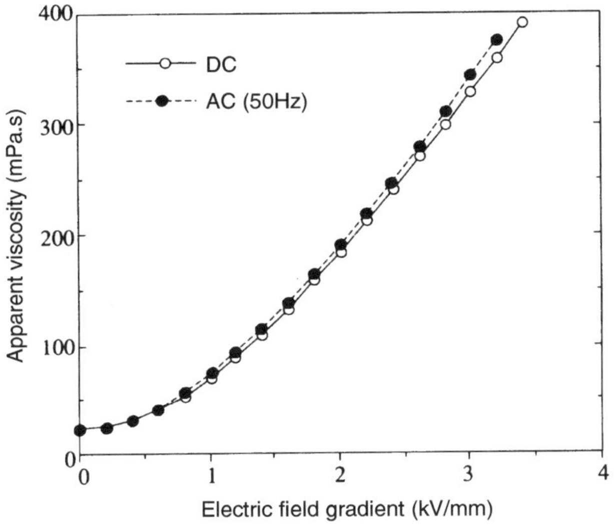

Fig. 4 Electric field gradient dependence of apparent viscosity of a carbonaceous material-filled ER fluid.

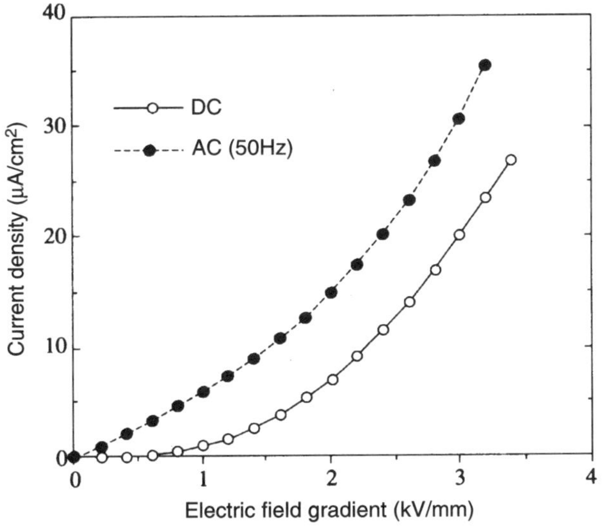

Fig. 5 Electric field gradient dependence of current density of a carbonaceous material-filled ER fluid.

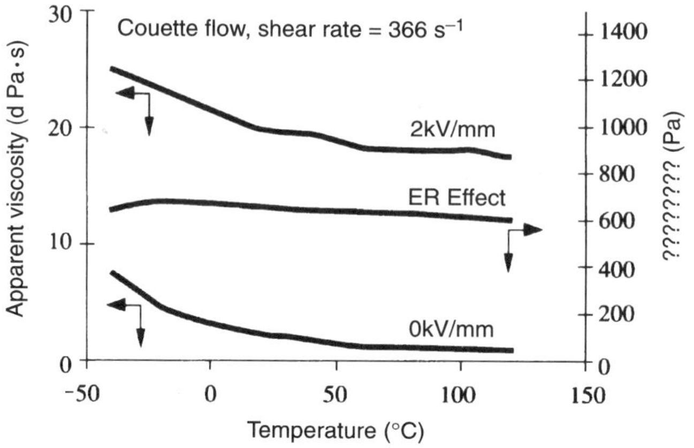

Fig. 6 Apparent viscosity and temperature dependence of carbonateous material-filled ER fluid.

and show excellent stability. However, even for this cabonaceous ER fluid, problems remain in the area of parts design before it can be used in automotive parts.

For ER fluids, further improvements in performance, stability and reliability are necessary. Furthermore, achieving these goals and decreasing costs is another important consideration.

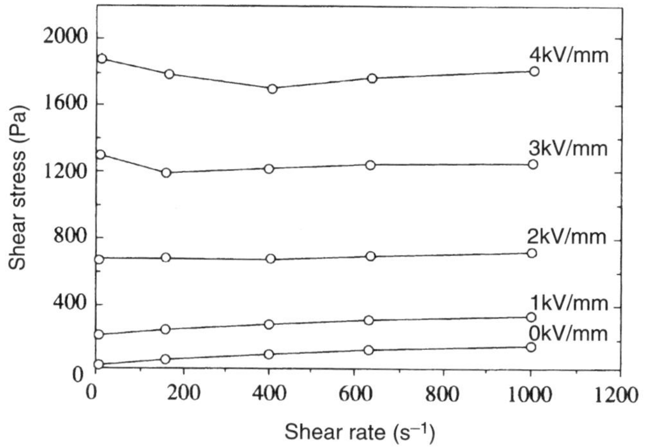

Fig. 7 Shear rate dependence of shear stress of carbonaceous material-filled ER fluid under various electric field gradients.

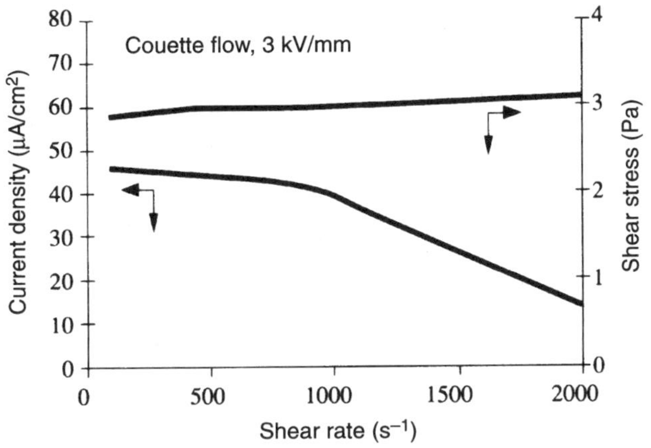

Fig. 8 Shear rate dependence of current density of carbonaceous material-filled ER fluid.

The ER fluids can convert an electric signal directly into resistance and its strength can be controlled by voltage. As shown in Fig. 11, two approaches are being evaluated. In one an ER fluid is passed between two fixed electrodes and in another one of the electrodes is moved. For the latter, shear resistance is generally used while the distance between the electrodes is kept constant. The device can also be used in tension and compression, although the operation is limited to a special case where only small deformation is necessary.

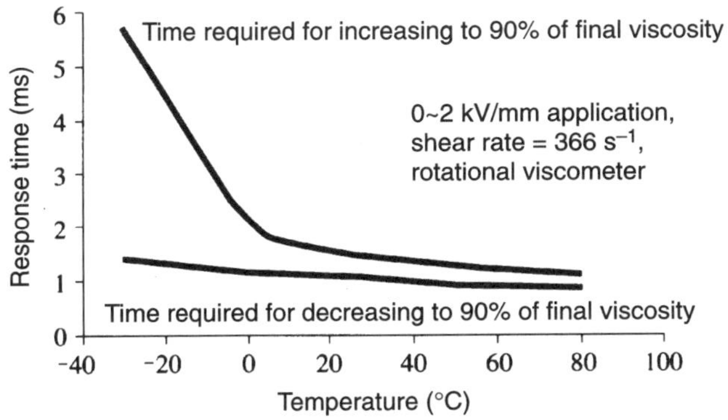

Fig. 9 Temperature dependence of response of carbonaceous material-filled ER fluid.

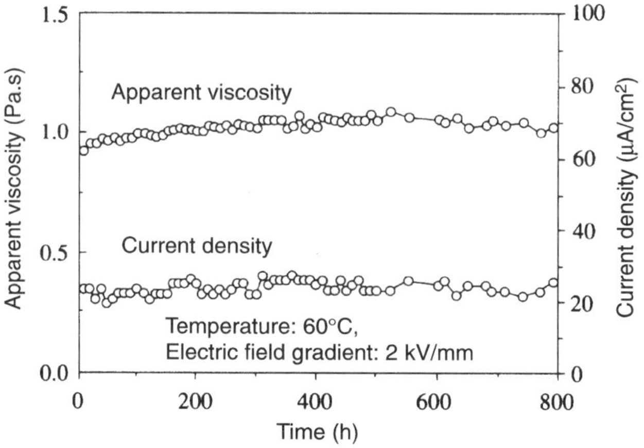

Fig. 10 Stability of apparent viscosity and current density of a carbonacous material-filled ER fluid.

Automotive parts are the most attractive application area for ER fluids. Applications for a controllable engine mount or semiactive controllable damper whose properties can be adjusted depending on driving or road conditions, respectively, are likely areas. Furthermore,

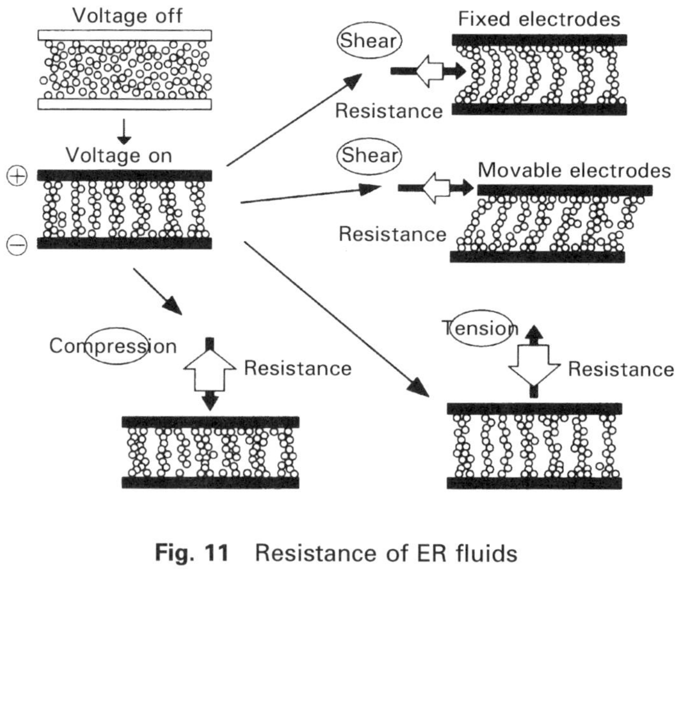

Fig. 11 Resistance of ER fluids

transmission applications are also attractive because design freedom increases and performance improves. Figures 12 and 13 illustrate engine mount and damper, respectively.

Other than for automotive parts, applications for semiconductors, electronics, vibration damping for manufacturing machines, and industrial impact absorption devices are also possible. Commercialization in these areas is expected to occur in the near future and these areas may be greatly expanded. Applications in other areas, including robotics, home appliances and construction and civil engineering, are also being evaluated.

For ER fluids to be used industrially in the aforementioned areas, it is necessary to satisfy the initial property requirements for instruments, such as viscosity without electric field, the ER effect, current value, and responsivity, as well as dispersion stability and permanence of the ER effect. In the following, examples using a carbonaceous material to explore these requirements will be described, with the focus on optimization of nonwater-containing microparticles.

[Example 1]. A carbonaceous material in which silica or alumina microparticles are homogeneously dispersed shows a good ER effect.

[Example 2]. Carbonaceous microparticle dispersed composite particle. A silica composite particle in which carbonaceous material is dispersed also shows a good ER effect.

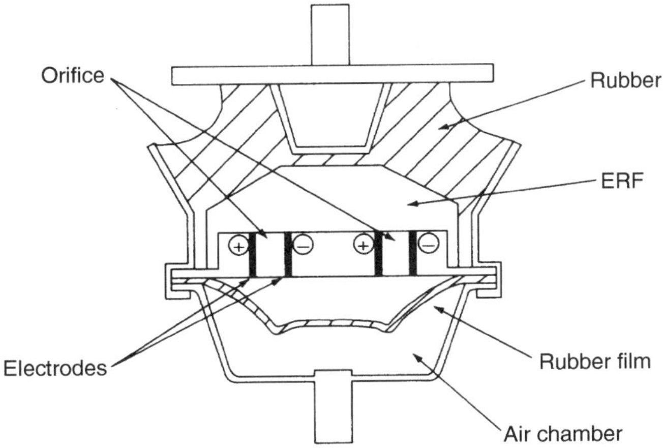

Fig. 12 Structure of ER fluid-controlled engine mount.

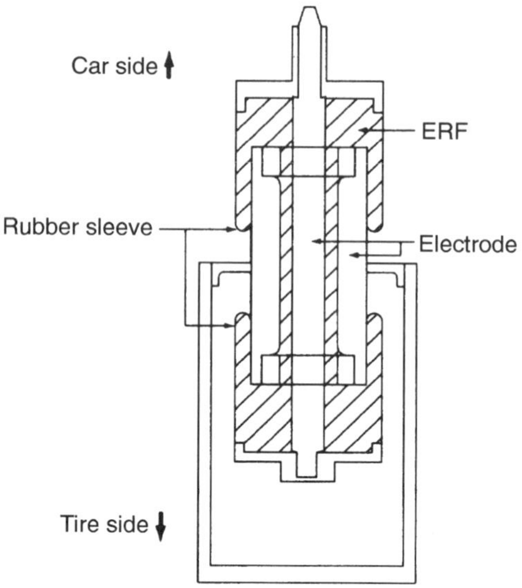

Fig. 13 Structure of ER fluid-controlled semiactive damper.

[Example 3]. Gradient functional composite particles.

Silica microparticles are dispersed at higher concentration near the surface and at lower concentration at the core of a carbonaceous material. These gradient-composite particles showed an extremely high ER effect. Figure 14 shows this microparticle.

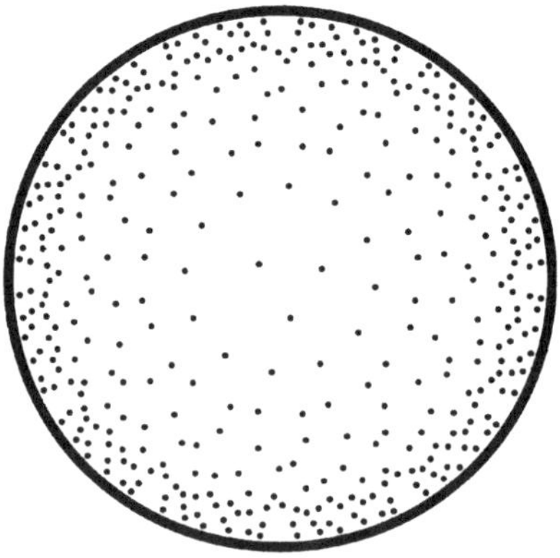

Fig. 14 A model for a gradient-functional composite particle.

[Example 4]. Upon intercalation of acrylonitrile monomer into clay gallery, it was polymerized and subsequently carbonized. These composite particles showed good ER effect with a low current.

From these results, an ideal ER particle should have a sufficiently high electrical conductivity within the particle in order to obtain high polarization. Because the particle does not charge near the surface, minimal current will pass between particles. It is also desirable to have minimum particle size without sacrifice to the ER effect, uniform internal structure with low time variation, good antifriction properties, and appropriate hardness to avoid electrode abrasion.

To achieve commercialization, the product, including a high voltage source, must be reliable and of low cost. There are many problems to be overcome in both development of fluid and design of products. However, judging from the recent development speed, nonwater-containing systems are expected to be industrially utilized in various machines and to contribute to improved performance and miniaturization.

8.3 MAGNETIC FLUIDS

TOSHIHIRO HIRAI

8.3.1 Introduction

In general, once a material has been magnetized, both the magnitude and direction of magnetization depend on the material used. As is well known, iron can be strongly magnetized in a direction along the magnetic field and thus it is attracted to a magnet whereas copper can be magnetized only slightly and will actually repel a magnet.

Magnetization of iron is called ferromagnetic and that of copper is labeled antiferromagnetic. Among materials that are attracted by a magnet, some materials are paramagnetic (i.e., they exhibit low magnetization), others have linearly proportionate magnetization, and for some there is no magnetization in the abasence of a magnetic field. An antiferromagnet, like a paramagnet, remains magnetized only in the presence of a magnetic field and the direction of magnetization is opposite. Accordingly, there are then various types of magnetization.

If this knowledge is reorganized from the viewpoint of gels, it is possible to obtain interesting insights on material development even though there are only a few examples of such studies at this time. In this subsection, the discussion will be limited only to the properties and application examples of magnetic fluids.

A fluid in which the microparticles of a ferromagnet are stably dispersed is called a magnetic fluid [91]. Microparticles range from several nm to 100 nm in size. In this particle size range, a ferromagnet shows a property called superparamagnetism. Superparamagnetism is the property in which paramagnetism occurs in the absence of a magnetic field. Such a magnetic property is observed in microparticle suspensions that consist of a single magnetic domain of sufficiently small size.

8.3.2 Properties of Magnetic Fluids

8.3.2.1 Structure and preparation of magnetic fluids

As already described, a magnetic fluid is a stable colloidal dispersion of solid ferromagnetic particles with subdomain sizes in a dispersant. The concentration of particles in a dispersant can be as high as $10^{23}/m^3$. Historically, materials similar to magnetic fluids were prepared by dispersing a ferromagnetic powder, such as iron carbonyl, with diameters (d) of

0.5–40 μm in mineral oil [92]. This suspension which aggregates under a magnetic field, has had possible uses examined. However, the magnetic fluids discussed here have particles more than three orders of magnitude finer than these coarse particles (particle diameter of ≈3–15 nm). Such magnetic fluids exhibit practically no degradation or separation as a function of time if properly prepared. Furthermore, the fluid responds reversibly to the presence of a magnetic field and no residual magnetization appears. Hence, it exhibits the properties of superparamagnetism.

Particles in a magnetic fluid are always attracted towards the direction of an applied magnetic gradient. This process competes with the diffusion of the particles due to thermal fluctuation. According to Boltzmann statistics, the maximum diameter of a particle in which thermal fluctuation overcomes the aggregation force of the magnetic field is expressed by the following equation:

$$d \leq \left(\frac{6}{\pi} \cdot \frac{kI}{\mu_0 MH}\right)^{1/3} \tag{1}$$

where k is the Boltzmann constant, T is absolute temperature, μ_0 is the permittivity of a vacuum, M is magnetization, and H is magnetic gradient.

Satisfying this condition means that, in a monodispersive particle suspension, concentration fluctuation remains in the average concentration range. For example, when $k = 1.38 \times 10^{-23}$ nm/K, $T = 298$ K, $H = 1.59 \times 10^6$ A/m, and M of the domain is 4.46×10^5 A/m, from Eq. (1), $d < 3.0 \times 10^{-9}$ m. The value derived, 3.0 nm, corresponds to the lower limit of particle size of magnetic fluids. Due to the volume occupied by the particles, the concentration change will be more limited than shown here. The ratio between the gravity acting upon a particle and the magnetization is given by

$$\text{Magnetization/gravity} = \frac{\mu_0 M|\nabla H|}{g\Delta\rho} \tag{2}$$

where g is the acceleration due to gravity and ρ is the density of magnetic particles. Under extreme conditions such as high magnetic field gradient in magnetic fluid seal, this ratio can sometimes reach 1.5×10^5. This is an extremely high value. In many colloids that are stable under gravity, they do not behave like magnetic fluids.

Two overall preparation methods for magnetic fluids will be summarized here. In one, microparticles are made using coarse particles and in the other, fine particles are obtained by chemically precipitating them. Thus, smaller particles can be obtained by spark erosion, electrochemical means, or wet grinding. Both decomposition of metal carbonyl and coprecipitation from a salt solution are used to derive smaller particles electrochemically.

The wet grinding method is best used when a low viscosity liquid and appropriate dispersion are possible. At the refining stage, the concentration of magnetic particles will be increased by solvent exchange and the excess dispersion agent will be eliminated from the solution. Under appropriate conditions, it is possible to exchange the surfactant on the particle surface [93]. Furthermore, eliminating the solvent through evaporation or diluting the particle concentration with addition of more solvent allows particle concentration to be adjusted. Figure 1 depicts an electron photomicrograph of magnetic fluid microparticles that were obtained by coprecipitation [94].

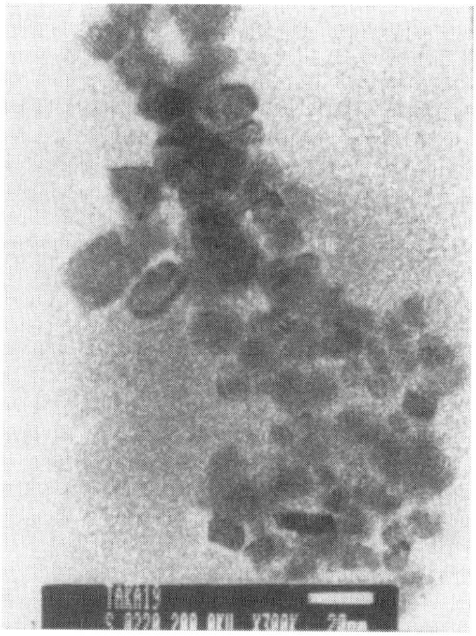

The bar in the photograph corresponds to 20 nm.

Fig. 1 Transmission electron photomicrograph (TEM) of magnetic particles.

A broad range of solvents can be used, including water, glycerin, paraffin, aromatic solvents, esters, halogenated carbons, and silicones. The requirements for the surfactant used as a dispersion agent are that it have a *head* that can adsorb onto the particle surface and a *tail* (approximately 2 nm long) that is compatible with the solvent. A polymer with an appropriate adsorbing group (anchoring group) is the most suitable stabilizing agent, even though spatial occupancy is too high. Compatibility between the tail and the solvent can be estimated by a solubility parameter. However, accurate interaction and dispersion stability must be determined experimentally.

8.3.2.2 Stability of colloidal dispersion systems

Among dispersed magnetic particles, four different interparticle interactions exist—these are the van der Waals forces, magnetic attraction, steric repulsive forces, and electric repulsive forces. Van der Waals forces or London dispersion forces originate from the interaction between orbital electrons or induced vibrating dipoles. For the equivalent two spherical particles, Hamaker's equation holds [95]. This force is strong only within short distances.

When particle diameter *d* is less than the critical value of several tens to several hundred nanometers, the magnetic particles will be single domain. Hence, the particles in a magnetic fluid are considered homogeneously magnetized single-domain particles. In this case, the potential energy within a magnetic particle pair can be accurately described using the equation for dipoles. This potential reaches a long distance and changes gradually as a function of distance. When there is no magnetic field, thermal motion reduces the orientation of dipoles and attraction energy also reduces.

Steric repulsion appears when long, flexible molecules adsorb on the magnetic particle surface. It has already been mentioned that these molecules have a polar anchoring group as the head and a simple chain portion as the tail. The long tail forms a loop. Except for the anchored head, the adsorbed molecules undergo thermal motion. When the second molecule approaches the already adsorbed molecule, the portion of the molecular chain will be spatially (entropically) restricted. For solvent molecules that are solving the adsorbed molecule, it is also necessary to overcome this energy of solvation. This latter, enthalpic influence works in both directions. When polymer molecules are significantly self-associating there will be less repulsive force or rather attractive force. This indicates that steric stabilization can be greatly influenced by solvent composition. Hence, it is difficult to calculate the repulsive energy of adsorbed polymers, or the calculated values are inaccurate. However, it is possible to estimate such entropic effect of short polymer chains used for magnetic fluids. Although electric repulsive forces are not usually considered, this might become important with some preparation methods. Electric repulsive forces between particles are the coulombic repulsion between charged surfaces. The charged state changes depending on the desorption of ions from the surface or adsorption onto the surface. The repulsive forces are known to reduce by the screening effect of the surrounding ions. The net potential curve is determined by the sum of these attractive and repulsive forces.

8.3.2.3 Magnetic properties

A system in which particles of magnetic moment m are dispersed in a liquid is similar to a paramagnetic gas. Dipoles that are in equilibrium under an external magnetic field fluctuate somewhat due to thermal movement. By applying Langevin's classical theory, superparamagnetic properties can be explained if interparticle magnetic interactions can be ignored. Brownian motion becomes important here. Under different circumstances, when magnetic particles become too small, a single magnetic domain cannot be maintained. In this case, the response of the internal structure of the particle to the external magnetic field leads to the superparamagnetic property.

For real magnetic fluids, two more parameters that influence the magnetization curve must be considered. One is the distribution of particle size, which can be determined by electron microscopy. Another factor is the reduction of effective magnetic radius as a result of reduced particle diameter d_S. This is necessary when nonmagnetic surface layers are formed by the chemical interaction of the particle surface and the dispersion agent. Chain-like aggregation formation of colloidal magnetic particles and the influence of a homogeneous magnetic field on this process are worth attention. De Gennes and Pincus [96] and Jordan [97] assumed that this process could be expressed by the equation of state of dilute gases. The degree of deviation from the ideal gas corresponds to the attractive forces between magnetic particles. Other forces are ignored. By considering pairwise correlation, it was found that ferromagnetic particles form a chain structure along the parallel direction of a strong magnetic field. The average number of particles in the chain can be expressed as follows using the volume fraction ϕ of the magnetic solids,

$$n_\infty = \left[1 - \tfrac{2}{3}(\phi/\lambda^2)e^{2\lambda}\right]^{-1} \tag{3}$$

where λ is the nondimensional connection coefficient and the quantity $\lambda = \mu_0 m^2/4\pi d^3 kT$ is the measure of the interparticle interaction strength. Here, m is the dipole moment. When the second term of Eq. (3) is greater than 1, the approximation does not hold. In this case, clusters rather than linear chains are formed. Even if the external magnetic field is zero, if $\lambda \gg 1$, existence of chains with certain lengths is predicted. The average chain length n_0 is expressed by the following equation:

$$n_0 = \left[1 - \tfrac{2}{3}(\phi/\lambda^3)e^{2\lambda}\right]^{-1} \tag{4}$$

This chain length is shorter than the case with strong magnetic field and the chains are randomly oriented. Peterson and Krueger studied cluster formation of a magnetic fluid in a tube placed perpendicular to the magnetic field direction [98]. Cluster formation is notable in water-based magnetic fluids. These clusters redisperse by thermal fluctuation when the magnetic field is removed.

8.3.2.4 Viscosity of magnetic fluids

Magnetic fluids are materials that possess the properties of both magnetic materials and magnetic fluids. The fluidity of the system is maintained even under saturation magnetization. However, the existence of a magnetic field influences rheological behavior. In the following, the influence of the magnetic field on the viscosity of magnetic fluids will be summarized.

In the absence of a magnetic field, the system behaves like nonmagnetic colloids in which solid particles are dispersed in a liquid. In this case, a theoretical model can be applied. A theory derived by Einstein states that the strain of a flow field is perturbed by the presence of a sphere.

It is desirable for magnetic fluids to possess high particle concentration, high saturation magnetic moment, spherical particle with large diameter r, thin adsorbed layer with thickness δ, and large fluidity. Such requirements for δ or r conflict with the conditions required by a stable colloid. Thus, in a practical magnetic fluid, a compromise must be sought. If a magnetic field is applied to a magnetic fluid under shear stress, the magnetic particles try to maintain their orientation along the magnetic field. As a result, the velocity field around the particles becomes larger than without the presence of particles. Rosensweig *et al.* studied the effect of a perpendicular magnetic field on the viscosity of a thin horizontal layer of a magnetic fluid under a homogeneous shear stress along the horizontal direction [99]. Upon dimensional analysis, the following relationship was derived:

$$\eta_H/\eta_S = f(\Gamma), \Gamma = \gamma\eta_0/\mu_0 MH \tag{5}$$

where η_H is the viscosity under magnetic field, η_s is the viscosity of the magnetic fluid in the absence of magnetic field, η_0 is the viscosity of the dispersing fluid, M is the magnetization of the magnetic fluid, γ is the shear rate, and H is the applied magnetic field. The relationship between the relative viscosity η_H/η_s in Eq. (5) and torque modulus Γ is anti-sigmoidal and it is roughly divided as follows. In the $0 < \Gamma < 10^{-6}$ region, relative viscosity maintains a maximum value. In the $10^{-6} < \Gamma < 10^{-4}$ region, viscosity depends on the magnetic field and shear stress. And in the $10^{-4} < \Gamma < \infty$ region, the viscosity is constant independent of the magnetic field. Furthermore, when viscous flow and magnetic field are parallel for the dilute dispersion of single domain spherical particles, the particles freely rotate and the magnetic field will not influence viscosity. Such a system obeys Einstein's relationship. On the other hand, when they are perpendicular to each other, the influence of the magnetic field on the viscosity is the greatest and this is expressed as follows:

$$\eta_H = \eta_0\left(1 + \frac{4\alpha + \tanh\alpha}{\alpha + \tanh\alpha}\phi\right)(\Omega\tau_B \ll 1) \tag{6}$$

where $\alpha = \eta_0 mH/kT$. The equaiton is for the $\Omega\tau_B \ll 1$ case and the rotation is somewhat restricted (Ω is the rotational speed of the liquid and τ_B is the relaxation time of rotation by Brownian motion). When α is small, this equation is reduced to Einstein's equation, $\eta_H = \eta_0(1 + 2.5\phi)$. However, when α is large, η_H approaches $\eta_0(1 + 4\phi)$. The particles are fixed physically by the magnetic field and rotation is restricted. In the fluid flowing in a capillary, theoretical prediction can be confirmed.

8.3.2.5 Fluid dynamic properties and various behaviors

There are many interesting and important reports on the fluid dynamic properties of magnetic fluids. In the following, characteristic behaviors will be introduced.

Corn-shaped meniscus Neuringer and Rosensweig found that a magnetic fluid creeps up to an electric wire when it is placed perpendicular to a fluid surface. The observed shape is illustrated in Fig. 2 [100]. The height of the meniscus Δh is a function of magnetic field H and is expressed as follows.

$$\Delta h = h - h(\infty) = \mu_0 \bar{M} H / \rho g \tag{7}$$

The magnetic field H has the following relationship with current I and the radial distance from the electric wire r_d,

$$H = I / 2\pi r_d \tag{8}$$

where $\bar{M}$ is the average magnetization (see Eq. 7).

Bulging of interface by magnetic field As shown in Fig. 3, if external magnetic field is applied perpendicularly to a magnetic fluid interface, the interface bulges by Δh. The magnetic field above and below the interface is H_2 and H_1, respectively.

$$\Delta h = h_2 - h_1 = \frac{1}{\rho g}\left(\mu_0 \bar{M} H_1 + \mu_0 \frac{M^2}{2}\right) \tag{9}$$

This relationship was derived from Bernoulli's equation. Berkovsky and Orlov analytically investigated many problems of the free interface of magnetic fluids [101].

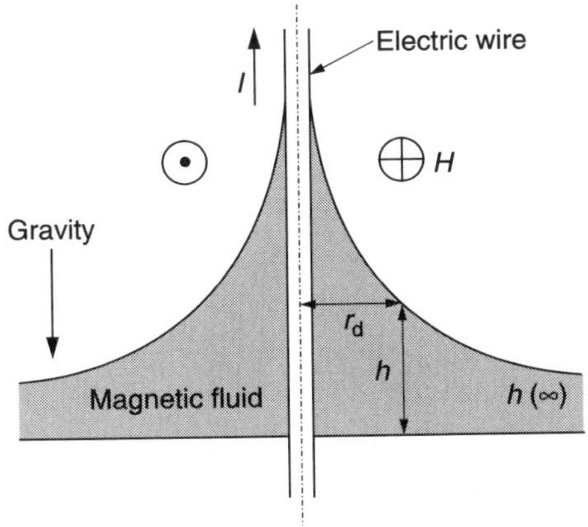

Fig. 2 Meniscus formation of a magnetic fluid by magnetic field.

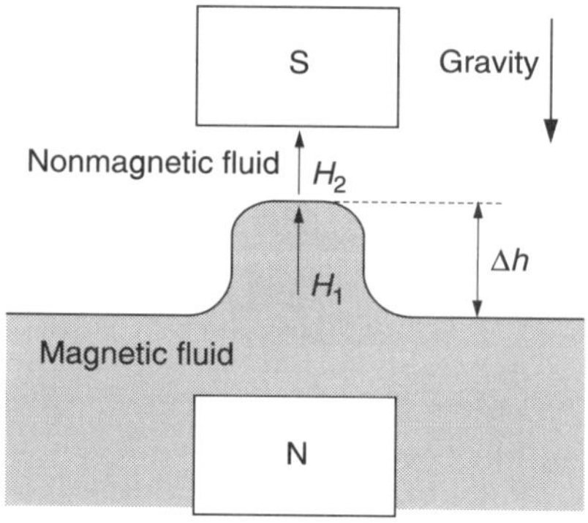

Fig. 3 Bulging of magnetic fluid interface.

Magnetic levitation It has been pointed out theoretically that an isolated collection of charges cannot be levitated stably by a static magnetic field. However, antiferromagnetic materials or superconducting materials are not restricted by the Earnshaw theory. As shown in Fig. 4, a nonmagnetic material levitates in a magnetic material under an applied magnetic field. Even if there is no external magnetic field, a permanent magnet self-levitates in a magnetic fluid [102].

8.3.3 Applications of Magnetic Fluids

As already described, magnetic fluids show various properties. There are application examples using such properties.

Seal bearings Sealing, which separates areas of differing pressures, is one of the most successful examples of magnetic fluid applications. A bearing using a magnetic fluid utilizes levitation and, unlike with a ballbearing, it produces almost no sound and has nearly no friction.

Dampers A damper using a magnetic fluid transfers kinetic energy into thermal energy using fluid viscosity and dissipates energy as heat. Applications range from delicate instruments to electromagnetic transport of materials.

Transformers A transformer is a device that changes one physical quantity to another physical quantity. Magnetic fluids possess various properties that are required for transformers. These transformers include a sound transformer, a pressure generator, a position detector, a current

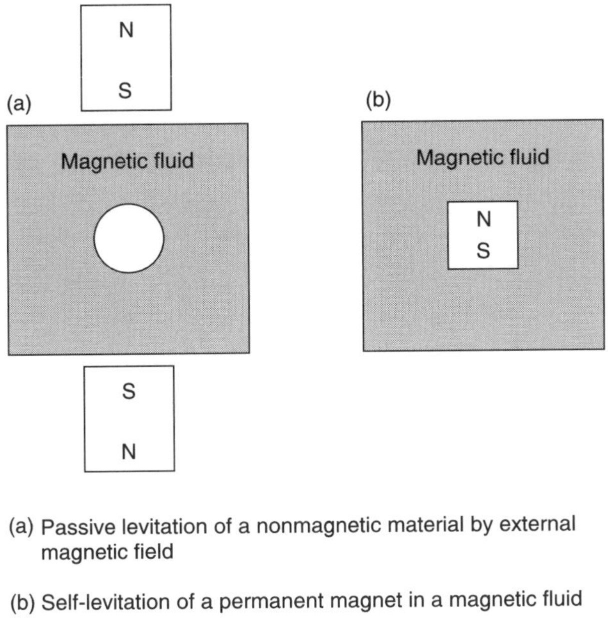

Fig. 4 Magnetic levitation phenomena by a magnetic field.

detector, an accelerometer and a water position detector. Furthermore, the properties of magnetic fluids can be applied to capillary collection, magnetic detection, extensometers, as well as display and optical shutters.

Actuators A mechanical actuation example will be introduced. Two fixed points are connected by a bag, which contains a magnetic fluid. A solenoid coil is installed in the bag. In this condition, if current is passed through the coil, it shrinks in the radial direction and stretches along the axial direction [103]. The advantage of this system is lack of fatigue during actuation (see Fig. 5).

8.3.4 Magnetic Fluids and Gels (Magnetic Deformation of Magnetic Fluid-Containing Gels)

Because magnetic fluids possess many interesting properties there have been attempts to use these properties in gels. The first method incorporates a magnetic fluid into the gel networks by inclusion fixation. The other

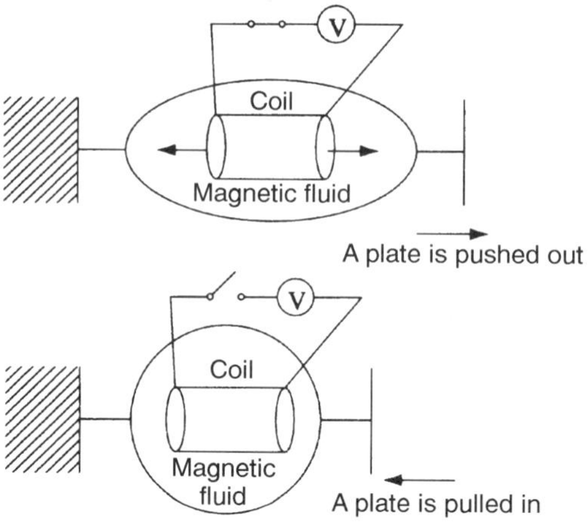

Fig. 5 An actuator utilizing a magnetic fluid.

approach fixes a magnetic fluid at the point at which dispersion of the magnetic fluid is destabilized. The polymer used for gelation was poly-(vinyl alcohol) (PVA) and the solvent was water or dimethylsulfoxide (DMSO). The degree of polymerization of the PVA used was 1700.

8.3.4.1 Fixation of a magnetic fluid into a gel

The magnetic fluid, which is coated and stabilized by oleic acid, is mixed with a PVA aqueous solution. After repeated freezing and thawing, a hydrogel was prepared. By changing composition, magnetic fluid content can be controlled. At a PVA concentration of <8%, the mechanical strength of the gel weakens significantly and the gel cannot be used [104].

Fixation of DMSO gels The same magnetic fluid as the one described in the hydrogel section was mixed with a DMSO solution of PVA. The gel obtained was a DMSO/H_2O one at a low temperature. The DMSO/H_2O ratio and the magnetic fluid content can be changed. In this case, even if the PVA concentration is 4%, gelation is achieved. Such a gel is called a DMSO gel. In any case, gel modulus can be changed by altering the concentration of magnetic fluid or polymer. Generally speaking, the modulus increases when the polymer concentration or its degree of polymerization increases but the modulus decreases with increased magnetic fluid content.

8.3.4.2 *Morphology of magnetic fluid-fixed gels*

Observation of the surface of a magnetic fluid-fixed hydrogel showed the same macroscopic networks as in a freeze-thaw PVA gel. In the case of the magnetic fluid-fixed gel, the networks become only slightly larger. Similar results are also observed in the case of the DMSO gel. Immersing these gels into acetone and subsequent drying to make a xerogel will highlight the structural differences. According to the SEM pictures shown in Fig. 6, the dried hydrogel shows large and clear network structures whereas the dried DMSO gel shows aggregates of microparticles. Thus, the gel structures appear to have some differences. Although these SEM images do not reflect the original gel morphology, they nonetheless maintain enough information to study the structural differences.

(a) (b)

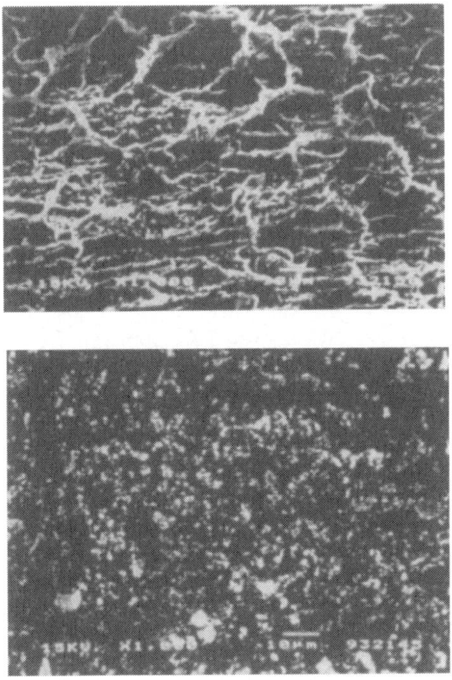

(a) from hydrogel, (b) from DMSO gel
The bar on the photograph corresponds to 10 μm

Fig. 6 Scanning electron photomicrograph (SEM) (cross section) of a dried gel in which magnetic fluid is fixed.

8.3.4.3 Magnetic field-induced deformation of magnetic fluid-fixed gels

In the following, how these gels show deformation behavior under an applied magnetic field will be discussed [105]. Although magnitude of magnetic deformation is small, it is characteristic for the hydrogel to expand along the magnetic field direction and the DMSO gel to shrink. In terms of rate, it is approximately the same as for electrodeformation. Ninety percent of all deformation occurs in 0.1 s. When the magnetic material content is constant, the strain depends on the magnetic fluid content, polymer content, and gel modulus. This remarkable difference between the hydrogel and DMSO gel is caused by the magnetic fluid property in the gel. From morphological observation, it was found that a magnetic fluid is forming aggregates in the DMSO gel and PVA is coating and connecting them (see Fig. 7).

On the other hand, in the case of the hydrogel, both the PVA and the gel that does not contain magnetic fluid have similar morphology. No obvious aggregation structures of the magnetic particles can be observed. Such differences suggest that either the magnetic fluid forms separate domains in the hydrogel and during xerogel preparation it is lost or it is homogeneously distributed in the PVA matrix. According to optical microscopic observations, the hydrogel showed a homogeneous distribution and the DMSO gel showed an aggregated structure of the magnetic particles. Further observation by TEM indicated that ferrite particle size in the magnetic fluid was 5–20 μm.

Considering the aforementioned fundamental properties of magnetic fluids, when a magnetic fluid forms a stable dispersion with such particle sizes, a spiking or bulging structure is formed in a magnetic field. Thus, if the magnetic particles fixed in the gel still maintain the properties of magnetic fluid, they will deform along the direction of the magnetic field. On the other hand, in the case of a DMSO gel, in which macroscopic aggregates are dispersed, the association of the aggregates accelerates by magnetic field application and the gel shrinks along the magnetic field direction.

8.3.4.4 Structural changes accompanying deformation of magnetic fluid-fixed gels

It is useful for material design of magnetic field-responsive gel actuators to elucidate the structure of the magnetic fluid in the gel or the gel structure itself in a magnetic field. For this purpose, structural changes of gels have

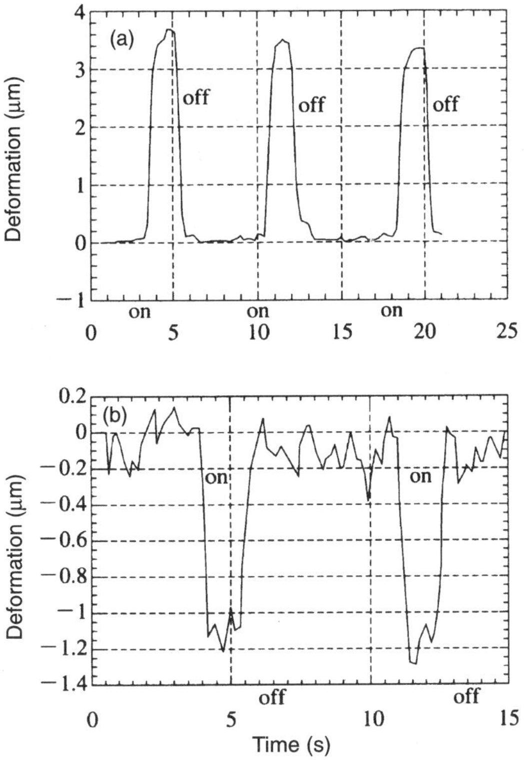

(a) Hydrogel: PVA 8 wt%, magnetic fluid 75 wt%
(b) DMSO gel: PVA 4 wt%, magnetic fluid 50 wt%

Fig. 7 Actuation of a magnetic fluid-containing gel by magnetic field.

been studied by small angle x-ray scattering [106]. The magnetic fluid contents of the hydrogel and DMSO gels used were both 5% and 25%. Unlike the sample used in the actuation study, the magnetic fluid contents were low due to the high x-ray absorption of the magnetic particles. This measurement became possible for the first time with a high-intensity x-ray from a synchrotron source. The measurement range in terms of scattering vector was $0.06 < q_s < 2$. In this scattering, there is a shoulder around $q_s = 0.75\ \text{nm}^{-1}$ which corresponds to approximately 8.0-nm particle size. This shoulder reflects relatively weak nearest neighbor particle interactions. This shoulder is more clearly observed in the hydrogel than in the DMSO gel; correspondingly dispersion stability is better in the hydrogel due to a surfactant. From the TEM photograph, the magnetic particle size is ~10 nm, showing good agreement with a value of 8.0 nm. Because particle size distribution broadens by particle aggregation, this shoulder becomes unclear in the DMSO gel.

To obtain further detailed knowledge, inverse Fourier transformation was applied to this scattering pattern and the distance distribution function is summarized in Table 1. It is shown that magnetic field influence is different for the hydrogel than for DMSO gels. The maximum vector length within the particle is 50–60 nm. This differs from the TEM observation, which indicates that a similar aggregation state is observed in both gels. For the 25% gel in which actuation is observed, this value increased in the hydrogel but decreased in the DMSO gel upon application of the magnetic field. This change is clear and corresponds to opposite direction deformation when they were actuated.

Table 1 also lists the Porod slope value α_P [107]. This is the slope of the log-log plots of the scattering function $I(q_s)$ and q_s. This slope provides knowledge on the fractal structure of the scattering materials. Here, slopes in the range $0.22 < q_s < 0.45$ are shown. The slopes in the range $-4 < \alpha_s < -3$ correspond to a surface fractal, although the actually observed value is $\alpha_P = -3.2$. The q_s range studied corresponds to particle sizes of 14–29 nm. Hence, the observed Porod slope corresponds to the interface with 14–29-nm fractal roughness. In the gel, which contains 25% magnetic fluid, the α_P value increases for the hydrogel whereas it decreases for the DMSO gel. The DMSO gel tends to tighten the aggregates under magnetic field. The momentum radius R_G based on the Guinier analysis is also listed in the table [108]. To evaluate the aggregation structure, relative changes in radii were compared. In the gel that contains 25% magnetic material, R_G changed significantly in the hydrogel whereas in the DMSO gel only a slight change was observed.

Table 1 Structural changes of the hydrogel and DMSO gel in which a magnetic fluid is fixed.

	Magnetic field (T)	Porod slope	D_{max} (nm)	R_G (nm)
Hydrogel	0	−3.255	50.3	16.2
	1.4	−3.146	58.0	17.5
DMSO gel	0	−3.152	59.8	19.7
	1.4	−3.168	55.8	18.4

Magnetic material content = 6 wt%, PVA content = 8 wt%.
DMSO content in the DMSO gel = 41 wt%.

The structural changes induced by the magnetic field depend on the state of the magnetic particles in the gel. Such structural differences are well reflected on the mode of actuation.

8.3.5 Conclusions

Magnetic fluid studies have been actively pursued in electronic engineering but only very little in polymer science. This trend might be related to the recognition that magnetic materials are inorganic. However, by viewing these materials as hybrids of inorganic materials and organic polymers, new developments may be expected. If a magnetic fluid can be incorporated into a polymer in a stable manner while fluidity is maintained, it will be possible to apply these materials as magnetic field-induced deformation materials. Magnetic fluid-fixed polymer gels showed different behaviors depending on the gel preparation methods. This suggests that it is possible to develop materials that maintain morphology and also exhibit characteristics of a magnetic fluid.

REFERENCES

1 Minami, H. (1978). Polyelectrolyte, Committee on Polymer Experiments, Soc. Polym. Sci., Jpn., ed., Kyoritsu Publ.

2 Hanai, T. and Asami, K. (1991). *Lecture Series on Experimental Chemistry 9: Electricity and Magnetism*, 4th ed., Maruzen Publ.

3 Komatsu, N., Nitta, T., Miho, T., Gong, J.P., and Osada, Y. (1996). *Polymer Preprints, Jpn.* **45**: 366.

4 Kwak, J.C.T., and Hayes, R.C. (1975). *J. Phys. Chem.* **79**: 265.

5 Oka, S. (1954). *Dielectricity Theory*, Iwanami Publ.

6 Oka, S. and Nakada, O. (1960). *Solid-State Dielectricity Theory*, Iwanami Publ.

7 Chen, P., Adachi, K., and Kodaka, T. (1992). *Polymer* **33**: 1813.

8 Chen, P., Adachi, K., and Kodaka, T. (1992). *Polymer* **24**: 1025.

9 Chen, P., Adachi, K., and Kodaka, T. (1993). *Polymer* **25**: 473.

10 Hashimoto, H., Inamori, I., Chiba, A., and Tajitsu, Y. (1995). *Proc. 7th Symposium on Polymer Gels, Japan.*, p. 23.

11 Gong, J.P. and Osada, Y. (1995). *Chem. Lett.* **6**: 449.

12 Tajitzu, Y., Ogura, H., Chiba, A., and Furukawa, T. (1987). *J. Appl. Phys., Jpn.* **36**: 554.

13 Furusawa, H., Kimura, Y., Ito, K., and Hayakawa, R. (1993). *R.P.P.J.* **36**: 55.

14 Furusawa, H., Kimura, Y., Ito, K., and Hayakawa, R. (1995). *Proc. 7th Symposium on Polymer Gels, Japan.*, p. 17.

15 Ito, K. and Hayakawa, R. (1993). *Fundamentals on Polymer Properties*, Chapter 1, Section 4, Kyoritsu Publ.

16 Pathmanathan, K., and Johari, G.P. (1990). *Polym. Phys.* **28**: 675.

17 Gong, J.P., Nitta, T., and Osada, Y. (1994). *J. Phys. Chem.* **98**: 9583.

18 Osada, Y., Umezawa, K., and Yamauchi, A. (1989). *Bull. Chem. Soc. Jpn.* **62**: 3232.

19 Miyano, M.K., and Osada, Y. (1991). *Macromolecules* **24**: 4775.
20 Sawahata, K., Gong, J.P., and Osada, Y. (1995). *Macromol. Rapid Commun.* **16**.
21 Miyano, M., and Osada, Y. (1990). *Polym. Preprints, Jpn.* **39**: 620.
22 Ohnishi, S., and Osada, Y. (1991). *Macromolecules* **24**: 6588.
23 Duff, A.W. (1896). *Phys. Rev.* **4**: 23.
24 Quinke, G. (1897). *Ann. Phys.* **62**: 1.
25 Herzog, R.O., Kudar, H., and Paersch, E. (1934). *Phys. Z.* **35**: 446.
26 Shiga, T., Okada, A., and Kurauchi, T. (1993). *Macromolecules* **26**: 6958.
27 Koyama, K. (ed.) (1994). *Development and Application of ER Fluids.*
28 Winslow, W.M. (1947). US Patent 417,850.
29 Winslow, W.M. (1949). *J. Appl. Phys.* **20**: 1137.
30 Block, H., and Kelly, J.P. (1988). *J. Phys. D.: Appl. Phys.* **21**: 1661.
31 Inoue, A. (1992). *J. Rheol. Soc., Jpn.* **20**: 67.
32 Otsubo, Y., and Edamura, K. (1994). *J. Colloid Interface Sci.* **168**: 230.
33 Gamota, D.R. and Filisko, F.E. (1991). *J. Rheol.* **35**: 399.
34 Block, H., and Kelly, J.P. (1989). *Proc. First Int. Conf. on Electrorheological Fluids*, English Publ., p. 16.
35 Inoue, A. (1990). *Proc. Second Int. Conf. on Electrorheological Fluids*, Technomic Publishing, p. 176.
36 Otsubo, Y. and Watanabe, K. (1990) *J. Rheol. Soc. Jpn.* **18**: 111.
37 Klass, D.L. and Martinek, T.W. (1967). *J. Appl. Phys.* **38**: 67.
38 Kondo, T., and Sugimoto, A. (1973). *Report on the Industrial Safety Institute*, p. 1.
39 Otsubo, Y. and Sekine, M. (1990). *Proc. 38th Rheology Symposium*, p. 129.
40 Boissy, C., Atten, P., and Foulc, J.N. (1995). *J. Electrostatics* **35**: 13.
41 Takimoto, J. (1992). *J. Rheol. Soc. Jpn.* **20**: 95.
42 Takimoto, J. (1991). *Proc. Int. Conf. Electrorheological Fluids*, p. 15.
43 Yang, I.K., and Shine, A.D. (1992). *J. Rheol.* **36**: 1079.
44 Inoue, A., Maniwa, T., Sato, T., and Taniguchi, K. (1993). *Proc. 41st Rheology Symp.*, p. 73.
45 Honda, T., and Sasada, T. (1977). *Jpn. J. Appl. Phys.* **16**: 1775.
46 Sasada, N., and Honda, T. (1980). *Machine Res.* **32**: 983.
47 Inoue, A. (1995). *J. Appl. Polym. Sci.* **55**: 113.
48 Fukumasa, M., Yoshida, K., Ohkubo, S., and Yoshizawa, A. (1993). *Ferroelectrics* **147**: 395.
49 Tanaka, K., Fujii, A., and Koyama, K. (1992). *Polym. J.* **24**: 995.
50 Uemura, T., Minagawa, K., Takimoto, J., and Koyama, K. (1995). *J. Chem. Soc. Faraday Trans.* **91**: 1051.
51 Duff, A.W. (1896). *Phys. Rev.* **4**: 32.
52 Winslow, W.M. (1949). *J. Appl. Phys.* **20**: 1137.
53 Winslow, W.M. (1947). US Patent 2417850.
54 Koyama, K. (1994). *Development and Application of ER Fluids*, CMC.
55 Havelka, K.O., and Filisko, F.E. (1995). *Progress in Electrorheology*, New York: Plenum.
56 Sugimoto, A. (1985). *Junkatsu* **30**: 859.
57 Sugimoto, A. (1977). *J. Soc. Mechanics, Jpn. (Part 2)* **43**: 1075.
58 Klass, D.L., and Martinek, T.W. (1967). *J. Appl. Phys.* **38**: 67.
59 Martinek, T.W., and Klass, D.L. (1968). US Patent 3412031.
60 Martinek, T.W., Klass, D.L., and Folkins, H.O. (1968). US Patent 3367872.
61 Stangroom, J.E. (1978). US Patent 4129513.

62 Stangroom, J.E. (1991). *J. Stat. Phys.* **20**: 859.
63 Izutsu, Y., and Yoshiwa, M. (1974). *Tokkyo Kokai*, 5117.
64 Sugimoto, A., Omura, T., and Inoue, H. (1991). *Tokkyo Kaiho* 200897.
65 Fujii, Y., Nakamura, E., Sato, H., and Kanya, T. (1983). *Tokkyo Kaiho* 33459.
66 Kobayashi, H., Asako, Y., Shimomura, T., and Sano, T. (1989). *Tokkyo Kaiho* 180238.
67 Kobayashi, H., Asako, Y., Shimomura, T., and Sano, T. (1989). *Tokkyo Kaiho* 180239.
68 Asako, Y., Kobayashi, H., Shimomura, T., and Sano, T. (1990). *Tokkyo Kaiho* 35933.
69 Asako, Y., Okada, I., Aoki, M., Ono, T., and Kobayashi, M. (1992). *J. Soc. Rheology, Jpn.* **20**: 61.
70 Asako, Y. (1994). *Development and Application of ER Fluids*, CMC, p. 56.
71 Asako, Y., and Itoh, K. (1994). *Kino Zairyo* **14**: 24.
72 Weiss, K.D,. Coulter, J.P., and Carlson, J.D. (1990). In *Recent Advances in Adaptive and Sensory Materials and Their Applications*, C.A. Rogers and R.C. Rogers, eds., Lancaster, PA: Technomic, p. 605.
73 Ishino, Y., Maruyama, T., Ozaki, T., Endo, S., Saito, T., and Goshima, K. (1993). *Proc. 41st Rheology Symp.*, p. 112.
74 Kingenberg, D.J., and Zukoski, C.F. (1990). *Langmuir* **6**: 15.
75 Itoh, K. (1994). *Yuatsu & Kukiatsu* **25**: 824.
76 Konishi, M., Kawakami, T., Aizawa, R., and Asako, Y. (1995). *Proc. 4th Polym. Mat. Forum*, p. 199.
77 Winslow, W.M., (1947). US Patent 2417850.
78 Stangroom, J.E. (1978). US Patent 4129513.
79 Block, H. *et al.* (1987). European Patent 4687589.
80 Catalogue from Bayer Co.
81 Ishino, Y., and Ozaki, T. *et al.* (1991). *Tokkyo Kaiho* 47896.
82 Inoue, A. (1988). *Tokkyo Kaiho* 97694.
83 Edamura, K., and Otsubo, T. (1995). *Tokkyo Kaiho* 26284.
84 Recent Topics, (1994). *Kobunshi*, 236.
85 Kurachi, I., and Fukuyama, Y. (1992). *Tokkyo Kaiho* 227996.
86 Kurachi, I., and Fukuyama, Y. (1992). *Tokkyo Kaiho* 227996.
87 Kurachi, I., and Saito, T. (1992). *Tokkyo Kaiho*, 227796.
88 Fukuyama, Y., and Kurachi, I. *et al.* (1991). *Tokkyo Kaiho*, 252498.
89 Ushijima, T., Takano, K., and Noguchi, T. SAE Tech, Paper No. 880073.
90 Ishino, Y., Saito, T., and Goshimo, K. (1994). *M & E* **74**.
91 Rosensweig, R.E. (1971). *Ferrohydrodynamics* Encycl. Dictionary Phys. Suppl. **4**: 411.
92 Rabinow, J. (1949). *Franklin Inst.* **248**: 155.
93 Rosensweig, R.E. (1975). US Patent 3620584
94 Shinoiizaka, J. *et al.* (1976). *J. Chem. Soc., Jpn.* **6**.
95 Kruyt, H.R. (1952). *Colloid Science*, vol. I, New York: Elsevier.
96 de Gennes, P.G., and Pincus, P.A. (1970). *Phys. Kondens Mater.* **11**: 188.
97 Jordan, P.C. (1973). *Mol. Phys.* **25**: 961.
98 Peterson, E.A., and Kruger, D.A. (1978). *J. Colloid Interface Sci.*
99 Rosensweig, R.E., Nestor, J.W., and Timmins, R.S. (1965). Mater. Assoc. Direct Energy Conversion, Symp. *AIChE-I, Chem. E. Ser.* **5**: 104.
100 Neuriinger, J.L., and Rosensweig, R.E. (1964). *Phys. Fluids* **7**: 1927.
101 Berkovsky, B.M., and Orlov, L.P. (1973). *Magnetohydrodynamics (English Transl.)* **38**.

102 Rosensweig, R.E. (1978). *Thermodynamics of Magnetic Fluids*, B. Berkovsky, ed., Washington, D.C.: Hemisphere, p. 195.
103 Sabelman, E.E. (1972). *NASA Jet Propulsion Laboratory, S/N 235*, 295, Pasadena, CA.
104 Hirai, T. *et al.* (1993). *Polym. Preprints, Jpn.* **43**: 3126.
105 Takamizawa, T., Hirai, T., Hayashi, S., and Hirai, M. (1994). *Proc. Int. Symp. Fiber Sci. and Technol.*, (ISF'94 in Yokohama), p. 293.
106 Hirai, T. (1995). *Zairyo Kagaku* **32**: 59; Takamisawa, T. *et al.* (1995). *Proc. 7th Polym. Gel. Symp.*, p. 69.
107 Porod, G. (1953). *J. Polym. Sci.* **10**: 157.
108 Guinier, A. (1939). *Ann. Phys.* **12**: 161.

Section 9
Shape Memory Properties

YOSHIHITO OSADA

9.1 INTRODUCTION

Shape memory generally means the recovery or restoration of deformed material to its original shape. It appears as though the material remembered its original shape. Around 1964, Ni-Ti alloys were found to possess shape memory properties. In the 1970s, the properties of shape memory alloys were clarified. They possess ordered structure and a thermoelastic martensitic polymorph. Since then, shape memory capability has been an important consideration in the search for functional materials. Along with the development of processing techniques, various machine parts and devices that use shape memory are being developed.

Polymers are viscoelastic materials in which deformations that are caused by an external force are time-dependent. Hence, the shape of the material is influenced by the previous mechanical history and it exhibits the memory phenomenon [1]. A typical shape memory aspect of polymeric materials is that of rubber elasticity. A rubber returns to its original shape even after terrific deformation. On the other hand, plastics return to their original shape slowly after they have been stretched or bent. There are fibers or stretched films that shrink suddenly at a certain temperature. In a broad sense, polymeric materials more or less possess shape memory effects. It is possible to enhance the memory further and add the shape memory function to a polymer by purposefully controlling both the molecular and the supramolecular structures. A shape memory polymer can exhibit shape memory after being exposed to heat, chemicals, and other physical forms of energy. Hence, a broad range of applications is possible.

The mechanisms of the shape memory effect in polymeric materials will be summarized here. Furthermore, examples of shape memory gels [2–5] and their properties will be described. These shape memory gels are soft and wet materials that are similar to biomaterials. Accordingly, there is potential for application either in an actuator or as implanted artificial organs that are light, do not corrode or break, are easy to process, and which possess the properties of polymers.

9.2 SHAPE MEMORY OF POLYMERS

For polymer materials to show shape memory properties, it is necessary for them to be chemically (crosslinking agent) or physically (entanglement of polymer chains, ionic bonds, etc.) crosslinked [2]. Here, crosslinking indicates formation of networks by fixing relative positions of polymer chains through chemical or physical bonds. Shape memory of polymeric materials can be classified as follows.

9.2.1 Use of Glass Transition Temperature (T_g) of Polymers

This method involves deformation of a polymer above the glass transition temperature, fixing of the deformation below T_g, and reheating to recover the original shape (see Fig. 1). Here T_g is the temperature at which polymer main chains begin micro-Brownian motion. Below this

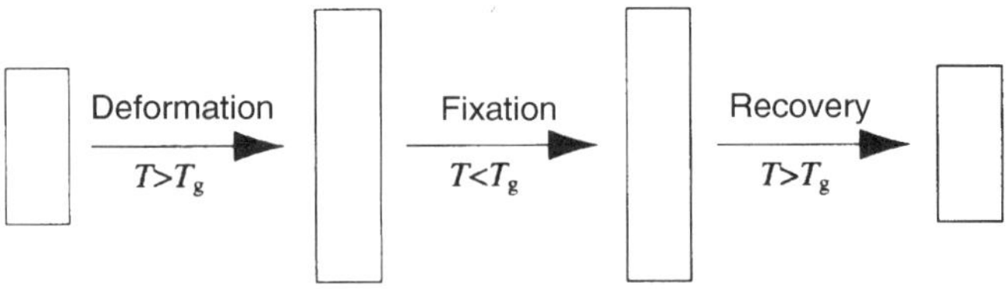

Fig. 1 Shape memory behavior by T_g.

temperature, polymers are hard and above this temperature they are as soft as rubbers. In this method, a polymer softens when it is heated above T_g and then freely deforms. If the polymer is cooled below T_g while this deformation is maintained, the polymer main chains stop movement and the deformation is fixed. When the deformed material is reheated above T_g, the polymer chain moves towards the direction of the increased entropy (entropy elasticity). As a result, the material returns to its original shape by itself.

Polymers that show shape memory at T_g are polyethylene and polynorbornene (see Fig. 2). Because polynorbornene has a very high molecular weight of 3 million, it is crosslinked by physical crosslinks [3]. Its T_g is 35°C and it shows a shape memory effect around this temperature. Polynorbornene disentangles above 150°C (viscous flow condition). Thus, a new shape can be memorized by thermally pressing the material at around this temperature.

Another method is to cold press below T_g, anneal above T_g for a short period, and then fix the deformation. Subsequently, the polymer is heated above T_g to recover the original shape (see Fig. 3). If a polymer is cold-pressed the interatomic or intermolecular potential energies of mainly bond angles change (change of the internal energy), which leads to shape change. If the sample is heated above T_g while maintaining this deformation, the stored internal energy is transformed into entropy changes of the polymer main chains (annealing). If the polymer is cooled to below T_g while the deformation is maintained, the polymer chains freeze and the deformation will be fixed. When the deformed material is reheated, the polymer main chains move towards the direction of entropy increase. As a result, it returns to its original shape by itself. Polymers that show such behavior include polycarbonate (see Fig. 2), poly(vinyl chloride), and poly(methyl methacrylate).

The only difference between the two methods that use T_g is the actual deformation temperature (in Fig. 1, it is $T > T_g$ whereas in Fig. 3, it is $T < T_g$). A shortcoming of the method in Fig. 1 is that slight recovery of the initial deformation occurs even if it is below T_g. This is caused by the recovery of the residual strain that is easily relaxed even at a low temperature. On the other hand, the method shown in Fig. 3 allows homogenization of the polymer chain entropy by annealing to eliminate the portion that relaxed easily. With this method, residual strain can be completely fixed.

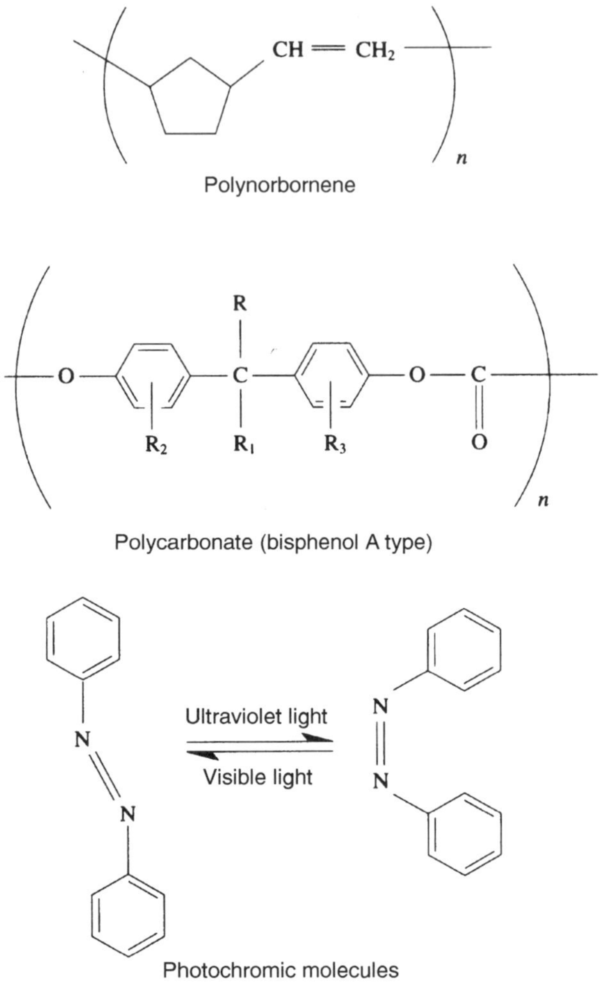

Fig. 2 Examples of shape memory polymers.

9.2.2 Utilization of Crystallization Temperatures (Main Chain and Side Chain) of Polymers

This method involves deformation above the melting temperature T_m, fixation of the strain below the crystallization temperature T_c, and finally reheating to recover the original shape (see Fig. 4).

In general, melting is a phenomenon in which an ordered arrangement turns into a random structure. In order to melt, the polymer main

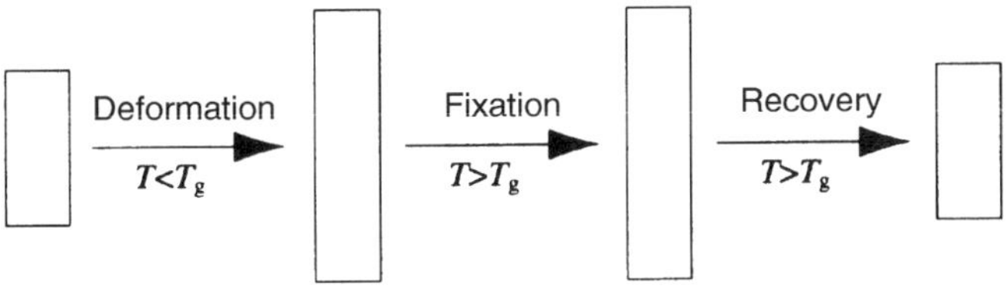

Fig. 3 Shape memory behavior by annealing.

chain must obviously be able to move around. Hence, the melting temperature T_m is higher than T_g. When a sample is heated above T_m, it softens and deforms freely. When the melt is cooled below T_c, the movement of polymer main chains is restricted by crystallization and the shape is fixed. When this material is reheated above T_m, the material recovers its original shape by entropic elasticity. Materials that show such properties include fibers and stretched polymer films.

9.2.3 Utilization of Chemical and Physical Stimuli

This method utilizes pH, chelate formation, oxidation-reduction reaction, or light to deform polymers isothermally and reversibly (see Fig. 5). Unlike the methods described thus far, the main characteristic of this method is reversible shape changes. For example, poly(acrylic acid) fibers that have the shape memory property with changes in pH possess dissociated charges. These charges repel each other and the polymer chains are stretched. On the other hand, in a low pH range, there are no charges and the polymer chains shrink. Another example is partially phosphated poly(vinyl alcohol) film, which responds to chelate formation. If Cu^{2+} is included, the film shrinks by formation of crosslinks.

When a stronger chelating agent, ethylenediaminetetraacetic acid, (EDTA) is added to this film, Cu^{2+} ions are removed from the film and it

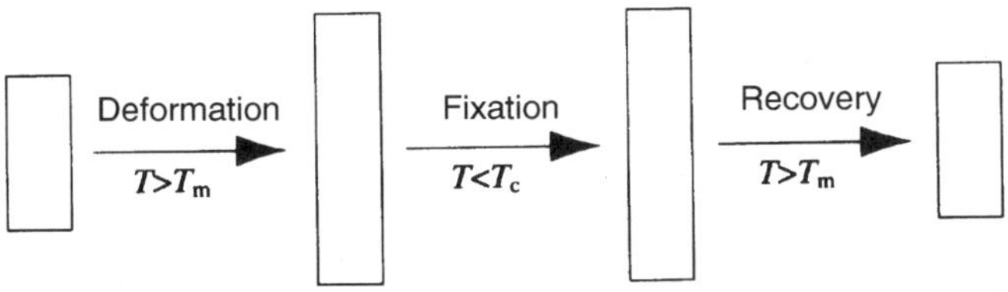

Fig. 4 Shape memory behavior utilizing T_m.

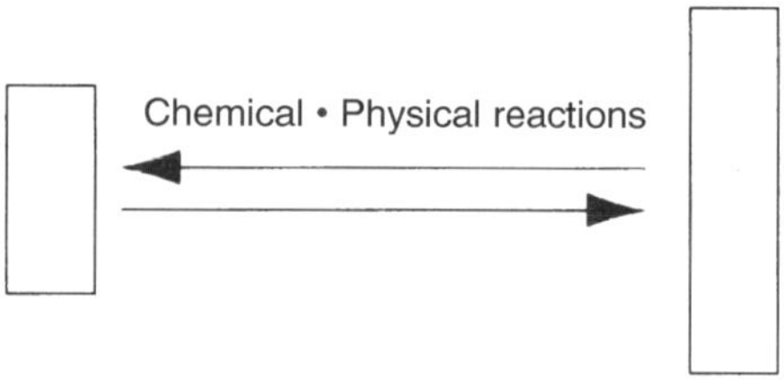

Fig. 5 Shape memory behavior by chemical and physical reactions.

recovers its original shape by crosslinking loss. Further examples are photochromic molecules such as azobenzene (see Fig. 2), which shrink and stretch reversibly due to the cis-trans transformation of polymer chains upon irradiation of light.

Accordingly, in the shape memory function that appears upon external thermal, chemical and physical stimuli, all deformations are based on changes in the stereoscopic positions of the polymer main chains. Hence, unless the relative position of the main chain is not fixed, the polymer will not show shape memory properties.

9.3 SHAPE MEMORY POLYMER GELS

Macroscopic morphological changes of a gel accompany reversible changes of swelling and shrinking. Because the diffusion rate of a solvent into the networks is the limiting step, deformation rate of the gel caused by an external stimulus is small. A shape memory gel with structural regularity on the molecular and supramolecular levels and fast response time was developed recently. It is made of a crosslinked copolymer, poly(SA-co-AA) of hydrophilic acrylic acid (AA) and hydrophobic stearyl acrylate (SA). Although it contains water, the SA side chains form a regular structure by hydrophobic interaction [4,5]. This gel shows crystal-line-amorphous transition at 49°C. The mechanical properties also change with this transition. The moduli of swollen gels are plotted in Fig. 6 as a function of temperature [4–6]. At a low temperature, the gel is hard due to the regular structure. Upon heating, the SA side chains melt and the Young's modulus of the gel suddenly decreases, softening the gel. Such changes of modulus can be performed reversibly (order-disorder transition).

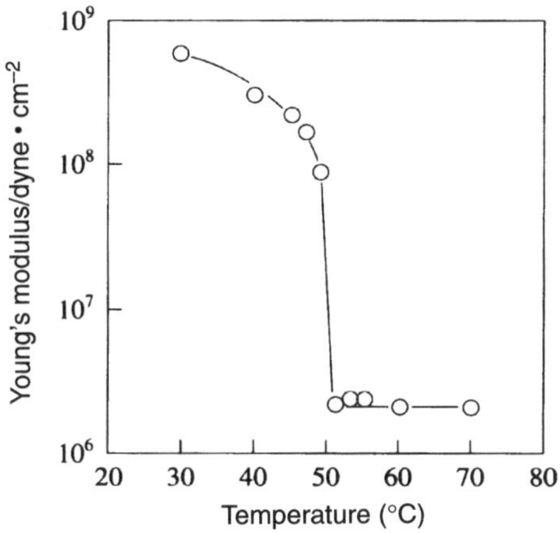

Fig. 6 Temperature dependence of the Young's modulus of poly(SA-co-AA).

Figure 7 depicts a shape memory behavior of a swollen gel whose mechanical properties drastically change [6, 7]. A stick-shaped gel softens above 50°C and can be wound into a coil shape. When this is cooled to room temperature, the gel hardens and the coil shape will be maintained even after removal of the external stress (0 s). Upon heating this coiled gel above 50°C, the gel softens and recovers the original shape by itself.

In general, the shape memory of rubbers is due to the maintenance of random coil structure by crosslinking. Hence, stable shape by chemical crosslinking such as vulcanization is determined at the time of synthesis and the memory is added. On the other hand, poly(Sa-co-AA) gel exhibits shape recovery by the mechanism shown in Fig. 8. The glass transition temperature T_g of the main chains of the swollen gel is −68°C when it is swollen with water. Shape memory effect of this gel appears only when the gel contains water. The T_g of dry poly(acrylic acid) is 108°C, which reduces drastically upon swelling by water. Under this condition, the mechanical property of the gel is controlled mainly by the SA crystals. In its dry state, this mechanism does not exist and thus the shape memory effect disappears. At a transition temperature of < 50°C, the SA side chains associate with each other and form SA crystalline domains. Thus, the main chains of the gel cannot move. However, if the gel is heated above the transition temperature, the SA side chain crystals melt and the main chains start moving freely. By applying external stress, the gel readily deforms. At this time, if the gel is coded below the transition, the

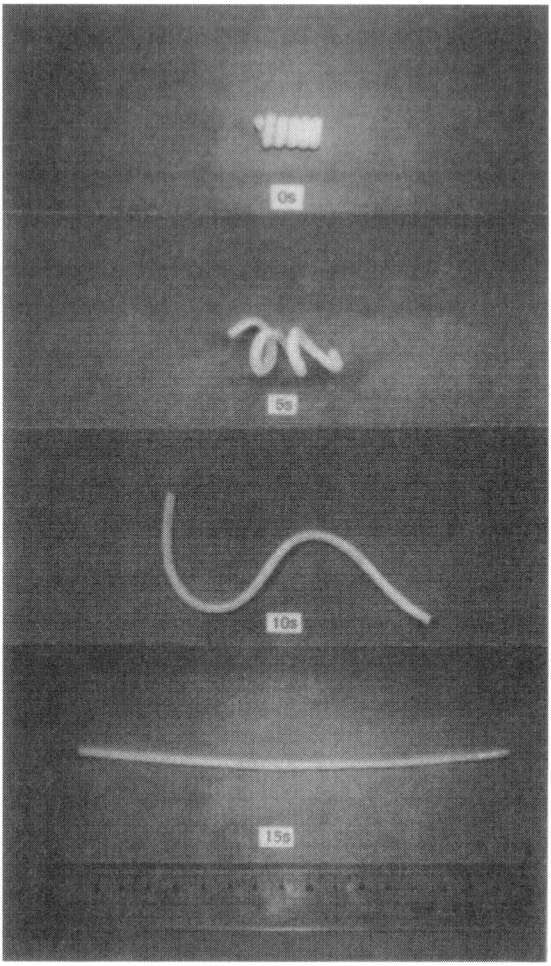

Fig. 7 Shape memory behavior of a poly(SA-co-AA) gel.

SA side chains recrystallize and form physical crosslinks. In this situation, the main chains are restricted in their conformations and fixed at a reduced entropic state. When the gel under such conditions is reheated above the transition temperature, the SA side chain crystals melt and the main chains move towards the direction of increased entropy (entropy elasticity). The sum of the movement of each main chain becomes the driving force for the recovery of the macroscopic shape. The hydrophobic SA of this gel forms crystalline domains, which are incompatible with water. Therefore,

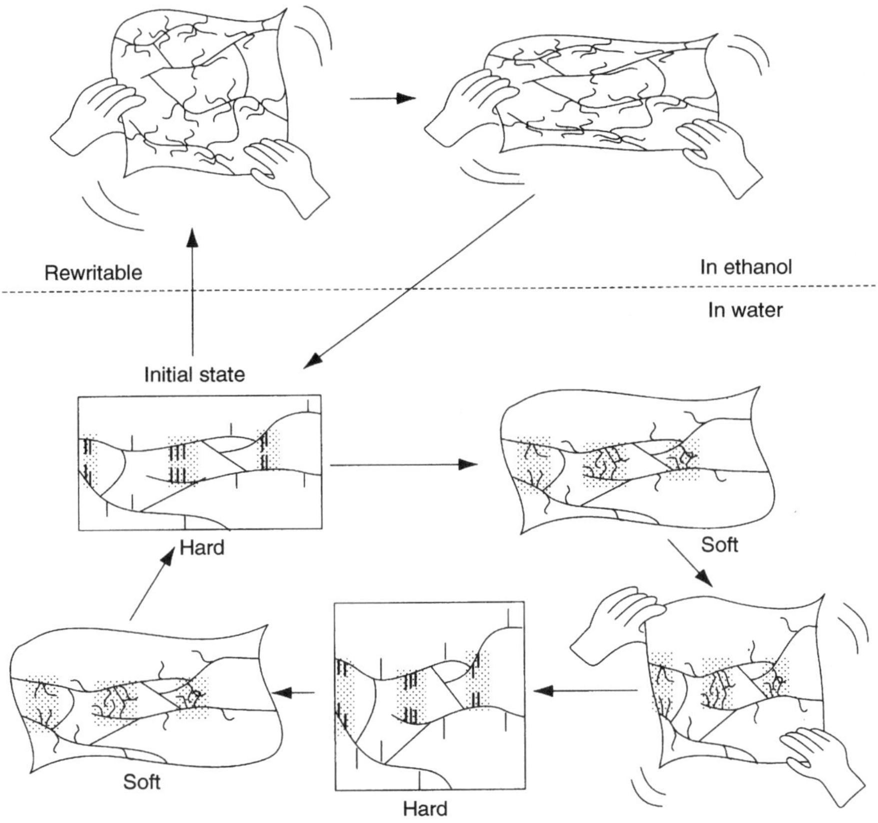

Fig. 8 Mechanisms of shape memory behavior.

even above the transition temperature the mechanical integrity of the gel will be maintained due to the hydrophobic interaction. However, if this gel is swollen by amphoteric ethanol at room temperature, the SA domains readily dissociate. Immersing a gel in this condition into water allows new SA domain formation and thus new memory to be inputted. Accordingly, this gel is an erasable shape memory material.

A gel made of a thermoplastic styrene-butadiene rubber (SBR) to which asphalt is added is also reported to be a shape memory material [8]. A small amount of SBR forms 3D networks in which a large amount of asphalt forms a phase separated structure (both components form co-continuous phase). When this gel is deformed at a low temperature, the plastic deformation of asphalt, which is much harder than SBR, dominates the process and maintained the deformed shape. Upon heating, the asphalt turns quickly into a fluid and the rubber elasticity of SBR becomes a main driving force to recover the original shape.

9.4 CHARACTERISTICS OF SHAPE MEMORY MATERIALS

Table 1 lists the characteristics of shape memory alloys, shape memory polymers, and shape memory gels. The principle of shape recovery in shape memory involves a martensite polymorph (the crystal lattice changes by temperature), whereas in polymers and gels it is the changes in stereoscopic conformation of polymer main chains (entropy, elasticity, etc.). These polymers show shape memory properties not only by heat but also by chemical and physical stimuli such as pH and light. Shape memory alloys are soft (almost like soft aluminum) at low temperatures. When they are heated and returned to their original shapes, they harden (3×) to overcome the external force. In contrast, shape memory polymers and gels are harder at low temperatures. Upon heating, they recover the original shapes through a rubbery state. The response time for shape

Table 1 Comparision of shape memory alloys, polymers and gels.

	Shape memory alloys	**Shape memory polymers**	**Shape memory gels**
Principles of shape recovery	Martensite polymorph	Entropy elasticity	
External stimuli	Temperature	Temperature, pH, light, etc.	Temperature, pH, light, chemicals
Change of hardness from low to high temperature	Soft-hard	Hard-soft	Hard-soft
Response time	ms order	Several s order	Several s order
Recovery force	Several tens kg/mm^2	1 kg/mm^2	—
Recovery ratio	7%	400–500%	400–500%
Solvents	Dry	Dry	Swollen with water
Price/kg	Several hundred thousand yen	5000 yen	
Density	6.5	1	
Phase transition	Crystal-crystal	Glass state-rubbery state	Order-disorder

memory alloys is on the order of milliseconds whereas those of polymers and gels are several tens of seconds [9].

The recovery forces of shape memory alloys are several tens kg/mm^2 whereas those of polymers are approximately 1 kg/mm^2. The shape recovery ratio of the alloys is 7% at the maximum compared to those of polymers and gels, at an amazingly high 400–500%. The shape recovery temperatures of shape memory alloys can vary by as much as or more than 100°C by several percent variations in composition. This is in contrast to nearly constant recovery ratio in polymers and gels, which depends on the type of materials used. The price of typical Ni-Ti alloys is several hundred thousand yen/kg, which is much more expensive than the several thousand yen/kg that shape memory polymers and gels cost. Alloy density is ~ 6.5 whereas polymer and gel density is ~ 1.

9.5 APPLICATION OF SHAPE MEMORY GELS

Table 2 lists the requirements and application examples of polymeric shape memory materials [10]. Application of shape memory gels basically follows Table 2. Among those application areas, commercialized products are mostly daily use items. However, following improvements in the shape

Table 2 Applications of shape memory polymers [10].

1. Insert from a narrow entrance—expand inside and fix
   Rivet, filling material of gaps in a house, lining materials for repair of old pipes, opening or blockage of blood vein, coating materials of printer roles
2. Carry in a folded shape—recover upon use
   Baby bath, a pool for children, various portable goods (camping goods), shoes for traveling
3. To make ones, favorite accessories easily and change their shape at will
   Flower vases or vessels, accessories (earring), toys (body of dolls), stationary, eyeglass frames (to fit face properly or to exchange lens easily)
4. Change easily into desired shapes—maintain the shape for a long period
   Replica of body parts (for making shoes), rehabilitation tools, sport protectors, chairs for physically impaired, helmets, mold for ice sculptures
5. Shape change during use—recovery
   Tires, belt for Japanese kimono, core materials for shoes, cotton for futon, artificial hair, folding of clothes, car bumpers
6. Correct into desired shape—maintain the shape
   Lady's underwear (bras), corrective materials for tooth
7. Temperature change—modulus changes significantly
   Sensors, automatic choking of engines
8. Improved efficiency and ease of difficult works
   Connection of pipes with different diameters, U-shaped pipe (elbows)

memory function of polymeric materials, further and broader application is underway.

REFERENCES

1 Nakagawa, T., and Kambe, H. (1959). *Rheology*, Misuzu Shobo, p. 369.
2 Shimizu, K., Irie, M., and Suiki, T. (1986). *Memory and Materials*, Kyoritsu Publ.
3 *Nikkei Mechanical*, (1984). Jan., p. 20.
4 Matsuda, A., Sato, J., Yasunaga, H., and Osada, Y. (1994). *Macromolecules* **27**: 7695.
5 Matsuda, T., Kagami, Y. and Osada, Y. (1995). *Polymer Preprints, Jpn.*, 469.
6 Tanaka, Y., Kagami, Y., Matsuda, A., and Osada, Y. (1995). *Macromolecules* **28**: 2574.
7 Osada, Y., and Matsuda, A. (1995). *Nature* **376**: 219.
8 Fukahori, Y., and Mashimo, N. (1996). *J. Rubber Soc., Jpn.* **69**: 608.
9 *Nikkei New Materials*, (1988). Nov., p. 40.
10 *Development and Application of Shape Memory Polymers*, CMC.

Section 10
Viscosity Enhancement and Flow Properties of Microgels

KENZO ISHII

10.1 MICROGELS

Microgels are internally crosslinked polymer microparticles and, in a narrow sense, they are defined as ultrafine particles with < 100 nm in diameter. In microgels, the structural characteristics of polymer gels, such as 3D crosslinked structure, etc., are maintained. Nonetheless, they dissolve or colloidally disperse. These properties are the same as those molecular properties seen in branched polymers. They are sometimes called intramolecularly crosslinked macromolecules (see Fig. 1) [1].

Microgels have been known for many years as the intermediates that are produced during synthesis of polymer gels. Unfortunately, macroscopic gelation during polymerization was unavoidable and made it difficult to obtain microgels. Thus the development of microgel synthesis methods was indispensable for characterization of microgel properties. In this subsection, the process of such development will be introduced.

Staudinger polymerized divinylbenzene under ultradilute conditions and predicted the formation of microparticle polymers [2]. The obtained solution had extremely low viscosity. Bobalek *et al.* [8] and Solomon and Hopwood [9] stopped the reaction immediately preceding the macroscopic gelation to synthesize microgels. When they synthesized an alkyd resin at

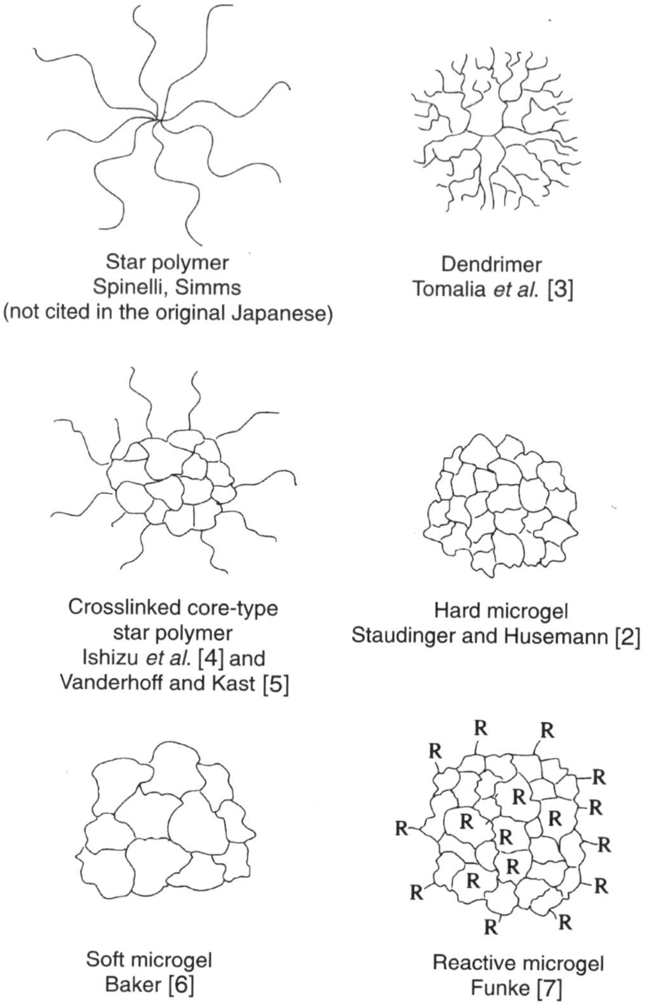

Fig. 1 Polymers having three-dimensional structures.

a composition that eventually led to gelation, they found that microgels 0.2–1 μm in diameter were formed immediately prior to gelation. Upon studying the melt viscosity of the alkyd resin that contains these microgels, they found that the viscosity was lower than the theoretical values predicted from the molecular weight. Funke synthesized a reactive microgel, which contains double bonds by emulsion polymerization [10]. The microgels obtained were superfine particles with $< 100\,\mathrm{nm}$ in

Table 1 Comparison of the viscosity of polystyrene and a microgel synthesized from divinylbenzene.

Polymer	Molecular weight	Solvent	Measurement temperature	[η] (ml/g)
Microgel	100,000	Benzene	25°C	10
Polystyrene	100,000	Benzene	25°C	50

diameter and colloidally dispersed in a solvent. This hard, nonswollen microgel dispersion solution had extremely low viscosity at concentrations over 40%. It also had extremely low viscosity when compared with polystyrene solutions (see Table 1) [2].

Since then, microgels have become available in a stable and reproducible form. Active studies have been undertaken especially in the area of coating. The ICI group commercialized a microgel of 200–300 nm in diameter by nonaqueous dispersion (NAD) polymerization using 1,2-hydrostearic acid, which has a terminal reactive double bond, as a dispersion stabilizer [11, 12]. Japan Paint also succeeded in commercializing ultrafine particles of < 100 nm in diameter by emulsion polymerization using an oligosoap in which amphoteric ionic groups serve as the emulsifier [13, 14]. Today, these materials are widely used for viscosity conditioners, mechanical property enhancers [15], water vapor passing membranes [16], and low profile agents. In particular, in the application of viscosity control functions, the flow properties of concentrated microgel dispersion solutions or mixtures with other polymers were studied. These materials were found to exhibit nonNewtonian behaviors such as plastic or quasiplastic flow, or thixotropy. These properties are used as viscosity controlling agents in liquid paints, inks, and adhesives.

10.2 PROPERTIES OF MICROGEL DISPERSED LIQUIDS

10.2.1 Solvent Swelling of Microgels

When microgels are colloidally dispersed in a good solvent, they absorb the solvent in the same manner as other crosslinked polymers and swell. The degree of swelling can be determined by particle diameter measurements or volume fraction determination from dilute solution viscosity measurement [17].

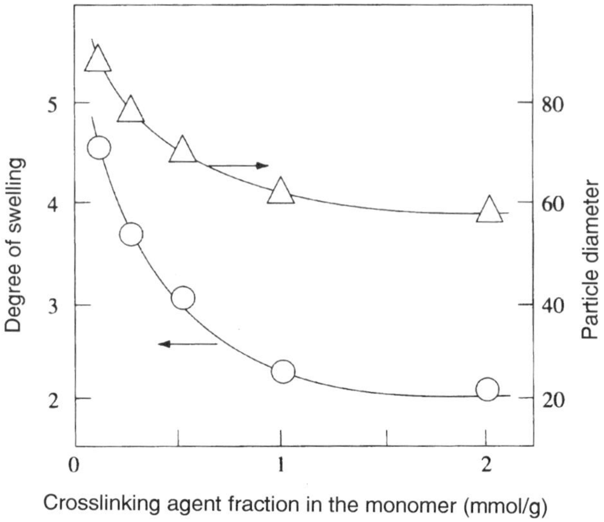

Fig. 2 Influence of the degree of crosslinking of styrene-acrylate copolymer microgel on the degree of swelling (○) and particle diameter (△).

The degree of swelling is influenced strongly by the degree of crosslinking of the microgel. Figure 2 illustrates the influence of the degree of crosslinking on the degree of swelling and particle diameters of a microgel dispersed in xylene. The microgel was synthesized by emulsion polymerization and the degree of crosslinking is adjusted by ethylene glycol dimethacrylate. The lower the degree of crosslinking, the higher the degree of swelling. The particles showed large diameters [13]. In an aqueous system, a polymer with a carboxyl group is used as a flow property-controlling agent after swelling the particles via base neutralization [18].

10.2.2 Flow Properties of Microgel Dispersed Liquids

A microgel-dispersed liquid is a Newtonian fluid with low viscosity when the particle concentration is low. However, at a high concentration it behaves as a pseudoplastic fluid in which there is an apparent field stress. Figure 3 shows shear-rate shear-stress curves of a relatively high concentration of 25% microgels dispersed in xylene. The influence of the degree of swelling is compared. The microgel dispersed liquid with a log degree of swelling showed Newtonian behavior whereas with high degrees of

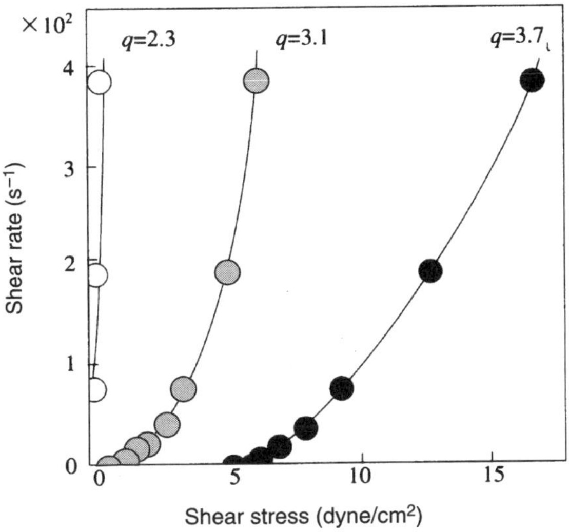

Fig. 3 Shear-rate shear-stress curves of a 25% styrene-acrylate copolymer microgel in xylene at various degree of swelling (*q*).

swelling it showed plastic or pseudoplastic behavior with high yield stress [13].

The fluidity of the microgel-dispersed liquid depends on the concentration of the microgel. At any degree of swelling, the fluid showed Newtonian behavior at low concentrations whereas at high concentrations it behaved as a pseudoplastic fluid with a yield stress. At further high concentrations, the liquid becomes a gel, which occurs quickly in a narrow concentration range. The concentration at which this fluidity change occurs is lower with a highly swollen microgel (see Fig. 4). The pseudoplastic behavior of the liquid relates to the state of particle packing. A highly swollen microgel has a high, occupied volume and thus becomes close packed at a low concentration, leading to a plastic or pseudoplastic fluid [13].

When the dispersate is a poor solvent, microgels cannot be dispersed stably and precipitate. In a mixed solvent system, fluidity changes significantly depending on the solvent composition. Figure 5 shows the viscosity of a microgel dispersed liquid when the alcohol (butyl cellosolve—a poor solvent) concentration is changed. When the polarity of the dispersate increases the fluid becomes viscous and deviates from Newtonian behavior. Viscosity changes drastically as a function of temperature.

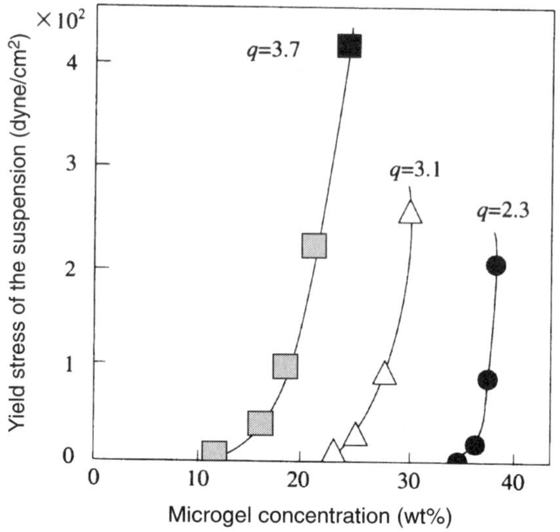

Fig. 4 Change of liquidity of styrene-acrylate copolymer microgel-dispersed xylene with various degrees of swelling and microgel concentration.

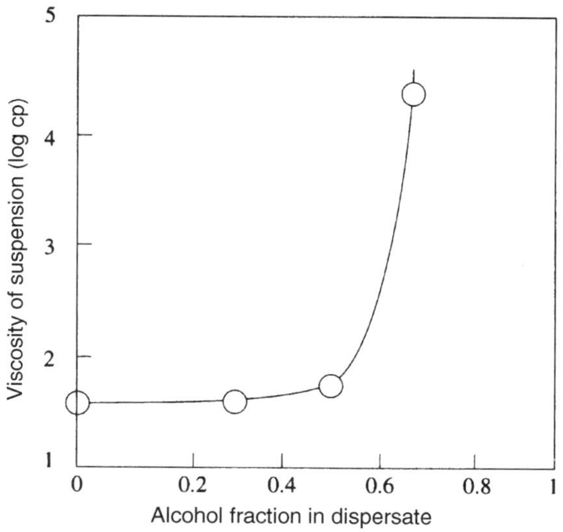

Fig. 5 Viscosity changes of styrene-acrylate copolymer microgel dispersed at 40 wt% in xylene-butyl cellosolve mixed solution as a function of the composition.

At high temperatures, the viscosity reduces and the fluid again shows Newtonian behavior (see Fig. 6). This temperature dependence is reversible. It is probably due to the interaction between the microgels and the dispersate [19].

10.2.3 Structure Formation of Microgels

When a microgel is used as a viscosity controlling agent for a paint, a concentration that is much lower than close packing will show plastic or pseudoplastic flow. This is probably due to network formation by microgel aggregates. This structure is destroyed by shear but will reform when the shear is removed, showing a reversible nature.

Figure 7 shows apparent particle diameter and yield stress of suspension measured by light scattering technique in a melamine resin solution, which accelerates microgel aggregate structure. When the apparent particle size increased due to aggregate formation, the yield stress appeared. This suspension was studied by electron microscopy using the freeze replica method, confirming a network-like aggregate structure of the microgels (see Fig. 8) [20].

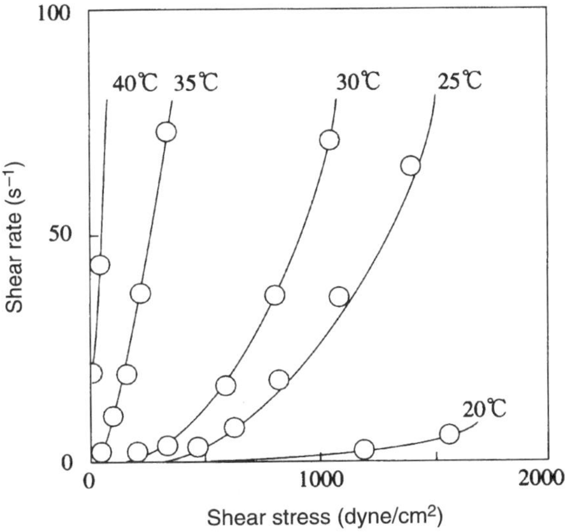

Fig. 6 Viscosity changes of styrene-acrylate copolymer microgel dispersed at 40 wt% in xylene-butyl cellosolve mixed solution as a function of temperature.

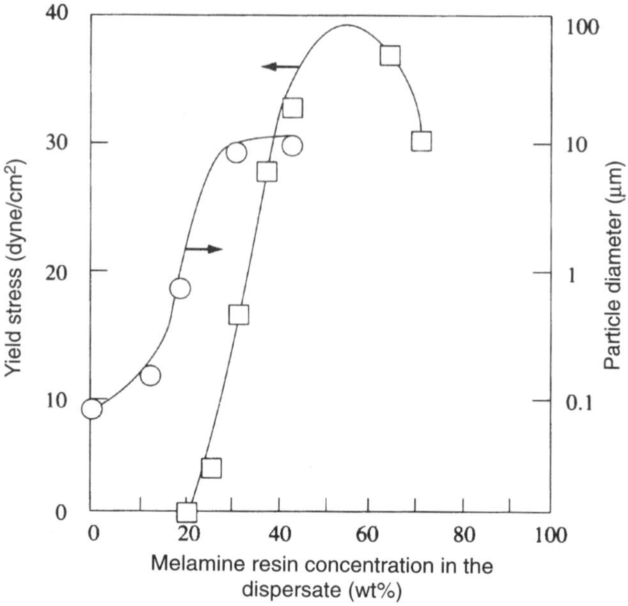

Fig. 7 Particle diameter of the microgels and viscosity of the suspension of styrene-acrylate copolymer microgels in melamine resin/xylene solution as a function of the melamine concentration.

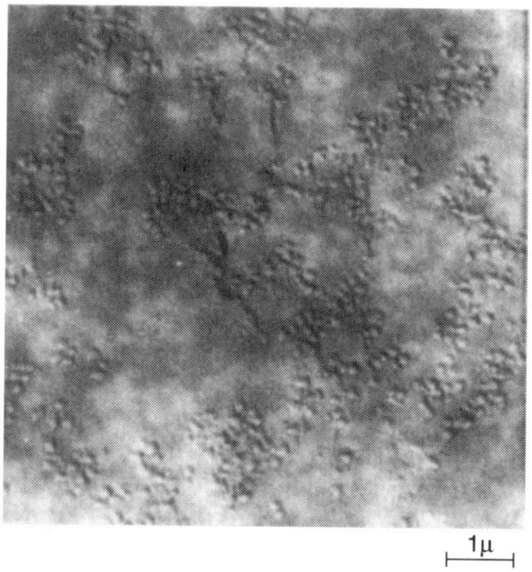

Fig. 8 Aggregate structures of microgels in a paint (electron photomicrograph of the frozen replica of the paint).

10.3 APPLICATIONS OF MICROGELS

Fluidity of liquid compositions such as paints, inks, and adhesives influences handling during manufacturing and use, emulsification and stability of product shape, stickiness and appearance, and the properties of the final product. Microgels are widely used as viscosity controlling agents.

For microgels the particle shape is stable in solvent of under heat, and thus the viscosity controlling function can be maintained utilizing the interface properties; and for synthetic polymers, solubility, reactivity, and charges can be controlled depending on the requirements. Application examples will be introduced here.

10.3.1 Improvement in Dripping Properties

When a liquid paint is painted onto a vertical object, dripping causes the appearance of the coating to suffer. To avoid this dripping phenomenon, it is necessary to paint a thin coating repeatedly and this results in reduced productivity. If a microgel is mixed into the paint, it becomes a pseudo-plastic fluid with yield stress and the limiting thickness that causes dripping increases signficantly (see Fig. 8) [14].

High-solid paints contribute to prevention of air pollution and conservation by reducing organic solvent evaporation during painting. To decrease the viscosity of paint, low molecular weight polymers are used. However, this also leads to thinning of limiting thickness for dripping. The high-solid paint to which a microgel is mixed exhibits high viscosity in the shear rate range of dripping, whereas within the shear rate to influence the leveling, it is low viscosity [14]. It is, therefore, possible to design a paint that has superb dripping characteristics and appearance [21, 22].

10.3.2 Orientation Control of Light-Reflecting Paints

A metallic paint uses flake-like aluminum or mica powders to add shine. When a flake-like powder aligns parallel to a substrate, the coating shows beautiful reflection or interference. Orientation of these powders is influenced by the viscosity of the paint. A paint with a microgel can be sprayed at a low viscosity and with low mist size and uniformity. Upon coating, the viscosity increases and improves the orientation of the flaky powder. The coating has a high reflection and exhibits a flip-flop property (change of brightness at changed angles of observation) [11, 23].

10.3.3 Edge Coating

Cationic electronic coating is superior to corrosion protection and is thus used for automotive undercoating. Electrocoating utilizes water-soluble, charged paint, which is water dispersable. A metallic coating object, which is immersed in the paint solution, is used as an electrode. By passing through direct current, the paint adsorbs onto the metallic substrate. This method has the advantage of achieving uniform coating on a complex object. However, the coating thickness of the edge is thinner due to the surface tension of the paint. This insufficient coating thickness can lead to corrosion of the substrate.

By adding a microgel to the paint, the coating thickness at the edge is improved because of the increased viscosity of the paint during the baking process [24].

10.3.4 Prevention of Pigment Sedimentation

If a microgel is added to a paint that contains a high density pigment (such as titanium dioxide or iron oxide), the sedimentation of the pigment can be inhibited and the shelf life of the paint improves greatly. In addition to the viscosity controlling effect, adsorption of the microgel onto the pigment surface adds charges and improves dispersion stability in a water-based paint [25].

10.3.5 Shape Maintenance of Photosensitive Print Materials

A flexible print material, which can be developed by a water-based developer, has been commercialized. Flexible print material is a gel that contains a photopolymerizable monomer or oligomer. Upon photocuring, the uncrosslinked portion is removed by a solvent and a print plate is prepared. From both the safety and the environmental standpoints, the trend is to use a water-based developer. To improve development capability, a microgel with a hydrophilic surface is used [26]. The solid property, transportability, and handling are improved as a result of the addition of the microgel.

REFERENCES

1 Funke, W. (1988). *J.C.T.* **60**: 68.
2 Staudinger, H., and Husemann, E. (1935). *Ber.* **68**: 1618.
3 Tomalia, D.A., Baker, H., Dewals, J., Hall, M., Kallos, G., Martin, S., Roeck, J., Ryder, J., and Smith, P. (1986). *Macromolecules* **19**: 2466.
4 Ishizu, K., Gammo, S., Fukutomi, T., and Karatani, T. (1980). *Polymer J.* **12**: 399.
5 Vanderhoff, J., and Kast, H. Unpublished results.
6 Baker, W.O. (1949). *Ind. Eng. Chem.* **41**: 511.
7 Funke, W. (1968). *Chimia* **22**: 111.
8 Bobalek, E.G., Moore, E.R., Levy, S.S., and Lee, C.C. (1964). *J. Appl. Polym. Sci.* **8**: 625.
9 Solomon, D.H., and Hopwood, J.J. (1966). *J. Appl. Polym. Sci.* **10**: 1893.
10 Funke, W.J. (1988). *J. Coat. Technol.* **60**: 69.
11 Backhourse, A.J. (1982). *J. Coat. Technol.* **54**: 83.
12 Bromley, C.W.H. (1989). *J. Coat. Technol.* **61**: 39.
13 Ishii, K., Kashihara, A., Kayano, H., Ishikura, S., and Midzuguchi, R. (1985). *Proc. ACS PMSE* **52**: 448.
14 Ishikura, S., Ishii, K., and Midzuguchi, R. (1988). *P. Org. Coat.* **15**: 373.
15 Kashihara, A., Ishii, K., Kida, K., Ishikura, S., and Midzuguchi, R. (1985). *Proc. ACS PMSE* **52**: 453.
16 Yagi, T., Saito, K., and Ishikura, S. (1992). *P. Org. Coat.* **21**: 25.
17 Crews, G.M., Widman, G.C., Grawe, J.R., and Bufkin, B.G. (1980). *J. Coat. Technol.* **52**: 33.
18 Clarke, J. (1994). *Surface Coat. Int.* **7**: 303.
19 Muramoto, H., Ishii, K., Miyazono, T., Ishikura, S., and Midzuguchi, R. (1987). *Proc. 13th Int. Conf. Org. Coat, Sci. Technol.*
20 Kojima, A., Ishii, K., Saito, S., and Ishikura, S. (1991). *Polym. Preprints, Jpn.* **40**: 4186.
21 Kurauchi, T. (1992). *J. Adhesion Sci., Jpn.* **28**: 287.
22 Kajino, T., and Uenoyama, K. (1990). *Toso Kogaku* **25**: 348.
23 Kuwajima, T. (1994). *TECNO-COSMOS* **6**: 42.
24 Ueda, R. (1991). *Toso to Toryo* **1**: 35.
25 Ishii, K., Ishikura, S., and Midzuguchi, R. (1988). *XIX FATIPEC KONGRESS Proc. IV*: 187.
26 Kanda, K. (1992). *TECNO-COSMOS* **2**: 74.

Section 11
Biocompatibility of Hydrogels

YOSHIHITO IKADA

11.1 THE HUMAN BODY AND GELS

To understand biocompatibility, it is first necessary to understand the human body. Here, we will treat both the human body and gels from a materials point of view.

The human body has hard systems that consist of hydroxyapatite and collagen and soft systems made of collagen and glycosaminoglycan (a mucopolysaccharide). The soft systems are typical hydrogels because their water contents range is 60–80%. Table 1 summarizes typical water contents of organs. From strictly water content, these cells could be regarded as hydrogels. However, here organs will imply connective tissues.

If the soft systems of the human body are hydrogels, should good artificial biocompatible materials also be hydrogels? At this time, the answer to this question is negative. Although it is difficult to explain simply, it is clear that natural hydrogels differ greatly from artificial ones.

Therefore it is difficult to use artificial hydrogels as replacements for natural hydrogels. First of all, if a natural hydrogel is compared with an artificial hydrogel, the natural hydrogel is superior to the artificial one in mechanical properties. Figure 1 shows stress-strain (S-S) curves of various blood vessels. There are no materials among artificial materials available today that show very low initial modulus and quite high strength against

Table 1 Water contents of various organs [1].

Organs	Origin	Water content (%)
Lens	Normal human	67–70
	Human with cataract	67–81
Vitreous	Human	99.0
Skin	Human	57.7–61.0
Tongue	Human	60–68
Interior of stomach	Human	77.6–79.0
Muscle of stomach	Human	80.6
Small intestine	Human	80.0
Kidney	Human	73–77
Thymus	Human	82.2
Vagina	Human	79.9
Cornea	Cow	77.9–81.0
Achilles tendon	Cow	62.9
Ligament (neck)	Cow	57.6
Skeletal muscle	Rabbit	77.4
Bladder wall	Rabbit	82.4
Wall of urinary tube	Rabbit	57.7

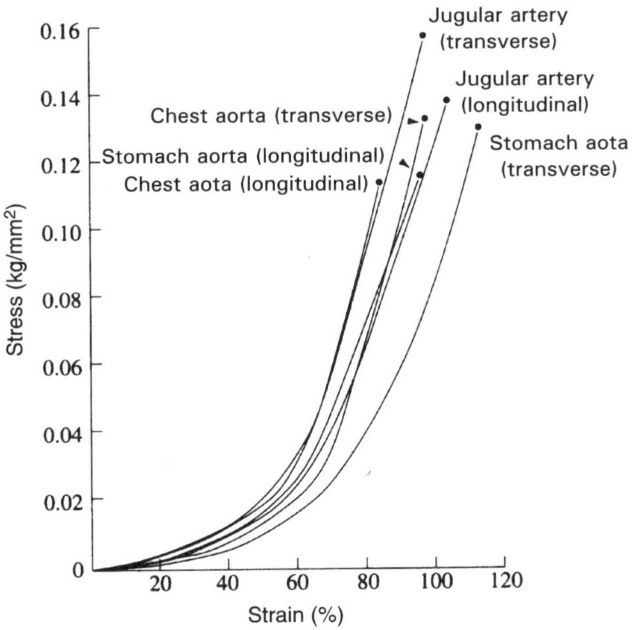

Fig. 1 Stress-strain curves of blood vessels of humans 20–29 yr [2].

breakage. The superior mechanical properties of biomaterials are inherent in their higher-order structures as shown in Fig. 2. No artificial hydrogels of higher order exist, as shown in Fig. 2. If an S-S curve as shown in Fig. 1 is desired, the material needs to be woven from fibers. The basic unit of the

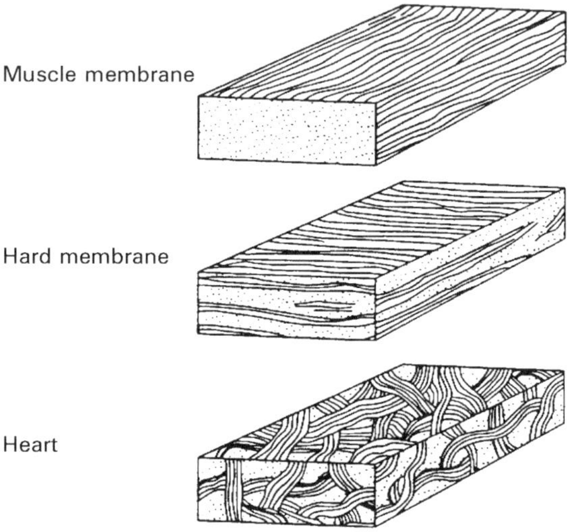

(a) Orientation of collagen fibers in biomembranes [3]

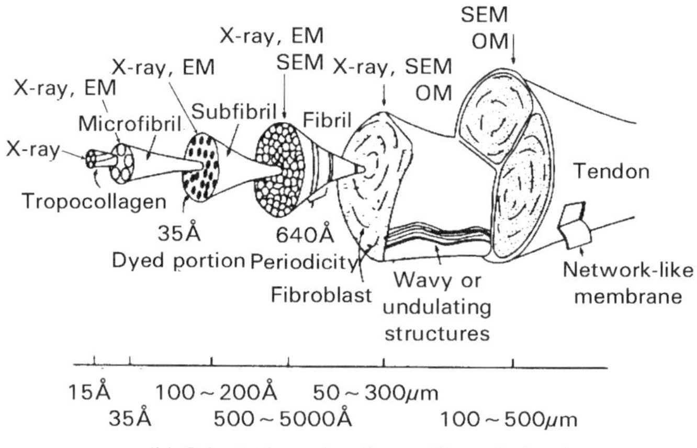

(b) Orientation of collagen fibers in tendon

Fig. 2 Higher-order structures of soft materials of organs [4].

hydrogel in Fig. 2(a) is collagen fibers and, even in this case, a fiber is the base structure.

Among the other characteristics of biomaterials, the polymeric constituents are digested by enzymes that are then absorbed by living organs and, additionally, fat and ions cannot necessarily penetrate freely into the natural hydrogels.

11.2 WHAT IS BIOCOMPATIBILITY?

Biocompatibility is a property of any medical material that is used in contact with a living body. A material that is obtained by artificially modifying a natural material will also be included as a biomaterial. For a biomaterial, the properties different from those of industrial materials will be required and these are summarized in Table 2. As can be seen in the table, the requirements for a biomaterial are nontoxicity, an ability to be sterilized, and biocompatibility. Biomaterials are also materials and thus possess their own unique functions. In order to maintain function, durability is also required. However, as this is a property generally required for any materials, it is not included in the table.

It is important to pay attention to the distinction between nontoxicity and biocompatibility in Table 2. Nontoxicity means biologically safe, that is, inflammation, both chronic and acute, bleeding, allergic responses and diseases like cancer are not caused or triggered by the biomaterial. This property is a minimum requirement for any biomaterial. Without satisfying this requirement, as well as the requirement that the material be sterilizable, it is not possible to obtain permission from the Ministry of Health in Japan. At this time, the toxicity of a material is due mainly to the soluble compounds found in the material or microbes that adsorb onto the

Table 2 Requirements for biomaterials.

Requirements	Examples
1. Nontoxic (safe to use)	Will not cause chronic or acute inflammation, bleeding, allergy or cancers
2. Ability to be sterilized	Will endure radiation, ethylene oxide, dry heat, or autoclave
3. Biocompatibility	Will not disturb homeostasis. Will not cause unwanted rejection reaction, nor cause anomalous growth, absorption, or death of organ systems. Will adhere strongly with organs in some situations.

material surface. Thus, it is possible to meet regulations if the material is synthesized and processed in a sterile environment so as to exclude soluble materials. As long as the material does not contain live cells or easily denatured proteins, it will not cause serious problems.

The remaining problem is that of biocompatibility. It should be emphasized that this is a different property from nontoxicity. Unfortunately, even among biomaterials researchers, there are many cases in which nontoxicity and biocompatibility are not distinguished one from the other. Use of both of these words helps to clarify the meaning of biocompatibility.

Good examples are alumina and poly(tetrafluoro ethylene) (PTFE), which are not toxic but often exhibit poor biocompatibility. Alumina cannot bond strongly with bone structures and PTFE forms blood clots. Hence, neither one is biocompatible.

Biocompatibility, summarized in Fig. 3, can be divided largely into bulk biocompatibility, in which the bulk properties of materials are important, and interfacial biocompatibility, in which the surface of materials relates to the problem. In the following, the biocompatibility of hydrogels will be explained according to the classifications as shown in Fig. 3.

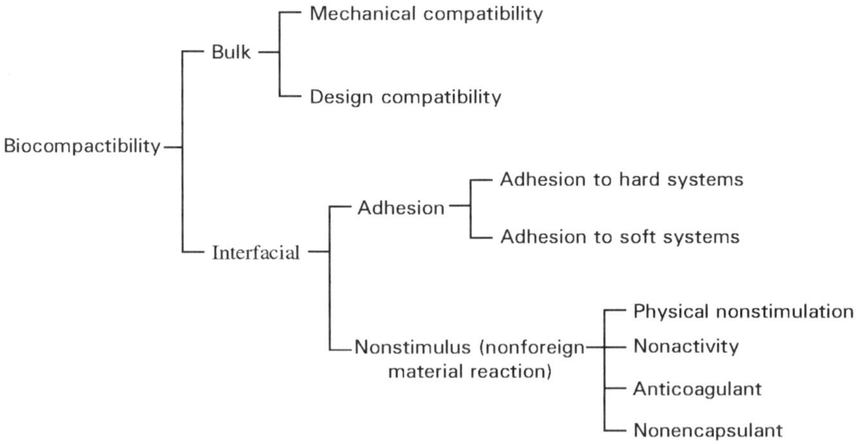

Fig. 3 Classification of biocompatibility [5].

11.3 BULK BIOCOMPATIBILITY

Bulk biocompatibility, also called mechanical biocompatibility, is considered as the ability of material to inflict a physiological mechanical stimulus on those systems that surround the material. This compatibility is not governed by strength but by the elasticity or modulus of a material. For example, when an artificial blood vessel is connected to a natural vessel, the artificial material inflicts excessive mechanical stress on the blood vessel when the artificial material is harder than the living material. Consequently, it leads to thinning of the blood vessel. This is a material hardness mismatch and is a typical example of poor mechanical biocompatibility.

For a material that requires good mechanical biocompatibility, it is desirable to use a material that has the same mechanical properties as the organs. Unfortunately, as described in the introduction here, artificial hydrogels that exhibit S-S curves similar to those of living systems have not to date been synthesized. If the modulus is reduced to the level of living organs, weakness results. If the strength is appropriate to the job, the initial modulus is too high.

Because strength is a more important requirement than modulus, hydrophobic high-strength materials rather than hydrogels are used as artificial blood vessels. Poly(ethylene terephthalate) fibers are mostly used as a woven cloth or twisted bundles to obtain flexible yet strong artificial blood vessels. If the diameter is ≤ 6 mm, PTFE can also be used, although it is far from having the properties of gels. For even smaller diameter veins, elastomers such as polyurethane have been investigated. Although hydrogels have the properties of elastomers, unlike hydrogels polyurethane is hydrophobic and water content is nearly zero.

Artificial breast implants require a flexible rather than a strong material; hence, the use of synthetic hydrogels seems appropriate. However, the material actually used is not a hydrogel but a silicone elastomer with zero water content. Using this hydrophobic material, a bag is made in which a silicone gel or saline solution is filled. From the mechanical point of view, a hydrogel could be used as a breast implant. However, it is not used because of the possibility of calcination by diffusion of Ca^{2++} or PO_4^{3-}. This is not a concern with hydrophobic gels. It is possible, however, that fat might diffuse into the silicone gel.

Accordingly, hydrogels are not used when there is a long-term need even if strength is not required for a particular application. When a soft material is needed, a woven cloth or silicone rubber is used.

11.4 BIOMATERIALS

Biomaterials will be explained in the following. As described earlier, the soft materials of the body are hydrogels. Nonetheless, materials of biological origin other than hydrogels are used for various applications as shown in Table 3. They are used because these materials exhibit excellent mechanical properties. If the materials originate from animals (other than human), it is necessary to remove all antigens. Only connective tissues are used and even these materials absorb compounds from the body. Thus, in order to reduce the absorption rate, they are generally crosslinked. This crosslinking also helps eliminate the antigens.

Good bioabsorption and biocompatibility are sometimes reasons for using biomaterials. It seems that this form of biocompatibility relates more to interfacial compatibility than to bulk compatibility.

Biodegradation, through which degraded biological systems are absorbed by the body, can be a shortcoming when a biomaterial is used. Crosslinking can suppress hydrolysis. It is also possible to use a nonbioabsorbing synthetic hydrogel. In general, strengths of synthetic hydrogels are lower than biological hydrogels. However, if crystallization

Table 3 Clinically used biomaterials.

Applications	Origins		Absorptivity
	Animals	Organs	
Replacement of blood vessel	Sheep	Conective tissues[a]	Nonabsorptive
Replacement of blood vessel	Cow	Jugular artery	Nonabsorptive
Replacement of blood vessel	Human	Ligament	Nonabsorptive
Replacement of blood vessel	Cow	Endosperm artery	Nonabsorptive
Replacement of heart valve	Pig	Artery valve	Nonabsorptive
Replacement of heart valve	Cow	Heart valve	Nonabsorptive
Replacement of hard membrane	Human	Hard membrane	Absorptive
Wound dressing	Pig	Skin	Absorptive
Bone replacement	Calf	Spongy bone (*substantia spongiosa*)	Absorptive

[a] Prepared by inserting a mandrel into a muscle

is propertly utilized, gels with high strengths can be obtained. A well-known example is that of a physically crosslinked poly(vinyl alcohol) that is prepared by low-temperature crystallization from a polymer solution.

The water content-tensile strength relationship is shown in Fig. 4. This PVA gel shows the highest strength among synthetic gels. For reference, the strength of human amnion, a natural hydrogel, is shown in Fig. 5. The fact that glutaldehyde (GA) does not influence strength is probably because the amnion is not homogeneous. Chitin, and its partially acetylated product, chitosan, also show high strengths (see Fig. 6). These high strengths are probably due to crystallization.

Even an amorphous material can produce a strong material by improving molecular orientation. Figure 7 illustrates the stretching ratio dependence of the mechanical properties of dried gelatin. It is obvious from Fig. 8 that such high strength was the result of molecular orientation. This dried and stretched gelatin has a higher strength than other hydrophobic materials as seen in Table 4. Even this high-strength gelatin is not used as an implantation biomaterial.

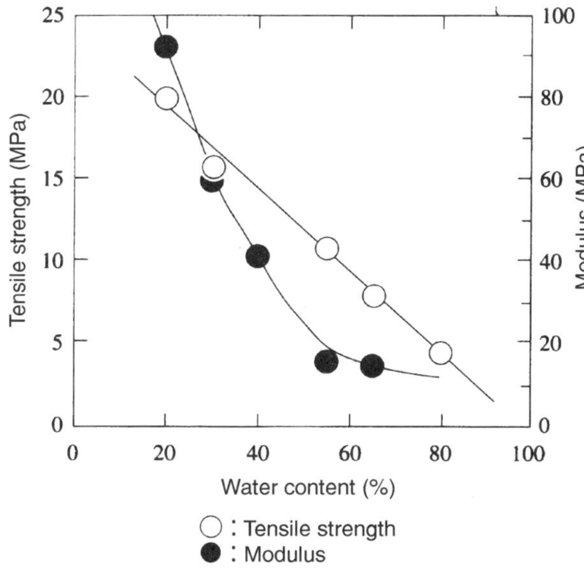

Fig. 4 Mechanical properties of a PVA hydrogel prepared by low temperature crystallization from a DMSO-water mixed solution as a function of water content (the degree of polymerization: 5000) [6].

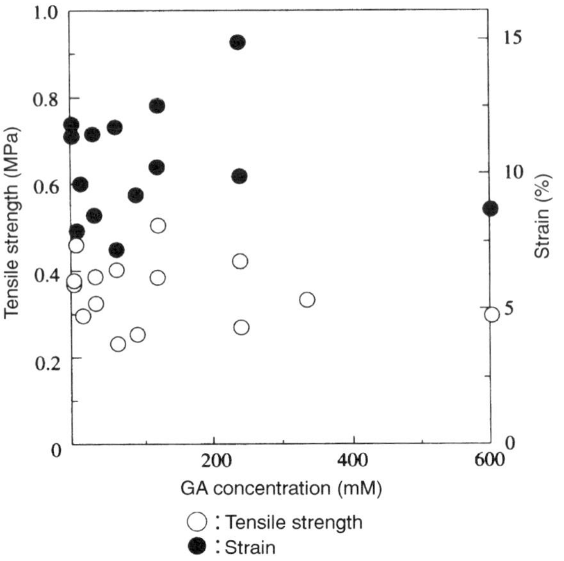

Fig. 5 Tensile properties of human amnion which is uncrosslinked and crosslinked by glutaldehyde (GA) under wet conditions [7].

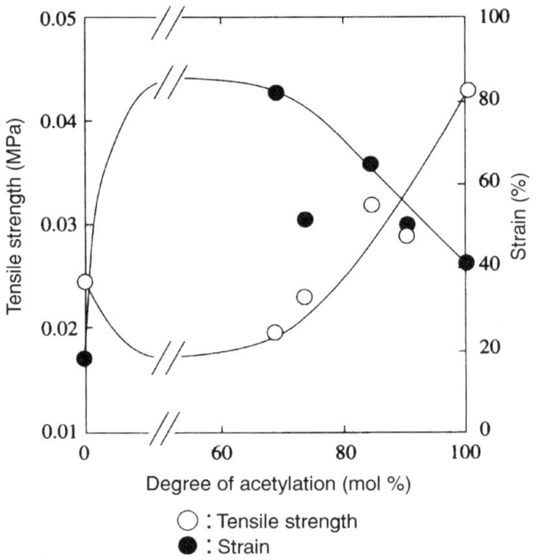

Fig. 6 Tensile strength of chitin and its deacetylated film under wet conditions (unacetylated material is chitin and 100% deacetylated material is chitosan) [8].

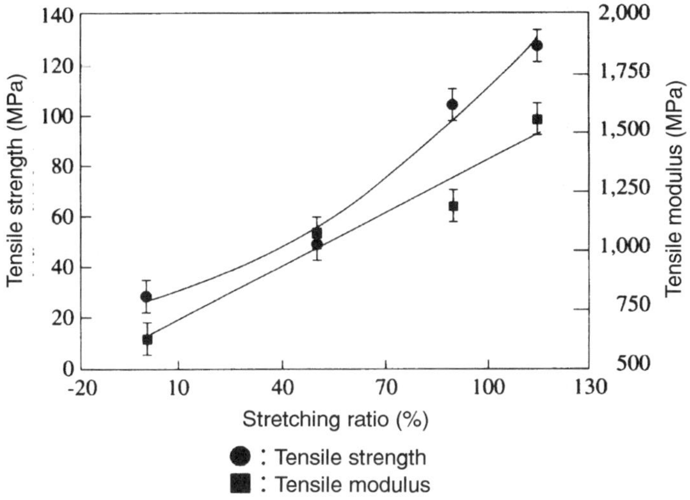

Fig. 7 Dry mechanical properties of a gelatin film that is axially stretched at 20–25 vol% water content [9].

One of the main reasons why nonabsorbing synthetic hydrogels are not clinically used is that their biosafety has not yet been confirmed. Calcination by C^{2+} diffusion may depend on where in the body the material is implanted. However, there has been no published research on positionally related calcination.

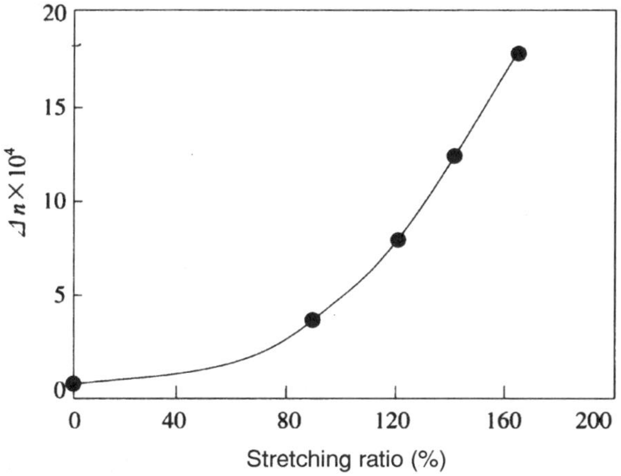

Fig. 8 Birefringence (Δn) of a uniaxially oriented gelatin film [9].

Table 4 Glass transition temperatures and dry mechanical properties of biodegradable polymers [9].

Polymers	T_g (°C)	Tensile strength (MPa)	Stretching ratio (%)
Starch	230	—	—
Cellulose acetate[a]	190	17 – 25	10 – 30
Poly(vinyl alcohol)	58 – 85	40 – 50	300 – 400
Poly(ε-caprolactone)	~ 60	21 – 31	600 – 1000
Poly(L-lactic acid)	50 – 59	50	3
Oriented gelatin	217	65 – 140	8–22

[a] Degree of substitution: 2.5

11.5 INTERFACIAL BIOCOMPATIBILITY

As shown in Fig. 3, there are many aspects of interfacial biocompatibility. In this subsection typical interfacial biocompatibility, blood compatibility, and tissue connectivity will be explained.

11.5.1 Blood Compatibility

Most extensively studied to date among biocompatibility issues is blood compatibility. However, only a few of the materials among those found to possess blood compatibility have been commercialized. This is probably because the biological changes that happen when the material surface comes in contact with blood are extremely complex. A very effective anticoagulant, heparin, is heavily used clinically. If the contact between blood and a biocompatible material is less than a day or so, heparin will prevent blood clots from forming at the material surface. However, when contact exceeds more than one week, heparin administration is dangerous as it might cause internal bleeding. Hence, the blood compatibility of the material must be considered.

Table 5 summarizes the blood compatibility of those material surfaces currently now in clinical use. In general, use of sustained drug release is limited by its effective period. On the other hand, improvements in material properties is too costly. For the quasi-intimation which is used for current artificial blood vessels, blood clot formation onto the material surface is rather needed and thus a special surface treatment is unnecessary. However, if the implantation period is long, the thickness of the internal membrane increases. Therefore, this approach cannot be used for artificial blood vessels with diameters of < 4 mm. However, this quasi-

Table 5 Blood compatibility methods for clinically used artificial materials.

Blood contact time	Methods	Applications
From several hours to one day	Administration of heparin	All external circulation
From several hours to one day	Fixation of a hydrophobic polymer	Artificial kidney membrane
Several days	Fixation of heparin	Artificial lung membrane
From several days to one month	Sustained release of heparin	Bypass type for operation
From several days to one month	Fixation of urokinase	IVH catheter
From several days to one month	Decalcination	Stored blood
More than one month	Quasi-intima formation	Artificial blood vessel
More than one month	Use of biosystems	Replacement blood vessel
More than one month	Administration of anti-coagulants and antiplatelet agents	Biovalve

intima formation might be an effective blood compatibility method when the distance between the surfaces is large (such as a blood pump in an artificial heart) [10].

11.5.1.1 Blood Compatibility of Biomaterials

Currently, clinically used blood contact-type biomaterials provide us with useful directions in terms of blood compatibility of artificial material surfaces. As shown in Table 3, such materials are already used for blood vessels and heart valves. Both are known to show excellent blood compatibility without forming quasi/intima for a prolonged period of time without the use of either the anticoagulant, heparin, or an antiplatelet agent [11]. With no epithelial cells on these materials, hydrophobic connective tissue is exposed. The water content of these materials is $\sim 80\%$, which is typical of hydrogels. The main component of the outermost layer, which is in contact with blood, is glucosaminoglycan. However, details are not yet known. If the excellent blood compatibility of the outermost layer surface is not due to the use of heparin, the hydrogel itself may have good blood compatibility.

11.5.1.2 Blood Compatibility of Hydrogels

In general, if the water content is as high as 80%, the surface of a homogeneous hydrogel that is in contact with water is considered to have the structure as shown in Fig. 9(a). This surface structure is the same as for the material on which hydrophilic polymer chains are grafted. This is also shown in Fig. 9(b). The great numer of *in vitro*, *ex vivo* and *in vivo*

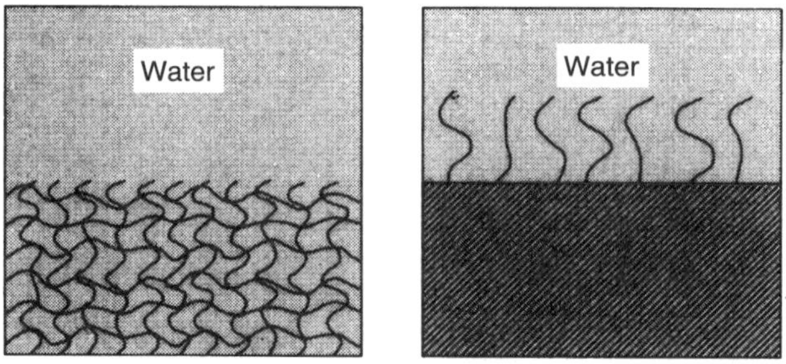

(a) Interface between hydrogel and water

(b) Interface between grafted surface and water

Fig. 9 Schematic diagrams of the interface between a hydrogel or grafted surface and water.

experiments that have been performed show that such grafted surfaces or hydrogels exhibit excellent antiplatelet properties. Some examples are shown in the following.

Figure 10 shows the water content dependence of platelet adsorption on a PVA hydrogel prepared with glycerin. Even at the same water content, adsorption differs depending on the gel's preparation conditions. This is probably due to the differing sizes and density of the surface amorphous regions as a function of the gel preparation methods as shown in Fig. 11. Figure 12 depicts the relationship between the number of platelets on various hydrogels and water contents in a phosphorus buffer solution. Interestingly, the number of platelets adsorbed exhibits a minimum at a 90% water content in all cases. As the water content increases, the number of platelets adsorbed increases again. Figure 13 illustrates the adsorption of platelets on these hydrogels in the presence of a protein. The obtained results are the same as those in Fig. 12. These results are due to the penetration of platelets inside the gel when water content significantly increases as shown schematically in Fig. 14. When a hydrophilic polymer is grafted on a hydrophobic polymer surface, the water content of the grafted layer cannot be accurately determined. However, this grafted layer affects significantly the adsorption of platelets and proteins. An example of such a study is shown in Fig. 15. As the graft content increases, the adsorption of the protein increased drastically. This is likely due to the penetration of the protein by the same mechanism as shown in Fig. 14.

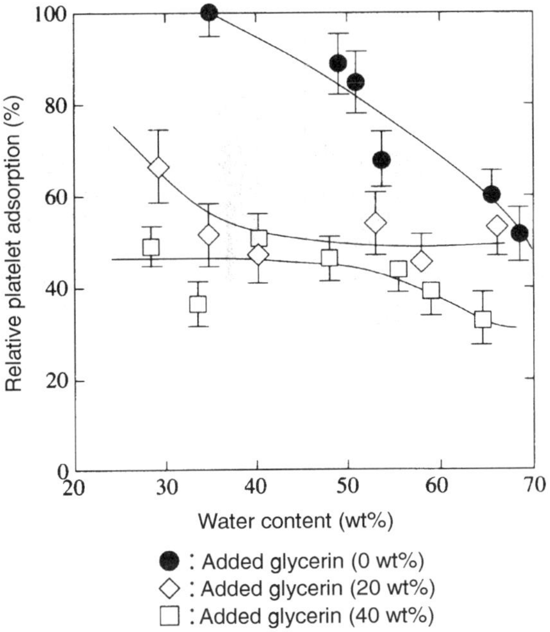

Fig. 10 Platelet adhesion onto glycerin-containing PVA hydrogel (glycerin in the gel was completely removed prior to the adsorption measurement) [12].

The *in vivo* results also indicated improved antiplatelet activity by grafting a hydrophilic polymer onto a material surface [15].

Even in a hydrophilic polymer, if surface density is high, a lubrication property is shown [16]. If the polymer has anionic or cationic groups, there will be ionic interaction with proteins or cells and this requires caution [17]. Adsorption of proteins or cells is difficult on a nonionic hydrogel. This has already been demonstrated by electrophoresis using a polyacrylamide gel, filtration of cefadex gel, and a cell culture on an agar gel.

11.5.2 Adhesion of Connective Tissues

Tendons and ligaments adhere strongly to bones. Teeth are also fixed strongly to the jawbones. Intestinal membranes adhere strongly to intestines and basement membranes adheres strongly with cells. Such strong adhesion is sometimes required between an artificial material and its

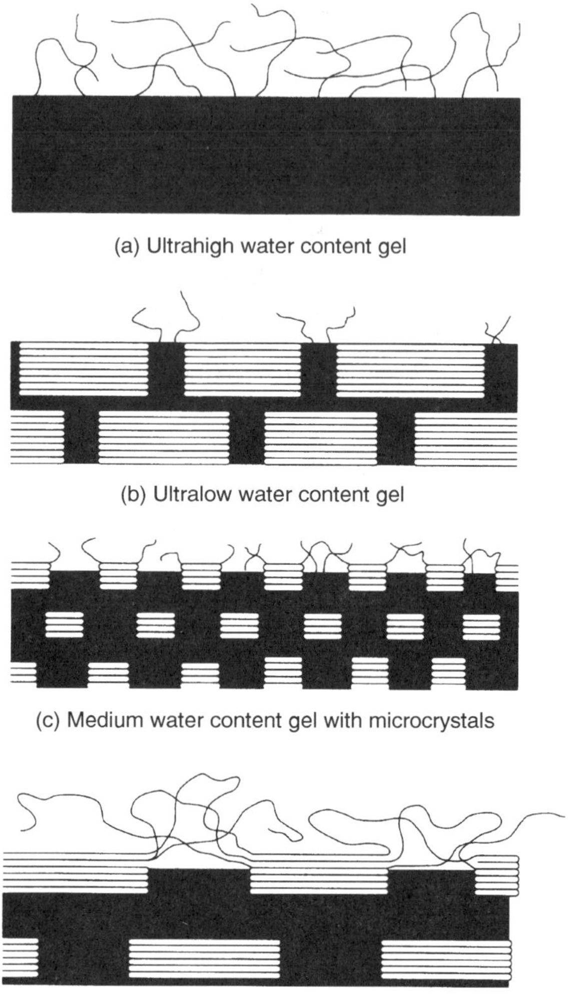

(a) Ultrahigh water content gel

(b) Ultralow water content gel

(c) Medium water content gel with microcrystals

(d) Medium water content gel with larger crystals

Fig. 11 A schematic diagram of the interfaces between PVA hydrogels with various internal structures and water [12].

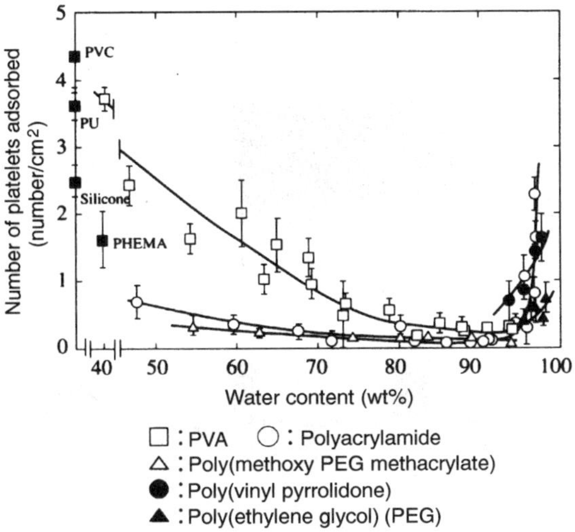

Fig. 12 Adsorption of platelets on various hydrogels in a phosphate buffer solution [13].

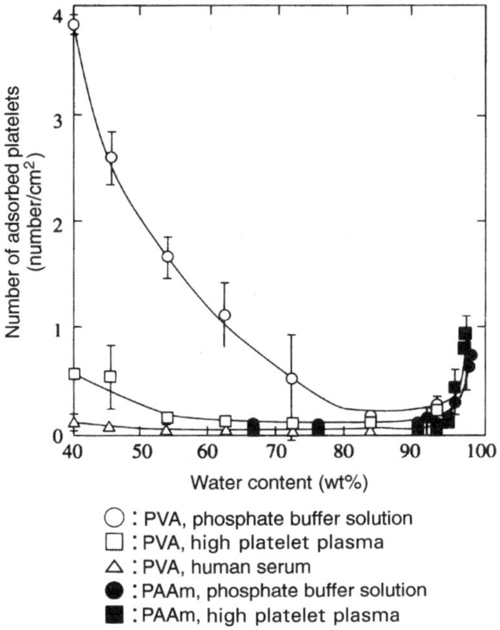

Fig. 13 Adsorption of platelets on PVA and polyacrylamide (PAAm) hydrogels in various media [13].

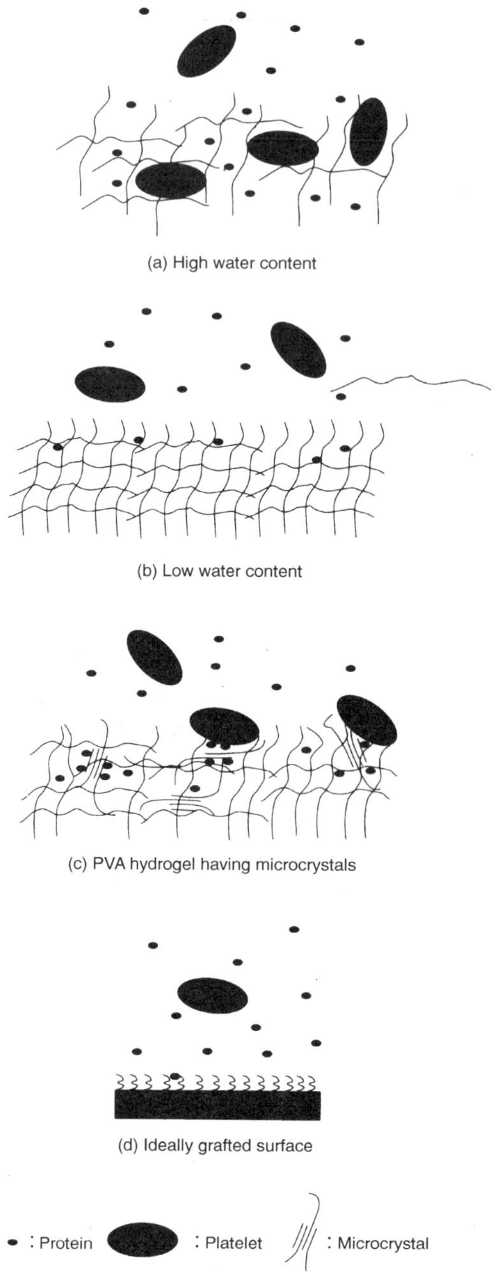

Fig. 14 A schematic diagram of adsorption of a protein and platelets onto a nonionic hydrogel having various water contents [13].

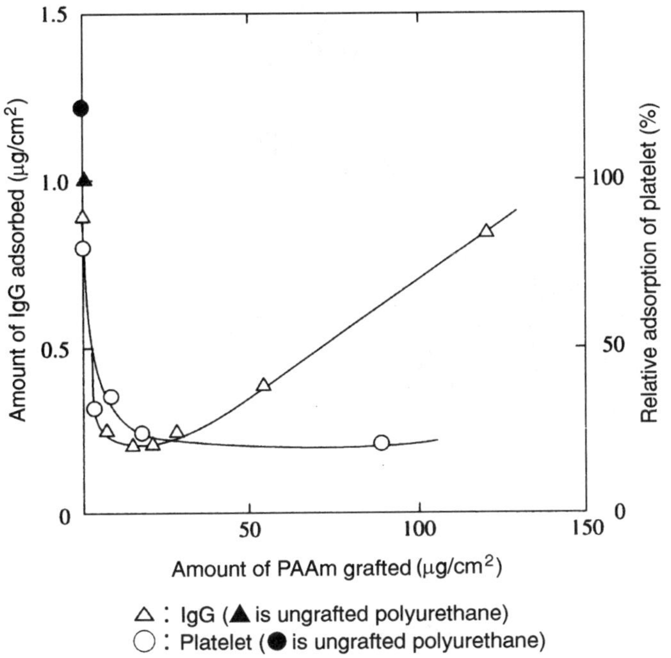

Fig. 15 Adsorption of γ-globulin (IgG) and platelets from concentrated human platelet solution onto a polyurethane surface to which polyacrylamide (PAAm) is grafted [14].

biosystem. An artificial cornea and the corneal system, a skin device and a transdermal system, and artificial valves and blood vessels are such examples. In these applications, surface treatments are required. It has been shown that soft tissue can adhere to a material surface by fixing a thin collagen hydrogel film [18]. This is probably related to the fact that the main component of the soft tissue is collagen and collagen is a protein that adheres to cells.

On the other hand, polymers that adhere well with bone systems are also known. One of them is a water-containing polymer reported by Radder *et al.* [19]. As already described, when a water-containing polymer is implanted, Ca^{2+} and PO_4^{3-} precipitate in the water-containing layer and calcium phosphate is formed. This inorganic crystal is highly bone-adhesive. We have demonstrated that, if a hydrogel made of a polymer having a phosphate group is fixed onto the material surface, the material bonds [20,21].

11.6 CONCLUSIONS

Despite the fact that soft tissues are made of typical hydrogels, namely high-water content gels, synthetic hydrogels are used very little as biomaterials except for use in soft contact lenses. The reasons why synthetic hydrogels are not used in implantation have been explained here. Synthetic hydrogels nevertheless remain useful biomaterials, depending on the application.

Finally, even if an ordinary gelatin is regarded as a weak material, it is worth emphasizing that a strong material can be prepared by increasing the orientation of polymer chains as in the example of the study on gelatin by Zhao *et al.* [9]

REFERENCES

1 (1979). *Biochemistry Databook [I]*, Biochemical Soc., Japan., ed., Tokyo Kagaku Dojin, p. 1617.
2 Ikada, Y. (1988). *Approaches for Biomaterials-Artificial Organs*, Nikkan Kogyo Shinbun Publ., p. 153.
3 Ikada, Y.T. (1994). *Biomaterial Science*, Sangyo Tosho, p. 234.
4 Ikada, Y. (1988). *Approaches for Biomaterials-Artificial Organs*, Nikkan Kogyo Shinbun publ., p. 154.
5 Ikada, Y. (1989). *Medical Polymeric Materials*, Kyoritsu Publ., p. 59.
6 Cha, W.-I., Hyon, S.-H., Oka, M., and Ikada, Y. (1993). *Biomedical Materials Research in the Far East* (I), pp. 145–146.
7 Burczak, K., Tomihata, K., Iwamoto, Y., and Ikada, Y. (1992). Proc. 21st Medical Polymer Symp., Kyoto, pp. 71–72.
8 Tomihata, K., and Ikada, Y. (1996). 211th ACS National Meeting, New Orleans, LA, March 24–28, 1996 Abstracts, BTEC, 015.
9 Zhao, W., Kloczhowski, A., Mark, J.E., Erman, B., and Bahar, I. (1996). *CHEMTECH*, March, pp. 32–38.
10 Menconi, M.J., Pockwinse, S., Owen, T.A., Dasse, K.A., Slein, G.S., and Lian, J.B. (1998). *J. Cell. Biochem.* **57**: 557–573.
11 Marois, Y., Boyer, D., Guidoin, R., Douville, Y., Marois, M., Teijeira, F.-J., and Roy, P.-E. (1989). *Biomaterials* **10**: 369–379.
12 Fujimoto, K., Minato, M,. and Ikada, Y. (1994). *Polymers of Biological and Biomedical Significance, ACS Symp. Ser. 540*, S.W. Shalaby, Y. Ikada, R. Langer, J. Williams, eds., pp. 229–242.
13 Kulik, E., and Ikada, Y. (1996). *J. Biomed. Mater. Res.* **30**: 295–304.
14 Fujimoto, K., Takebayashi, Y., Inoue, H., and Ikada, Y. (1993). *J. Polym. Sci. Part A: Polym. Chem.* **31**: 1035–1043.
15 Inoue, H., Fujimoto, K., Uyama, Y., and Ikada, Y. *J. Biomed. Mater. Res.* (in press).
16 Ikada, Y., and Uyama, Y. (1993). *Lubricating Polymer Surfaces*, Lancaster-Basel: Technomic Pub. Co..
17 Kato, K., Sano, S., and Ikada, Y. (1995). *Colloids and Surfaces B: Biointerfaces* **4**: 221–230.

18 Okada, T., and Ikada, Y. (1995). *J. Biomater. Sci. Polym. Edn.* **7**(2): 171–180.
19 Radder, A.M., Leendens, H., and Blitterswijk, C.A. (1996). *J. Biomed. Mater. Res.* **30**: 341–351.
20 Tretinnikov, O.N., Kato, K., and Ikada, Y. (1994). *J. Biomed. Mater. Res.* **28**: 1365–1373.
21 Kato, K., Eika, Y., and Ikada, Y. *J. Biomed. Mater. Res.* (in press).

INDEX

I

T

ISBN 0-12-394962-9
90182 >
9 780123 949622